TRADITIONAL CONSTRUCTION FOR A SUSTAINABLE FUTURE

Carole Ryan

Spon Press
an imprint of Taylor & Francis
LONDON AND NEW YORK

Published 2011
by Spon Press
2 Park Square, Milton Park, Abingdon, Oxon OX14 4RN

Simultaneously published in the USA and Canada
by Spon Press
270 Madison Avenue, New York, NY 10016, USA

Spon Press is an imprint of the Taylor & Francis Group, an informa business

© 2011 Carole Ryan

Typeset in Baskerville by
Swales & Willis Ltd, Exeter, Devon
Printed and bound in Great Britain by
TJ International Ltd, Padstow, Cornwall

British Library Cataloguing in Publication Data
A catalogue record for this book is available from the British Library

Library of Congress Cataloging-in-Publication Data
Ryan, Carole, 1948–
Traditional construction for sustainable new building/Carole Ryan.
  p. cm.
1. Building.   2. Sustainable construction.   3. Appropriate technology.
4. Vernacular architecture.   I. Title.
TH146.R93 2011
690.028′6–dc22   2020020901

ISBN13: 978–0–415–46756–8 (hbk)
ISBN13: 978–0–415–46757–5 (pbk)
ISBN13: 978–0–203–89550–4 (ebk)

This book is dedicated to the memory of
Clifford William Rose Ridout,
late of Stalbridge and Okeford Fitzpaine, Dorset,
who gave me his love and support for many years,
but who sadly passed away, just as this book had commenced.

CONTENTS

CONTENTS

ACKNOWLEDGEMENTS

There are many owners of traditional buildings who have contributed in their own way to the illustration of this book, especially in Shropshire and Dorset, over a career spanning 35 years.

Special thanks go to Rob Buckley of the Dorset Centre for Rural Skills, Farrington, Child Okeford, Dorset who has always been on hand to lend an ear and offer practical help, support, and encouragement. Long may his enterprise flourish. Also to the Weald and Downland Museum, Singleton, near Chichester, West Sussex, who have for many years been a source of inspiration and knowledge, with a special thank you to Diana Rowsell, Head of Learning.

To Amanda Hayhurst, grateful thanks for all her sterling work on the drawn illustrations.

THE NATURAL WAY OF LIFE AND THE TRADITIONAL HOUSE, ITS INCEPTION AND DEMISE

Ancient lifestyles and the use of natural materials

The meaning of vernacular/traditional

Vernacular building tradition is one based upon time-immemorial building techniques and the use of natural materials, but also upon the siting and location of buildings, the layout of their rooms, and room usage. It is passed on by word of mouth, and materials are used that are close to hand and do not require to be moved any appreciable distance. In addition its very essence is a format devised by the craftsman and occupier, acting in conjunction, and with function only in mind. Vernacular or traditional building is thus a response to a way of life, usually farming, passed on from generation to generation. That farming is now described as an industry is a very telling factor, as this way of life, still evident to our grandparents, has declined and with it the need for the buildings that catered for it.

Early lifestyles and shelters

Such vernacular buildings may have their origin in primitive buildings defined as temporary and constructed in flimsy materials. Today primitive buildings are hard to find, the most recent

Photo 1.1 A group of Herefordshire farm buildings indicative of a former way of life

that may survive include farm buildings constructed of sawn-down telegraph poles and clad in corrugated iron, a product of post-war Britain, when both skilled men and materials were scarce. No one can be entirely sure exactly how ancient man adapted natural materials to primitive shelters as the pre-emptor of this vernacular tradition, as such materials were by their very nature ephemeral, apart from rock shelters. Cave dwellings certainly provided a solid basis for shelter, and had the advantage of a temperature that was cool in summer and warmer in winter.

Give anyone an instruction to go into the forest and make a shelter and one can be sure that it will involve cutting equally sized timber poles, placing them in the ground in a circle or two lines, and tying them with vines at the summit. The cladding of this tent-like structure is then a matter of what is available, be it large leaves or bundles of organic material such as leaves or straw. The translation of this into the cruck frame of the earliest surviving houses, some of which bear more than a passing resemblance, being mostly roof and not wall, is a matter of pure speculation. The fact remains that this continued into historic periods with the shielings (summer camp for grazing) of Jura, Scotland, with branches of trees covered with turf in a tepee-like shape, reinforces the possible theory. So does the beehive hut discussed below (the Skellig Islands, County Kerry, Ireland), although these tent-like structures are expressed in stone, the forerunner of mass walling construction. The few remaining black houses of the Isle of Lewis have roofs constructed of branches clad with turf in a linear tent-like form, on stone walls packed with earth. They are the latest and most tangible reminder of tent-like structures, being mostly 18th and 19th century in date. Excavated examples are rare, possibly because they consisted entirely of a wooden branch frame clad with turf that rapidly returned to the soil when occupation ceased. All these examples are reminiscent of a tent-like precursor.

Another parallel tradition which has a very ancient origin can be seen in the presence of lintel and post structures in chambered tombs (dolmens) to be seen in great numbers in Malta but also in the British Isles, and even the structure of Stonehenge hints at this form of construction.

The post and lintel is the basis of the possible classical tradition of the ancient world of Greece. There many theories on the origin and reason for Stonehenge, but few dwell on the possibility of two ring beam structures, one inside the other, of post and lintel format. The inner ring beam

Photo 1.2 A rock shelter for animals on the island of Malta

Photo 1.3 Remains of a megalithic chambered tomb in Malta

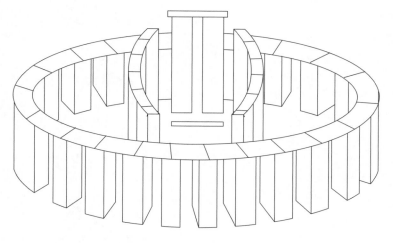

Diagram 1.1 Stonehenge, conjectural reconstruction (after Fletcher, 1921)

is higher, surely hinting at a possible use as a support for a hefty tree-like series of rafters, tied at the apex; in essence a giant megalithic tent. In some archaeological quarters such theories would be regarded as heresy, but the possibility cannot be discounted.

The influence of climate

Climate must have been the greatest factor influencing the development of buildings, and has a bearing also on the quality and productivity of their environment. Physiological and psychological tolerance by humans to extreme heat or cold and the capacity of the same to affect the ecology, particularly vegetation and other animals, would have determined the vernacular architectural response to shelter. The use of vines for binding timber, the quality of the timber itself which determines how it can be modified, and the nature of animal skins for

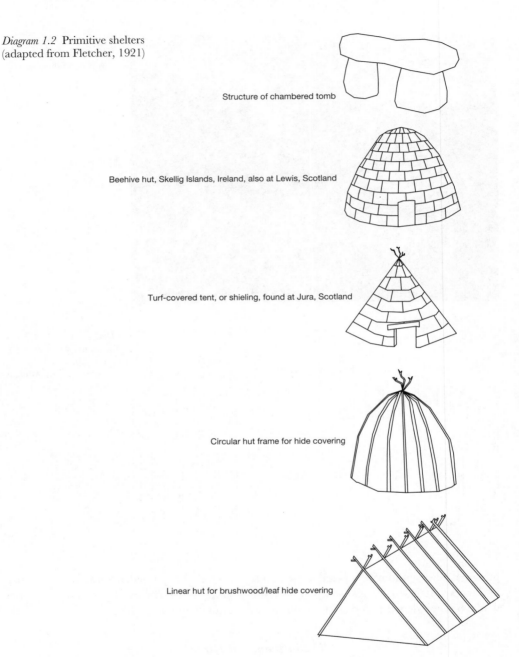

Diagram 1.2 Primitive shelters
(adapted from Fletcher, 1921)

Structure of chambered tomb

Beehive hut, Skellig Islands, Ireland, also at Lewis, Scotland

Turf-covered tent, or shieling, found at Jura, Scotland

Circular hut frame for hide covering

Linear hut for brushwood/leaf hide covering

possible cladding or roofs and walls, will all be determined by the effect of climate upon environment. An even greater influence of climate is on the roof form of buildings. Roofs need to respond to temperature, solar radiation, precipitation and winds. Steep roof slopes are used in wet-temperate and cooler zones, the most prime examples being surviving British farmhouses with long low sloping roofs facing down a valley, encouraging run-off long before water can reach the walls.

4

Photo 1.4 Long, low, sloping thatched roof deliberately contrived to protect the rest of the building from adverse weather conditions.

Hot dry zones, in comparison, utilise flat roofs that can be constructed in considerable depth to absorb the sun's rays long before the heat reaches the dwelling beneath. A more graphic reason why flat roofs are unsuitable for the British climate would be difficult to find, and yet they were constructed in great numbers in the post-Second World War period, indicating a loss of knowledge of traditional forms. In some cultures domed roofs were utilised to reflect the day's heat back into the night sky, because of their greater surface area, but this does not explain the format of the 5th-century early Christian huts on the Skelligs (Irish coastal islands, County Kerry) with their corbelled roofs. In this inhospitable location, the roof form may well be an idea imported from warmer climes, or simply a response to the only material being available. The naturally shaly rock would lend itself to forming a roof by corbelling inwards until the apex was reached.

Photo 1.5 Corbelled huts on the Skellig Islands

Climate certainly affects design for comfort levels in buildings, although it has to be remembered that our own comfort levels contrast greatly with those of our immediate ancestors, a factor that is rarely taken into account by most recent occupants of vernacular buildings. The builders of these were more intent on creating shelters from extremes of climate and a shelter for the fire, and their occupiers were content to endure greater contrasts between heat and cold then we are. They also placed reliance on clothing that functioned both internally and externally, with little differentiation. That the fireplace was essential for air quality was possibly recognised, but in general, ventilation and day lighting was rarely addressed. The former was catered for naturally by the many draughts, and in the choice of porous materials for wall and roof, the latter must have been more difficult unless tasks could be directly illuminated by the fire. Domestic tasks, often large scale and dirty, had more credence out of doors in natural light, but the British climate would certainly have militated against doing so. A good example of this is the Dorset covered rear loggia or shire, known to have housed a multitude of dirty tasks. Unfortunately these rarely survive, having fallen victim to the modern desire for a conservatory in this south-facing location, but its original format was doubtless originally designed to capture light and solar gain.

The nature of traditional construction demanded small openings for stability and the need to retain heat, the walls acting as a thermal store, so they did not allow for large windows. Conversely it is interesting to note that even until recent periods, elderly farmer's wives pursued their domestic tasks with the kitchen door wide open, even on the coldest day. Such a way of life would be unthinkable now. Another factor that characterises traditional buildings is the lime-washing of both interiors and exteriors. Widely thought to have its origins in the use of lime as a disinfectant and inhibitor of woodworm and fungal decay by virtue of its strong alkaline chemistry, the white interiors would have helped to reflect light to militate against the use of small windows. The external white coating would have kept the house cooler in summer, building up to a fine plaster to protect the masonry and joints, although the practice would have provided no solar gain in the depths of winter.

It is interesting to note that mass walling construction, which has its origin in the need to provide thick walls for stability, also results in walls that are capable of operating the thermal

Photo 1.6 A partially underbuilt Dorset shire for undertaking unpleasant tasks beneath its shelter

flywheel effect. The building is warm in winter, as the internal surfaces absorb the heat from the fire, reflecting it out when the fire dies down, but also keep buildings cool in summer by resisting the penetration of the sun's rays. It is known that walls need to be a minimum of 250 mm (10 in) to provide an eight-hour thermal lag (Oliver 1997a: 132), and walls in most 16th- or 17th-century farmhouses are at least 500 mm (20 in) and in some cases considerably more.

The location factor

The location of a farmhouse, in respect of its immediate landscape, is rarely examined today when making a decision to purchase (apart from a nice view), yet our ancestors would have taken many factors into account before deciding to build. The drainage of the ground would have been a key element, plus the presence of a shelter belt of trees, essential in mitigating wind-chill factor. Sadly the first thing that most recent occupiers do is to modify this landscape so that features like this are removed, because they block the view, giving little thought to the value of these trees not only in winter, but in providing essential shade for summer. By the same token the availability of water, a river or brook, was not only essential for human and animal life, but the amount of it falling from the sky, be it great or small, also determined the nature of the shelter. In areas where water is frozen, both the extreme cold and the difficulty of obtaining water will act together to deter settlement. The topography of the landscape also will determine the ease with which settlement can take place, particularly orientation.

Lowland areas, defined as lying at less than 300 m (1,000 ft) above sea level, and rich in vegetation, would always have been attractive to settlement and the vegetation itself a natural source of material for shelter. Upland zones and higher slopes, always more inhospitable in terms of access and materials for shelter, have traditionally been more sparsely populated except in times of war or stress, such as over-population of the lowland zone. Not withstanding this there were advantages, such as ease of defence, avoidance of disease, proximity to pastures, and land availability away from the crowded lowlands for poorer people. In addition, in warmer climes hilltop locations were more comfortable, being cooler in summer, but the British climate demands a hardier breed of occupant altogether. Hilltop towns, such as Bridgnorth in Shropshire, hint at more defensive reasons, and in all events are outcrops in a more lowland plain. It is interesting to note the connection between this form of settlement and the earliest

Photo 1.7
A Mediterranean style
farmhouse designed to
cope with very hot
weather

rock shelters. The cellars of hilltop towns, hewn out of the rock, may have provided the material for the rest of the dwelling. Also the void between the inclined hillside and the floor of the main dwelling provided an area for cold storage, taking full advantage of the thermal flywheel effect. In some instances the cut-outs in the rock were dwellings themselves, and until peremptorily closed by the Public Health Acts in the post-war period undoubtedly enjoyed all the advantages of being cool in summer and warm in winter. It is noticeable that the need for shelter from wind, as well as the need for warmth, encouraged the clustering together of dwellings. Deep eaves and sheltered balconies are the rule in more exotic locations such as Malta, providing shelter from winter storms and summer sun.

In British uplands there is more dispersal of settlement and ancient patterns in construction, but more pertinently a continuum in traditional lifestyle and customs, still visible today. This may be due to the slow change discouraged by difficulties of actual access and communication with other settlements. The south Shropshire Hills is an example where until quite recently one could still see a pastoral farming economy and a way of life unaltered by the passage of time.

The intimate connection with green spaces and the eventual loss of contact between man and nature

Settlement then is an essential part of what governed the nature and form of dwellings, but how did this settlement arise? At best it can be regarded as the colonisation of what started as vast tracts of forest covering large areas of the planet. In this vast 'spiritual landscape' ancestors of our many nations roamed at will, living on the fruits of the land, be they animal, vegetable or mineral, and moved when the seasons changed or resources became scarce. Migration of animals would have engendered migration of man. In this way humans gained an in-depth understanding of their environment and the sum of its components, in particular the effect of climate, the changing seasons, and the animals and plants influenced by these aspects. By doing so they formed an intimate relationship with their environment in a way that one can only dream about today, as one sits within an artificially created environment, such as an office block, staring into the distance at some green trees on the horizon. It is this disassociation from the spiritual landscape, which formed the whole territory as opposed to a territory delineated by defined boundaries, that is widely believed to be at the root of the general disorientation experienced by modern man. Home was the whole territory and spiritual landscape; now it is a small capsule, to be weighed, allocated inflated values, and treated generally as a piece of currency. When home was a hide tent, or a cave, or a wooden shelter composed of poles arranged in a tent formation and covered with hides, the immediate enclosure was only a small part of a much larger 'home' that is the landscape, as far as could be traversed or seen in the distance. From this may have derived the meaning of 'territory' as a tract of land, delineated by boundaries and administered by a social group. This implies settled agriculture, widely supposed to be the fostering influence upon settlement, and itself a result of scarce food resources and increasing population. 'Territory' may have other origins, akin to that of animals, who also jealously guard scarce food resources and who have a genetic link to their territory, expressed as 'hefting'. It is by this method that hill farmers rely on the innate knowledge of their flock not to stray beyond the boundaries of the farm. It may also be part of a general evolution. Throughout all these theories the possibility of the 'spiritual' landscape holds good, and despite the modern trends for mass migration about the British Isles and further afield, many people of rural origin still feel a pull back to the land of their birth.

Existing remote tribes in the Amazon may give an indication that the first settlements showed an attachment to an area rather than to a particular location within that area or place, and later nomadic cultures, such as the American Indians, had an attachment to 'place', or rather the spiritual landscape they perceive to be their surroundings. The evidence lies in the hearths, storage pits and evidence of shelters which, although a moveable feast, do so within a particular tract of landscape. The move from this to smaller enclosures defining the buildings themselves, such as walls, fences, hedges or trees, was another way of physically defining space, but always with regard to the larger landscape around it.

Such boundaries are thus only recognisable in terms of their use by the settlement and are mix of natural and man-made boundaries. In addition some are certainly designed to prevent entry, such as walls, whilst others are designed to be more psychological, such as trees and are more of a spiritual boundary, and again a link to the wider spiritual landscape. From this there appears to have been a gargantuan leap, admittedly over a long period of time, from spiritual boundary to the perception of boundary as a political entity, to be fought over and extended by force. This concept of 'ownership' and 'property' (now defined legally as the right to enjoy, and to exclude others from this enjoyment) is how people now view their relationship with the landscape, yet as late as the 18th century the wider landscape had significance. Parson Woodforde (1740–1803), in his *Diary of a Country Parson 1758–1802* (Beresford 1978: 162), described in graphic detail the beating of bounds of the Parish of Weston in Norfolk, some 12 miles and taking five hours.

The presence of high walls around buildings or settlements, either still standing as in some town walls or remaining archaeologically, is not always associated with the need to reinforce this political entity. In the 15th and 16th century, courtyard houses were not just a political statement of power. Their planform was a device to resist the penetration of wild animals, that essential link with the wider spiritual landscape, to be enjoyed for hunting, but whose inconveniences were not to be tolerated on an everyday basis.

The concept of tenure versus sense of place

Tied up with the concept of property was the eventual concept of tenure and inheritance. Thus 'place' became an asset to be passed on from generation to generation, a concept which usurps the meaning of the wider sense of place. Nonetheless 'sense of place' was not completely lost upon historical generations, and certainly not upon the lower echelons of society, farmers and farm labourers. The continuous association with the seasons via the farming cycle perpetuated that deep affinity with the seasons, as well as the vagaries of climate and the behaviour of animals and plants. In addition there was an inextricable link between the landscape and the materials that the immediate landscape provided, themselves a product of nature and the changing seasons. Stone, timber, and earth used as building materials for indigenous architecture are the materials of the landscape, and the buildings created from these materials, as well as responding to the local climate, became as one with the landscape, barely distinguishable from it. This is certainly true of British vernacular architecture, but is also true of, for example, Mediterranean and North African courtyard houses with their massive earth walls creating cool interiors and flat roofs designed to collect every scrap of rainwater. Whilst the examination of all types of indigenous or vernacular architecture is not possible in this text, it is interesting to note the commonality of the use of earth in both British and North African houses. Earth is as adept at keeping warmth in as it is at keeping heat out.

The development of the British vernacular house

It is a sad indictment of present-day society that traditional houses are viewed as some curious one-off, to be modified at will if they prove inconvenient, and with little thought of what it is they represent as a vast store of knowledge from a previous age. It is somewhat problematic that we cannot illustrate this latter phenomenon from its earliest inception. Archaeology hints at rectangular structures based on post holes and conjectural reconstructions hint at low walled structures with a wall plate of irregular shaped poles held in the notch naturally formed at the top of a young tree. From there it is a simple matter to conjecture a roof of slender pole rafters each coupled to its opposing neighbour by the device of tying with twine. The hipped roof is usually shown in such conjectures, and indeed many surviving thatched roofs are of this form in southern England. The preponderance of roof over wall is a natural function of the timber available but also has resonance with a possible tent-like precursor. Sir Banister Fletcher (1921), in his inestimable treatise on architectural history (first published 1896), once the staple of every student of architecture, illustrates two buildings that show precisely this phenomenon. Teapot Hall, Spilsby, Lincolnshire, described as probable 14th century, is a pure tent-like structure, the sloping sides clad partly in thatch, partly in large wooden boards on the lower slopes. The cruck-framed cottage at Didbrook, Gloucestershire, described as probable 15th century, on the same page shows the same tent-like structure, with the walls later raised by modification of the wall plate level, at least twice. The analogy between the two has always been striking.

Such early low walls could have been clad with horizontal or vertical boards, or even cob pressed into a wattle structure woven around the uprights. The latter may have its successors in 'mud and stud' structures still surviving in East Anglia today. Such buildings are thought to be the epitome of the Saxon (post-Roman) hall house, but with so little tangible archaeological evidence it is difficult to be precise.

One of the most pressing conundrums in understanding vernacular is how and why regional variations are found. Traditional building construction has many analogies to language: sadly, the degradation of vernacular buildings today goes hand in hand with that of local dialects, both due to post-war social and economic mobility. In an age of rather less mobility, ideas and the spread of cultural traits to a new place, or deposited on the way there, must have taken place, albeit slowly, and this can be referred to as cultural diffusion, but regrettably it has left no clear trace. It is our unhappy lot to try to reconstruct the diffusion processes by in essence working backward from the buildings as we now discern them. The problem is that we cannot

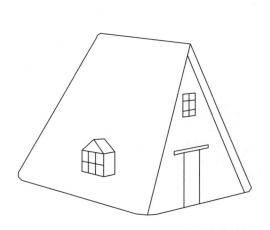

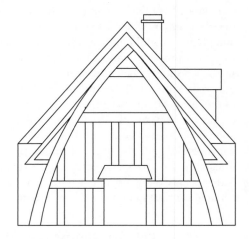

Teapot Hall, Spilsby, Lincolnshire, late 14th century?   Cruck Cottage, Didbrook, Gloucestershire, 15th century?

Diagram 1.3 Buildings with tent-like forms (after Fletcher 1921)

know what we have lost, and certainly there are many gaps in the progression. The issue of unknown and long-vanished buildings has been explained in terms of the Vernacular Threshold, that period in time when there is a clear emergence of buildings in materials permanent enough to survive. The problem is that we can never be sure that surviving buildings actually represent the Vernacular Threshold. Primitive buildings from early periods do not survive because they are built of flimsy materials and constructed to last a very short period. Vernacular buildings do survive because they are built of very sturdy materials, but their survival is also geared to social status and location. This is best examined through the medium of the main domestic building types and their location in the British Isles.

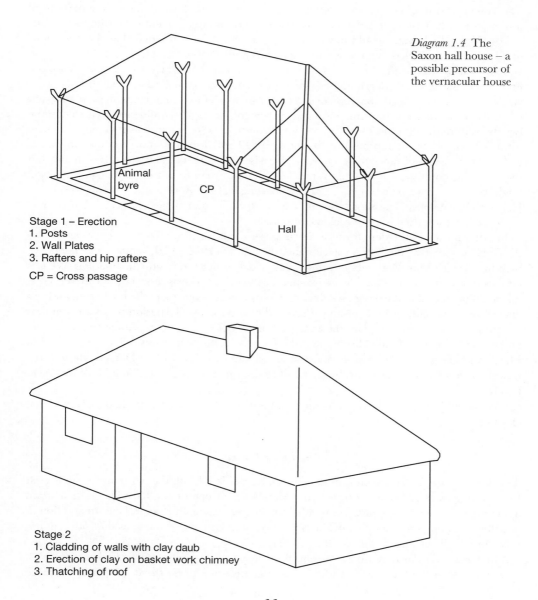

Diagram 1.4 The Saxon hall house – a possible precursor of the vernacular house

Animal byre

CP

Hall

Stage 1 – Erection
1. Posts
2. Wall Plates
3. Rafters and hip rafters

CP = Cross passage

Stage 2
1. Cladding of walls with clay daub
2. Erection of clay on basket work chimney
3. Thatching of roof

The main building types

The great house

The 'great house' has its origin the stone castles of the 12th century, and these in turn were replaced by the Elizabethan or Jacobean country mansion, yet there is little logic in the progression, more of a gargantuan leap. Many retained the vernacular tradition of using stone or timber, albeit that they tried to adopt classicism a somewhat clumsy way, described at 'artisan mannerism'. It is not until the late 17th/early 18th century that great houses adopted true classicism in the form of Baroque, Palladian, Neo-classicism, and Classical Revival, each following on from its precursor to either diminish the extrovert nature of the classicism or revert from a severe form of classicism to something more frivolous. These houses were bastions of power, just as the castle had been, and in the early periods little thought was given to comfort, although the thick walls would have acted as thermal stores provided that the enormous fireplaces were brimmed full of wood burning constantly day and night.

These Elizabethan and Jacobean great houses, the homes of the aristocracy, and their more humble counterparts, the large houses which succeeded the open medieval hall houses, were largely constructed by a combination of owner design, master mason or carpenter and itinerant or local contractors. The movement of such craftsmen was largely confined to an area of some twenty or thirty miles, others like Robert Smythson roamed over surprisingly large distances, from Wiltshire to Northamptonshire. (By the age of 43 Smythson had moved home some three or four times over a distance of 250 miles.) There is some early use of the word 'architect' but there is some confusion about what is meant, be it craftsman or patron, although Smythson might be said to have set a trend for drawing existing features (such as the window tracery for Henry VII's chapel) and passing these on to his son for utilisation in later projects. Here is some evidence at least for diffusion of architectural ideas.

In other instances the design was a much looser affair. Sir John Thynne, the Secretary to the Duke of Somerset, the Lord Protector after the death of Henry VIII, kept in correspondence with his steward John Dodd about progress at Longleat but rarely saw the building. What is clear is that he understood about such aspects as the need to season wood, but by the same token it is obvious that the design was based on only a vague concept of the final form, and was altered verbally as the project continued – a cardinal sin today. The period shows an emphasis on geometric forms for the plan, the triangle being a being used both as a conceit or device, and a nod towards the Holy Trinity and the Catholic faith, which was otherwise banned in Elizabethan England. Perhaps the most obvious is Lulworth Castle in Dorset, whose three round bastions set on a circular plinth stand as testament to the Catholic faith of the Weld family.

Occasionally some tentative plans survive, mostly sketch ideas rather than plans for final execution.

The large house

The 'large house' is that which is most commonly associated with the Norman war-lord's gift to their knights, who became Lords of the Manor. Their open halls, barn-like in their form, appear to have the most analogy to their possible forerunners, the Saxon Hall. In addition, it was the form apparently adopted by the earliest yeoman farmers, certainly by the 15th century, following the Black Death of 1348/9 and 1361.

The format was fairly standard: a central open space, floored at one or both ends, with a cross passage at the lower end, and a fire in a central hearth on the floor. Anything more

Photo 1.8 Smoke-blackened thatch from a more recent source, an 18th-century bakehouse

Photo 1.9 A medieval 15th-century roof, Gothelney, Somerset

designed to create draught it would be hard to find, the principal reasoning being to control the fire, the smoke from which ascended upwards into the cavernous space of the hall where it could disappear out of gablets in the hips, or a louvre-style box on the ridge, or just be absorbed by the thatch. Smoke-blackened thatch and roof timber are the main evidence for any surviving examples. The form predominated for the upper echelons of society until the Black Death.

The medieval hall planform, consisting in its fully developed form of an open hall with floored ends, one end containing a service with 'solar' (private retiring room, with a large window to gain sunshine) over and the other a parlour with 'solar' over, set the trend for the development of the traditional yeoman farmhouse (see small house below). The flooring over of the hall took place in stages, with first an internal jetty or canopy over the heads of those who sat at high

13

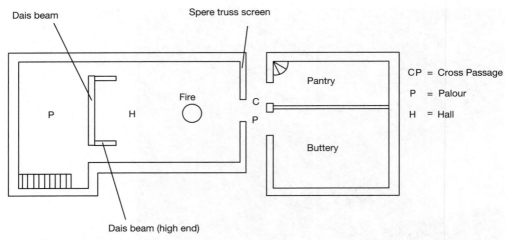

Diagram 1.5 Planform of a hall house c.1450 (Ryan, 1979)

table. This had the advantage of protecting the manorial lord and his family from the draughts that whirled about their heads. The incentive to elongate this into a partial ceiling creating another chamber above plus a definitive smoke bay to control the acrid fumes was soon realised, and eventually led to a full ceiling, creating a 'Great Chamber'. This in turn provided a plan that was the basis of the c.1600 purpose-built manor houses that followed.

The smoke bay became an enclosed fireplace with a timber framed smoke hood, the latter having obvious disadvantages regarding fire, and so was replaced by a solid stone or brick ingle and chimney breast above. Either would have acted as a vast central heat store.

The front elevation projections had numerous gables, with a particular emphasis on the projection forming the chamber over the porch and the chamber over the dais window.

Diagram 1.6 Purpose-built manor house c.1600, a typical elevation

Photo 1.10 A typical ingle fireplace

The enclosure of the fire and the need for a ceiling over the open hall has much to tell us about the lack of comfort endured before these improvements. These could in no way be described as sustainable houses in terms of heat conservation; the house was merely a shelter for the fire rather than a means of conserving the heat from the fire before the advent of the ingle enclosure.

The tiny chambers formed by the chambers over the porch or dais window and by the projecting bay windows in what are essentially very large houses are a source of some wonderment as to their function. Often glazed with large mullion windows and leaded lights, they might have acted as miniature sunspaces, a magnet for solar gain on a sunny day, as opposed to the cavern-like space of the Great Chamber behind, which in contrast must have been cold and draughty. There is an interesting analogy between these and the modern conservatory, which acts as both a space to sit and enjoy high light levels for reading or activities (needlework, etc.) and a sun space to attract solar gain. The heat would have been transferred into the stonework (thermal store), to be reflected out again at night. One can only speculate that such reasoning took place for their construction, as in true vernacular fashion, and despite these great and large houses being now described as supra-vernacular, verging on polite architecture, little was written down.

The small house

Following the demise of so many villeins (lowly village inhabitants) in the mid 14th century due to the Black Death, the lords of the manor appear to have found it difficult to perpetuate the feudal system of strip allocation in large open fields of ridge and furrow, and thus started the process of amalgamation of strips into fields and the practice of leasing manorial demesne land to the survivors. This first and tumultuous agrarian revolution saw the birth of the yeoman farmer, a man holding land to the value of 40 shillings. The title has its origin in the concept of military retainer, the freeholder who always held social prominence over his more servile neighbours.

Photo 1.11 'Bayleaf', a Wealden house at the Weald and Downland Museum

Photo 1.12 Boarhunt Hall, Weald and Downland Museum, a largely reconstructed possible post-Black Death house (photo courtesy of the Museum)

These early yeoman farmers were forerunners of an immensely important breed of social climber, who peppered much of lowland Britain with their regional small house/yeoman house types in the 16th and 17th centuries. Initially freeholders, yeoman farmers went on to hold land as copy hold (i.e. for rent for three lives). Accompanying this massive reorganisation were many other changes, and this partly accounts for our lack of understanding of earlier house types. Cottage dilapidation and consequent desertion of villages swept away the buildings that are essential to our understanding of the origin of the yeoman house, itself so distinctive in its surviving forms. Particularly early surviving yeoman houses of around 1400 seem to indicate that initially they followed the form of the large house of the lord of the manor, i.e. an open hall, with floored ends. The Wealden House, surviving in quite substantial numbers in Sussex, Kent and Essex, indicates this to be the case, although we can never be entirely sure that we have not lost examples of rather less substantial houses of early yeoman farmers.

Photo 1.13 A reconstruction of a pre-Black Death house at the Weald and Downland Museum

The starting point for any discussion of traditional building forms is thus the open medieval hall, enough of which survive in their original format to make sense of their modification by insertion of floors and chimneys. There would certainly appear to be some element of cultural diffusion down to the next social level of the yeoman farmer with their floored house-place and parlour with chambers over. In addition, in the south-west of England there is the retention of the cross passage. This is a primary element of the medieval open hall, but again its origin as a threshing floor, or animal passage (a function that it retained in the medieval Devon long-house until possibly the 19th century) in an even earlier medieval period, can only be surmised. Its function in the open hall, as discussed above, appeared to revolve around controlling draught to the fire situated centrally on the floor, although one suspects that rather too much draught was involved. That it should be perpetuated in the Dorset 'cross-passage house', a typical example of the 17th-century yeoman farmhouse with the central stack constructed to back onto it, but with no obvious use, is a matter for debate. Its original use, for either animals or controlling the draught to the fire, would have been long forgotten, although it may have perpetuated a clear distinction between upper and lower echelons of the household. Another distinctive but opposing trait in other parts of Britain is the blocking of the cross passage with a centrally placed chimney-breast to form the baffle entry house. This is a very strongly represented house type over much of lowland Britain. The central stack in both planforms acted as vast storage radiator, as home to the continuous fire, hence its construction in brick in the vast majority of cases, but its location blocking the cross passage is far more sensible in terms of sustainable living (creating a central heat store) than backing onto the cross passage. The question then arises as to why Dorset retained the cross passage. Notwithstanding this conundrum, the yeoman houses were and still are eminently sustainable both in their construction and in their planform. Thick walls, and thatched roofs, made for a fair degree of insulation, and a central core of heating, with one central room always well heated, meant a concentration of resources where they were most effective. These houses were and still remain the epitome of the natural and sustainable house.

That one can trace back the origin of such houses to the modification of the open hall is an endless source of satisfaction, but it leaves many other questions unanswered for the many variations of planform that abound in various regions of Britain. The illustration of the number of different planforms in the Parish of Westbury in Shropshire is a typical example of what can be found in one small parish in England.

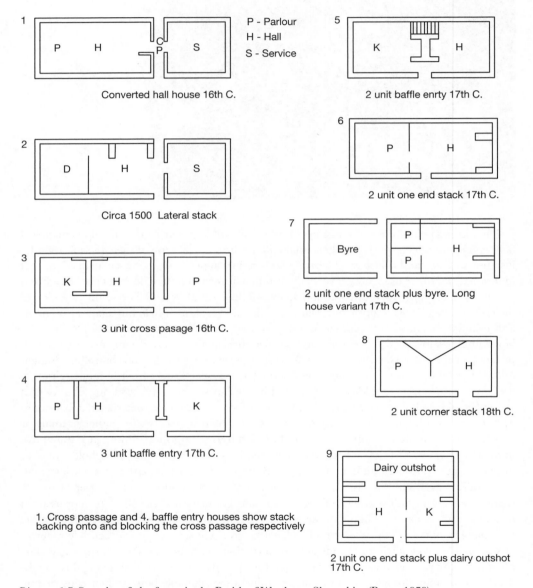

Diagram 1.7 Samples of planforms in the Parish of Westbury, Shropshire (Ryan, 1979)

The cottage

The social revolution which accompanied the end of the feudal period created other classes such as husbandmen and cottagers (and shopkeepers and artisans in towns) whose building forms must have been analogous to their villein predecessors, but whose primitive nature leaves little or no trace. It is difficult to comprehend that what is now understood by the word 'cottage' is actually 'estate agents' parlance for 'yeoman house'. True surviving cottages, the homes of landless labourers, whose villein precursors never made it to the rank of yeoman, rarely date before the 19th century and were either sub-divisions of yeoman houses (whose cellular format led to division down the truss (A frame) line into two or three cottages) or were somewhat fanciful creations of philanthropic landlords who swept away the hovels that may well have followed an earlier format of the vernacular house. Unfortunately these hovels were the very buildings needed to project forward from the flimsy homes of the villeins in order to construct any cultural diffusion, but they are long gone from our grasp.

The importance of planform or layout

The strong ethnographic basis for the dwelling is visible in the network of spatial relationships between rooms and the objects within them. Through the medium of probate inventories, itemisations of a deceased property by the immediate neighbours charged with the task, one reads the close relationship between the spatial quality of the room, and the everyday objects therein. The houseplace, always described as 'the house', had the principal fireplace with its andirons (fire dogs to support the upended log so it burns with air underneath) and gives an important clue to the everyday ritual behaviour within the environs of the ingle fireplace, 'the room within a room'. The concept of the ingle has its origin in the need to enclose the fireplace, to prevent smoke drifting into the chambers above the hall, during its many modifications to a completely ceiled room. Here was the source of all heat, light and nourishing hot food. The cultural significance of this most important area and of other spaces such as the parlour, used as a private retiring room, sleeping or work/store room, depending on social status, is at the very heart of what we understand as a traditional and natural way of life, passed down through many generations. It is as close as we can get to understanding the evolution of house types from the medieval period onwards, but also to understanding a way of life, geared to the seasons and the land, and passed down by word of mouth to succeeding generations. Even though a whole raft of precursors is lost to us, particularly those 14th-century yeoman houses that may have immediately post-dated the Black Death, it is possible to postulate that they were simplified versions, with an open hall divided into inner and outer rooms plus the central open hall as at Boarhunt Hall, Weald and Downland Museum. This is building is largely a conjectural reconstruction, with partitions of a simple wattle format.

Early classical influence (the Renaissance) and the start of the demise of the natural house

The great houses of the period around 1600 were precursors to the stately homes of the 18th and 19th centuries that followed.

The period around 1600 was characterised by the shift in emphasis of graceful living from ground floor to first floor, the latter having lofty rooms and tall windows. The ground floor hall was relegated to an occasional use of its original intention, for the mingling of the social classes, but more often than not as an entrance hall housing the grand staircase. The upper hall or great chamber in manorial households was used for the manorial court administration (or a separate courthouse was constructed), and in the great houses the most superior upper chamber or stateroom was allocated for visiting royalty. The creation of these so-called prodigy houses was the pre-occupation of their courtier occupiers. Elizabeth I was well known for her penchant for utilising the expenditure of others for her amusement rather than inflict her costly lifestyle on her own private household resources. These house types are also marked by a general increase in comfort with the introduction of numerous fireplaces to parlours and upper chambers, plus panelling and fireplace over-mantles, all to make draughty rooms cosier, together with decorative plaster ceilings. These 'improvements' were confined to the highest social status rooms, such as the great parlour and the great chamber. Such improvements at first sight appear to be counteracted by the floor-to-ceiling mullion and transom windows, 'more glass than wall' as at Hardwick Hall in Derbyshire 1590–96, designed by the famous Robert Smythson. Filling a whole wall with glass would certainly have resulted in a degree of discomfort in the English winter, but one also has to realise that such buildings usually formed on side of a courtyard, providing wind exclusion and something of a micro-climate.

The concept of the courtyard had many possible derivatives, including possibly an influence of ancient cultures, but one of the main drivers was that the location of these great houses was frequently in areas of wild landscape infested with wolves and outlaws. It should also be noted that the dissolution of the monasteries that took place from 1539 onwards had an influence on the retention of the courtyard plan. The typical religious establishment had its principle ranges of dorter, refectory, abbot's lodging and church arranged around a courtyard.

When these magnificent stone buildings were sold off as country residences, to courtiers who were holding the offices of state previously held by ecclesiastics, they required houses of a suitable status. In particular such men required a large house within a day's riding of the Court. The existing religious courtyard was modified without too much intervention. Longleat in

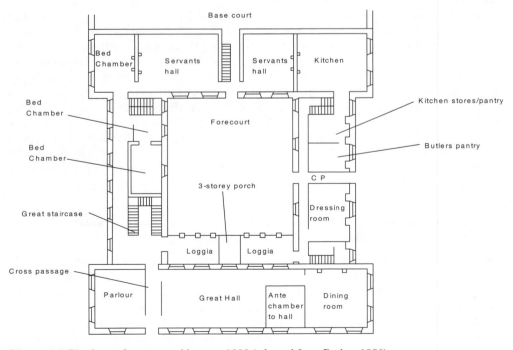

Diagram 1.8 Planform of a courtyard house c.1550 (adapted from Barley, 1980)

Photo 1.14 Woodsford Manor, Dorset – remains of a courtyard house

Wiltshire retained two courtyards, one of which probably represented the cloister garth of the ousted Augustinian canons. In addition, arranging the accommodation for household, servants, and visitors around the sides of courtyards (the forecourt being for the upper echelons and base court for the services) enabled cross-lighting of superior rooms, or lighting of the ranges that were two rooms deep. The loss of the courtyard tradition in the centuries that followed was very rapid, so that by the 18th century it was definitely out of fashion. A problem which besets large houses today reared its head, that of maintenance of such an extensive range of buildings. One can only bemoan the loss of this tradition, as wind-chill factor versus surface area of wall is a key feature of energy loss. It is similar to the loss of shutters in Georgian houses, important not only for security but also to prevent heat loss on cold nights. In general the loss of the courtyard house was a retrograde step, exposing those houses which demolished their base court service ranges (stabling, brewery, etc.), and often their forecourt as well (gatehouse, lodgings, etc.), to expose their main elevations to the wind-chill factor. The loss of the tradition also meant its loss as a cultural trait, which as a nation we would have done well to have retained in terms of energy conservation.

Courtyard houses inspired some of the finest supra-vernacular, yet traditional great houses of the period around 1600, such as Longleat in Wiltshire and Wollaton Hall in Nottinghamshire by Robert Smythson, followed in the early years of the 17th century by Hatfield House, Hertfordshire (in 1612), whose great staircase probably surpasses all others. Also worthy of mention is Hatfield's plan, with massive turreted blocks, whose only concession to an energy conscious way of life was individual apartments that could be occupied independently of the whole house. The sheer scale of the complete building, and the impossibility of heating it, must have inspired this apartment living and is a graphic example of early energy conservation. The blocks were conjoined with the central range housing the typical multi-tier porch, which had a hierarchy of orders (Tuscan, Ionic, and Corinthian), and the open-fronted loggia. This again is worthy of comment, in the context of sustainability. Such a feature was clearly designed with the Italian sunshine in mind, to provide shelter from the hot sun, and whilst it may well have functioned to keep wet, cold rain away from the ground floor walls, probably had little other real function in cold, wet England. This feature, together with the external 'stone wallpaper' in the form of 'pasted on' classical columns, crudely executed in terms of their classical proportions, graphically shows that these buildings were not part of a vernacular tradition. It is in these features that are typical of imported Italian renaissance, secondhand from Northern European sources (due to the discouragement of Catholic liaisons), that we see the first move away from traditional houses with a functional form. The move towards symmetry in elevations is also present, and the cross passage becomes subsidiary or is not present at all.

Although brick was used for the first time in great houses it did not displace the time-honoured popularity of stone, and in all events its seasonal production limited its use, dependent as it was on cold winters to break down the clay and warm summers to dry the bricks in open-fronted sheds. Timber framed structures did perpetuate a vernacular tradition, using a prodigious quantity of timber, but displayed the use of small curved boughs to make decorative panels. These were transformed into the decorative elements that infilled the small square panels creating an array of patterns such as bird's beak, quatrefoil, and diagonal bracing. Unwittingly this was a more sustainable use of timber, necessitated because large trees, which during the medieval period had been squandered to create cultivable land, were now in short supply, particularly due to ship building and charcoal burning. Examples can be found in Cheshire and Lancashire, such as Speke Hall in Liverpool.

In the first quarter of the 17th century there is an increasing acceptance of classical features such as corbelled-out columns executed in artisan mannerism, despite their rustication, as at Bolsover Castle, Derbyshire, and the introduction of the Dutch gable for pediments over roof projections. This style rapidly transposed itself into the Dutch gable topped with a triangular pediment, a classical format that was to have ever increasing significance. Internally other more significant changes were afoot. At Raynham Hall, Norfolk, 1622–23, the plan is a combination of the ground floor ceiled hall, with screens passages at both ends of the hall and smaller rooms surrounding it. From here it was a simple matter to transpose this into the plan of a Palladian villa by turning the hall through 90 degrees. Thus the Palladian ideal of a central room or saloon with small rooms grouped around was easily incorporated into the English system of the great hall with great chamber above, and with it the transference of external symmetry to the interior. Any concept of the traditional English house was subsumed in a veneer of classicism which was to dominate the high social status house for many generations to come. There had in fact been an even earlier purely classical house, the completely one-off Old Somerset House in the Strand, built by Protector Somerset in 1548–50. This was the first house in England of entirely classical design. Only the very haziest of illustrations survives, showing a house with a three-storey central narrow range, flanked with two-storey ranges, with a mixture of cruciform (transom and mullion) and sash windows. The latter must have been the earliest known of their type, with thick glazing bars, the sashes operating on a sliding mechanism, secured possibly with wedges. It would be the late 17th/early 18th century before such windows would be seen again.

Another fashionable innovation occurred at the very early date of 1653, at The Vyne, Hampshire. The two courtyards were demolished, and on the north side where the base-court had included dormitories for retainers, there was built a grand portico, in a style that is associated more with Palladian architecture of perhaps a hundred years later. In other great houses, the total enclosure of the base-court was in time reduced to a walled enclosure with at most a gatehouse. Throughout this tumultuous period of architecture for the great houses and even large houses, whose manorial owners emulated their aristocratic betters, the building of the traditional yeoman house continued apace, developing regional characteristics in their planform. Materials continued to be indigenous to the locality, as transport of materials any distance continued to be difficult and expensive. The same could be said of the great houses, their walls and roofs reflecting what could be obtained locally, even though their owners and designer/craftsmen may have travelled some distance. It was thus the yeoman farmers who enjoyed some degree of comfort on a winter evening, from the wood burning in their ingle fireplace, whilst their grand contemporaries possibly shivered in their opulent and larger great chambers.

Traditional building did not die out in the period up to 1700, it simply took on a less functional and fanciful disguise in the form of embellishment to façades of the houses of the rich and powerful, whilst the yeoman farmer perpetuated the functionality of the house. In Dorset the plank and muntin (stud) screen, which separated the house-place from the retained screens passage, was indicative of careful thought to the comfort of the occupants. So useful was this device, at all levels of the house, providing screening to first-floor corridors (a seemingly innovative feature as the rest of the country enjoyed little in the way or privacy at first-floor level) and for the basic division between upstairs rooms, that it was to continue into the 19th century for the same purpose. The methodology of construction changed into alternating boarding, then butt boarding against a horizontal member (ledged).

Photo 1.15 A newly discovered plank and muntin screen

The fashion or need for a fireplace for every room in the high social status yeoman house, including all upper chambers, made its appearance, although one suspects they were more of a status symbol than in universal use. Contemporary accounts in the 18th century indicate that they were only in use in times of illness, childbirth, and exceptionally cold weather (Beresford, 1978, quoting Parson Woodford, 1758–1802).

The 18th century – general trends

The upper echelons of society continued to move away from the traditional building, not so much in terms of its basic construction such as thick walls, acting as a thermal store, but in terms of its embellishment and house planning. At the same time it is clear that less thought was being given to constructing houses in tune with climatic demands.

The courtyard house, so attuned to the latter, suffered a dramatic fall from grace to be replaced by houses that appeared to rise out of the ground, the basement decorated with rustication, creating an illusion of light and shade by deep indentations between the stone blocks. This basement contained service rooms underneath stone or brick vaults, the function of which would have previously been in the base-court. Why or how this transition occurred is not really understood, although it is thought by some architectural historians to be a French influence. The trend was possibly influenced by the publication of *Vitruvius Brittanicus* (The British Architect) by Colen Campbell, which first appeared in 1715. This location for service rooms would in turn be replaced by service wings in the late 18th century. Doubtless food arrived at table somewhat chilled in either instance, hence the need for heated cupboards to be found in some 18th century ante-rooms adjoining dining areas.

How they were heated is often a mystery; it is possible that hot bricks from the cooking range were used, but their lead lining gives some clue as to their usage. The use of rooms generally in such extensive houses can at best only be speculated, as in many an instance the use changed at frequent intervals. Commentators of the period remark that there was a room for breakfast, another for supper, another for dinner, another for afternoon use, and the great arcade for walking. That rooms are named today by association with their furnishings, for example 'the

Photo 1.16 Heated cupboard at Milton Abbey School, 18th century

red room', is a source of some regret and not a little frustration, as usage has far more relevance to understanding than the colour of the walls or the many paintings in the room. A feature of late 17th- and early 18th-century houses, the apartment, was often broken up into other uses by the late 18th century, so that two or more adjacent rooms might be conjoined by means of demolishing their party walls, creating a large space for, say, the ubiquitous library. A library was the hallmark of any gentleman country house owner by the middle of the 18th century, whether he could read the books or not. This development can only be viewed as a reversal of sustainability. The only room which appears to remain unscathed during these 18th-century improvements of houses of 15th- or 16th-century origin is the long gallery. Peculiar to English country houses, because of the vagaries of the English climate for outdoor exercise, it maintained its use even in those houses that were remodelled in the late 17th-century Baroque or early 18th-century Palladian style, retaining its 17th-century panelling. The Vyne, Hampshire, is a prime example. Its primary use became that of a gallery for paintings and sculpture, although some house owners considered purpose-built galleries more to their benefit, as at Petworth, West Sussex.

What is obvious is that during the 18th century country houses became more attuned to the ostentatious display of paintings and sculpture, and thus simply evolved into gigantic galleries rather than addressing the functional basics of living, such as keeping warm and eating. This revocation from the cosy apartments of the late 17th/early 18th century in favour of lofty spaces for display is not documented as causing any particular personal grief, but it must surely have done so for the more elderly members of the household. The problem was largely overcome by retiring to the nearest fashionable town for the winter season to engage in socialising in a far more cosy townhouse, leaving the heating problem behind and the furniture in dust covers.

The obsession with grand staircases continued, although back stairs were always provided to allow servants to carry water for washing and fuel to bedrooms, and escape unseen with the chamber pot from the close stool.

Throughout the 18th century there was a change in the type of gentry who occupied a country seat, from the traditional land-based entrepreneur, dependent on agricultural rents, to the owners who depended on business, so that by the end of the 19th century the latter was the dominant force.

Photo 1.17 The Vyne, Hampshire, showing the early portico

By 1800 the country seat had become somewhat more modest, was often the home of the well-to-do merchant from the nearest town and had encompassed a more comfortable way of life. Gone were the trappings of state and the formal gardens with their parterres. Instead the principal rooms of the house on the ground floor were equipped with tripartite sashes, for extra light, or tall sashes with low cills or French doors for immediate exit onto the lawn, which now extended right up to the house. The exponents of this form of manicured setting, looking into the distance on a wild landscape, were Humphrey Repton and Capability Brown. The very rich engaged in fanciful exercises in creating follies, such as mock ruined castles, as eye-catchers. In some instances they embellished a naturally rocky landscape with an array of grottoes and obelisks, summer houses masquerading as temples, and romantic gothic cottages, even cages of wild animals that sadly did not live very long. Hawkstone Park in Shropshire and Stourhead in Wiltshire are two well-known examples, the former being the most flamboyant.

Internally, smaller country houses had more of interest to the reader concerned with traditional domestic life and its influence on traditional building construction. A typical c.1800 house is that of Pradoe, near Oswestry, Shropshire. A unique survival into the late 20th century, it presented a set of service rooms deliberately preserved in the late 20th century by its punctilious owner, Colonel Kenyon. It retained a servants hall, a butler's pantry, two kitchens, one of which contained a curry maker (a brick support for an iron basin reminiscent of our modern-day woks, possibly reminiscent of military activity in India in the 19th century) and large iron ranges. From the kitchen there was access to a game larder, bacon curing room, and general larders. From the back door could be reached a collection of external service rooms and areas, encompassing such activities as coal storage (a large rectangular low-walled enclosure), ash pit (cinders were riddled for paths, and in some households they were recycled on the fire, a perfect example of sustainability), and drains for kitchen waste, one of which exited into the pigsties adjacent. This close attention to utilising waste would put our recent efforts to shame. The range also included boot-boy's room, storage for paraffin and morning sticks (kindling), and servant's privies. A poultry run nearby has bee boles in the form of niches in the walls. A dairy room with intact tiled walls and cold slabs, cheese presses and butter churns, plus a 'boiling copper' for sterilising utensils, which also included a bread oven, and a drain in the floor, sat in the same grouping as the brewhouse and pigsties. The dairy house drain fed 'whey', the watery liquid left over when milk is curdled to make cheese, direct to the adjacent pigsties. The brewhouse was adjacent to the pigsties, to enable malted 'wort' (the leftover solids) to be easily fed to the pigs, with the pigsties adjacent to the slaughterhouse, all within easy reach of the rear service range, thus limiting effort spent on walking. The whole is the most stunning example of scrupulous attention to cutting down waste and wasted energy.

A hen house sat over the pigsties so that the pigs could defend the hens against marauding foxes. Nearby was a complete farm group, with a threshing barn, converted in the late 19th century to a three-door cowhouse to reflect the downturn in crop prices due to the importation of American wheat. The range included a granary, stables with 'bothy' over (accommodation for stable boy) and coach houses, plus a range of former open-fronted shelter sheds fronting onto a yard facing the winter sun. The walled garden enabled a continuous supply of fresh vegetables and fruit for the 'house' all year long, by means of a coal-fired hypocaust system set within the brick walls. The gardener's hut was an open-fronted stone enclosure where generations of workers had taken their lunch, hidden from view behind the artichokes. A carpenter's workshop, complete with a hoist door, glue pot and fireplace, was an essential component of estate building repair, together with an open-fronted timber store. All is indicative of total sustainability associated with a traditional way of life in traditional buildings, with animals and produce grown near to the house, processed in service buildings, cooked in purpose-built kitchens, with little or no waste. The message is clear. Even in houses built to impress, there was much attention to a sustainable way of life behind the scenes.

The obsessive attention paid to the importance of major rooms in the great and large houses of the aristocracy and local gentry in houses open to the public rather clouds understanding of the trappings of everyday life in their households as described above. It also overwhelms the rest of the story of traditional houses, which illustrate even more graphically that a sustainable way of life proliferated in every large or small village anywhere in England. That they are many and varied is without doubt. It was adequately illustrated in a lecture to the Vernacular Architecture Group by Machin (1994), who stated 'there is a bewildering variety of minor domestic architecture, as illustrated by a huge variety of building materials, plans, elevations, and dates'. One of the most important points made by Machin is that houses actually represent an enormous capital investment, and that means a diversion of funds from the purchase of stock or land. In addition, the primary build tends to have been influenced by tenurial history. Traditional houses of the 16th and 17th centuries that survive in great numbers in any one village may well have enjoyed freehold tenure, which put money into the freeholders' pockets rather than that of the lord of the manor, leaving a surplus for investment. Much also depends on whether villages were open or closed. In the former the landlords, often absentee, encouraged the settlement of poor craftsmen and labourers, housing them by dividing obsolete farmhouses into tenements, and extending some into irregular terraced rows. This was often done by simply infilling between existing free-standing yeoman houses. Dorset abounds in such villages, which at first sight make no sense to those skilled in understanding traditional buildings in other parts of the country. The village of Stalbridge, near the Somerset border, is such an example. This is in sharp contrast to the Midlands where closed villages abound, and where the lord of the manor and the wealthy yeoman farmers held complete sway. The clearly open village of Hazelbury Bryan, consisting of seven separate hamlets and also in Dorset, had an abundance of building trades as well as the usual range of tailors, shoemakers, chair and saddlery makers (Machin, 1994), such as would be found in any small market town. Machin describes it as 'an island of economic freedom in a sea of restrictive estates, the latter containing only opulent tenant farmers and their impoverished labourers.' A typical small house in the

Photo 1.18 The Knoll Bellypot Lane, Pidney, a striking example of entrepreneurial activity in the open village of Hazelbury, Dorset

hamlet of Pidney is illustrated but is sadly no more, such is the fate of key examples of houses which illustrate a way of life in an 'open' village.

Such a stark comparison is also hinted at in *Lark Rise to Candleford* (Thompson, 1945) where the norm pervades, that is, the concentration of artisans in the town of Candleford and the farm labourers in Lark Rise, with the exception of the former itinerant mason. Lark Rise, and its many equivalents, was one of those 'open' hamlets created by a demand for housing for poor labourers squeezed out of a nearby closed village, which those more wealthy shopkeepers in the town of Candleford exploited by erecting poorly constructed cottages to rent. That some of these were the successors of a yeoman house already on the plot, abandoned and let to a cottager, is also plain from this popular text. Yeoman farmhouses became the victims of engrossment, the practice of the reduction of many small farms into a few large ones. Landlords were particularly partial to this practice, and on the Bankes Estate at Kingston Lacy, near Wimborne, Dorset, Machin (1994) has discovered that as late as 1851, 26 freeholders survived. As soon a family died out or fell on hard times, however, because they had not diversified into by-trades, the Bankes Estate purchased the land and added it to larger farm. The house was either demolished or used to house a labourer's family. In the 1960s and 1970s, large numbers of the yeoman houses subdivided into cottages for farm labourers were once again united into a single house, their new owners convinced that they had always been cottages.

The import of this description of the state of traditional housing in the countryside by the 19th century is to emphasise how very precious is the survival of yeoman houses, and even more importantly the survival of 19th-century cottages. The latter are in fact an even more rare survival, as they were never very prolific, but how many times does one hear 'It is only 19th century so it is not so important!'

More pertinently, the process of engrossment continues with smaller farms being bought up to make larger farms and the house being left stranded, and this is a process that has continued throughout the late 20th century/early 21st century. In the mid 20th century, period houses were left to rot, and it is only with the inflation of house prices that the worth of traditional buildings has been seen. Even then, their essential features were often lost in the name of modernisation, and works were done that actually reduced the efficacy of their construction, such as lining out walls with foil-backed plasterboard. The latter effectively blocks the transmission of warm moist air outwards, creating zones of condensation in cold spots that

Photo 1.19 A rare surviving example of an untouched early 19th-century canal workers cottage

produce black mould. This is just one aspect of many that included removing traditional roof claddings such as thatch and replacing with concrete tiles, removing internal walls and making large draughty spaces, initially heated with cheap 1970s oil but now a hugely expensive way to heat, removing tight newel staircase and replacing them with staircases that took up all of a room, to name but a few.

The classical house and the demise of the traditional house

The movement away from the traditional house to the classical house by the upper echelons of society is widely attributed to one Inigo Jones. Born in 1573, Jones had a meteoric rise to fame from the humble son of a clothmaker and early training as a joiner. His skill at producing stage sets led to an appointment at Court for the same, and a trip to Italy with the Earl of Arundel. What followed was unprecedented: the meeting with pupils of Palladio, and becoming the first Englishman to make first-hand studies of Roman antiquities, including those by the Roman architect Vitruvius, was to set the seal for the advocacy of classical architecture in Britain. This was cemented by his appointment as Surveyor of the King's Works in 1615. What followed were some of the landmark buildings of classical architecture, such as the Queens House at Greenwich (begun 1616), containing a 12 m (40 ft) high cube in the centre, an open loggia at first-floor level, plus an internal balustrades walkway around the cube. This was rapidly followed by the Banqueting House at Whitehall, added to the Tudor Palace, as a purpose-built setting for Court entertainments. Also a double cube, its exterior was a masterpiece of two tiers of columns (composite set over Ionic) applied to rusticated masonry (deliberate emphasis of the arris of stone blocks to create light and shade), paired pilasters at each end, and the whole enriched with garlands (festoons) and masks. His contribution also included the first public square in London at Covent Garden, the Queen's Chapel at St James's Palace, the double cube room at Wilton House in Wiltshire and the innovative portico at The Vyne in Hampshire. These landmark buildings were to determine the future of classical architecture and Jones's influence made possible the careers of those who followed in the mid 17th century, including John Webb (his pupil), Sir Roger Pratt and Hugh May.

Classical orders and rustication became the most desirable features of buildings. Classical columns were surmounted by simple corbelled-out capitals (Doric), or a feature that resembled a pair of ram's horns or volutes (Ionic), or the much more elaborate Corinthian capital, with a cascade of acanthus leaves sprouting from the base of the capital. All was surmounted with an entablature decorated with metopes and tryglyphs, the latter resembling tassels, beneath a corbelled-out head, itself decorated with dentils. There was strict adherence to the proportions of the height of each element of the pillar. The rustication was created by extreme attention to light and shade, not just on quoins but also on the basement area of the building.

Pratt designed Kingston Lacy, near Wimborne, Dorset, and the ill-fated but stunning Coleshill House, near Swindon, in 1650, one of the finest examples of its type destroyed in a fire in 1952. Hugh May designed the royal apartments at Windsor Castle and Eltham Lodge in Kent in 1664, in the Dutch Palladian style. Essentially, however, the style began to epitomise what the Victorians later termed the Queen Anne House, a house of two storeys of nearly equal height, a double pile plan, a great staircase dominating the entrance hall, and a corridor running down the middle. The whole was surmounted by big dormers, a balustraded parapet and very tall chimneys. Windows had alternating heads of triangular pediments or segmental moulded arches, or flat heads (pulvinated) supported on console brackets. In addition the age of the portico was born, although it was not to come into any degree of prevalence on such major houses until the early 18th century.

Diagram 1.9 A typical Queen Anne façade with pulvinated and triangular headed windows

The age of Baroque

Before true classical architecture in its purest form in the early 18th century was to once again rear its head, it first had to go through a somewhat flamboyant phase – the Baroque. The terms was again coined in the 19th century to describe a style that evolved in Italy around 1620 as heavily influenced by the Roman Catholic religion. The word means misshapen or irregular, but was in fact the creation of sculptors rather than architects, three of whom in particular have the romantic sounding names of Bernini, Borromini, and Cortona. All were born in the late 16th century and active in the mid 17th century. They influenced the building of churches in particular on a vast scale, with curved forms. The whole was designed as an exercise in creating buildings that exhibited 'architecture of movement' and spatial innovation, accompanied by a complexity of richness in materials and dramatic effects of lighting. The use of colour and decoration on walls and vaults was used to create false architectural perspectives (trompe l'oeil), and the whole was an exercise in ingenuity and fantasy.

The form spread to France by the end of 17th century and became the essence of some of the great chateaux built at the time, and from then to Austria, Germany, Hungary, Poland, and Russia. In many of these areas it never ended, particularly for churches, but in England its progress was halted by true Palladianism due to the influence of Lord Burlington (see below). This was despite the best efforts of Nicholas Hawksmoor and Sir John Vanburgh, and also Sir Christopher Wren, who in the late 17th century led the movement for Baroque in England. Hawksmoor and Vanburgh created Castle Howard, a prodigious pile in the Yorkshire landscape later to be used for the famous television adaptation of *Brideshead Revisited*. Blenheim Palace, Woodstock, Oxfordshire (1705–24), was deliberately created as a monument, rather than a home, to the Duke of Marlborough, as grateful thanks from the nation following the Battle of Waterloo. Its original concept dwarfed its eventual execution and we are left with only two wings surrounding a great courtyard, although two such courtyards were envisaged. This was followed by Kimbolten, Huntingtonshire, Kings Weston, Gloucestershire, Seaton Delaval, Northumberland, and Grimsthorpe Castle, Lincolnshire. All these buildings exhibited a cyclopean grandeur utilising massive Doric pillars and rusticated masonry. They were not homes in the accepted sense of the word, more palaces; places to visit for the weekend house party rather than houses of traditional construction. They used gargantuan quantities of stone, and their layout was tortuous and as far removed from the functional working farmhouse as such houses ever could be. The local farming class must have been mystified by the sheer opulence of these landmark buildings.

Wren, who was in reality an astronomer/physicist, and certainly in the beginning an amateur architect, utilised the style for his famous rebuilding of St Paul's Cathedral (1665–1710) on the site of the medieval Cathedral destroyed by the Great Fire of London. His main contribution was the creation of a dome that no longer relied on stone, but was an octagon converted into a circle, which carried a light wood and plaster dome. He also reconstructed 51 London city churches, including St Stephen Walbrook, whose interior is embellished with Corinthian columns, each one of which bore the same baroque hallmarks of spatial quality and grandeur. His other notable works include Chelsea Royal Hospital, the south and east wings of Hampton Court, Kensington Palace, and Greenwich Palace. His Oxford Sheldonian theatre designed in 1663 was modelled on the Roman theatre of Marcellus. Associates included Robert Hooke, William Talman, and William Wynde, known more for their country houses, Talman in particular for the glorious Dyrham Park, Gloucestershire. The baroque country house formulae of giant pilasters, an attic storey with a flat parapet, and spreading pavilions was for a brief time the norm. These features were reflected in the work of other baroque architects, such as Thomas Archer (1668–1743) and James Gibbs (1682–1754), although the latter is more famous for his *Book of Architecture* (1728), and the notorious Gibbs surround, a deeply rusticated door surround with a large keystone (Calloway, 1991).

Other more parochial baroque architects included William Wakefield of Yorkshire, William Townend of Oxford, and Francis Smith of Warwick. They built numerous houses for the lesser gentry in their local areas, in an emasculated Hawksmoor–Vanburgh style, well into the mid 18th century and certainly into a period when true Palladianism was once more in vogue for the aristocracy. This is particularly true in the West Country where the Bastard Brothers of Blandford, Dorset, and Nathanial Ireson continued to reconstruct whole towns and churches in the baroque style, long after it had been forgotten in London.

Photo 1.20 Stock Gaylard, Kings Stag, Dorset – an unspoiled early 18th-century house of surprising simplicity

31

The result of this architectural innovation were churches and country houses whose very being was the antithesis of comfort and energy conservation; in short, they paid little or no heed to the rigours of the English climate. Vast church interiors, whose volume exceeded or equalled that of the minster churches, and double cube halls in country houses must have been icy cold in winter. Kingston Maurward, Dorchester, has such a double cube and a massive and opulently decorated fireplace, which must had a hard job to heat the space above it. Today, the space acts as a reception for a successful college, its comfort levels boosted by central heating. That some cognisance of the English climate was taken is evident by the 'port cochère', a massive projecting portico underneath which carriages could be drawn to shelter their occupants as they alighted to enter the building. In general, however, it is difficult to envisage what the temperature must have been like in the extremities of a cavernous room away from the fire. It would appear that the demands of fashion surmounted comfort levels to the detriment of the occupants, and more surprisingly they allowed it to do so.

The Palladian movement

The Palladian style of architecture, severe and purist classicism, was in reality a reaction by men of taste and influence against what they deemed to be the opulence and sheer extravagance of the Baroque style. It was a political move, the Whigs against the Jacobite Toryism that had begun to assume less importance with the death of Queen Anne in 1714. This is the date from which true Palladianism is deemed to have developed. The Whigs associated the new style with a 'natural' classicism as opposed to the false and artificial style of the Baroque. Thus Palladianism became associated with major Whig landowners of whom Lord Burlington and the Earls of Leicester and Pembroke led the field. Burlington became the premier figure, and was hailed as almost a high priest of something akin to a religion. His disciples were the long gone Palladio, Vitruvius, and Inigo Jones.

A series of new texts launched the movement, notably Colen Cambell's *Vitruvius*, Leoni's *Architecture of Palladio*, and William Kent's *Designs of Inigo Jones*. With the exception of Cambell's Burlington House in Piccadilly of 1715, the movement was one associated with the rebirth of the more severe and classically proportioned country house, such as that at Wanstead, near London (1714–20), whose giant portico was one of the first after the one-off at The Vyne, Hampshire. The most significant building was Chiswick House, a collaboration between Lord Burlington and William Kent in 1726. Its hallmark feature was the giant rotunda with lunette windows to light an octagonal hall below, surrounded by small rooms of varying shapes whose spatial quality reflected those found in the baths of Ancient Rome. The front had a giant portico and flanking staircases to reach the main floor above the basement service range. This magnificent centrepiece is flanked by only two tall windows. It was modelled on the Villa Rotunda by Palladio, and its design is every bit as dedicated to the hot Italian sun as this Italian villa, the limited windows creating a dark interior. The building was clearly nonsense in terms of the British climate, its interior resembling an icy tomb in the depths of winter, yet its stark exterior and Palladian window was to influence the building of great houses throughout the 18th century. It must rank as one of the least sustainable types of architecture ever invented.

The surprising element was the use of Baroque for interior decoration, the problem being that the visiting gentry had very little access to Palladio designed interiors, so only had their knowledge of Baroque to fall back upon. The real purpose of the building, and indeed of many of the Palladian great houses that followed, was to display a collection of paintings and sculpture plus the many illegal Roman artefacts acquired during the Grand Tour. The latter became the 'must have' of every son of a great landowner, as did somewhere to put the booty. Thus the house was in essence a large gallery, whose purpose was to invite visiting gentry to view the contents and at the same time be indoctrinated by the Whig message. Stourhead, Wiltshire, is sadly a much more limited example due to the re-arrangement of its interior in line with subsequent fashions, fires, etc, but still has a magnificent Palladian exterior, modelled upon Palladio's Villa Emo. With its perfect proportions and temple front it was the height of fashion. In addition, it had the wild and romantic landscape, the latter to reflect the natural Italian landscapes so adeptly painted by French artists such as Salvator Rosa and Claude Lorraine, hanging in the gallery. William Kent, as a garden designer, was much in demand to remodel existing formal parterres that had been popular for many of the earlier phases of great houses, into glades, wildernesses, and serpentine paths, complete with shell-studded grottos, obelisks, and urns. Any landowner blessed with a naturally undulating setting revelled in its potential to create such a utopia without undue cost, whilst those who did not enjoy this facility paid handsome sums for serpentine lakes to be constructed where previously there had been cow pasture.

Such garden reorganisation accompanied even modest houses, medieval in origin such as Rousham Park in Oxfordshire. Of considerably more impact and importance because of the political persuasion of the landowner, Lord Colke of Holkham, was Holkham Hall, which utilised the skills of Lord Burlington and William Kent to create a Palladian house of gigantic proportions in 1734. Its entrance hall, based on a rotunda, was on two levels, with an Ionic colonnade. Again, the interior was less severe than that of the outside, due no doubt to Matthew Brettingham being the architect. His contemporaries were Henry Flitcroft, Isaac Ware, and Roger Morris, all Whig-sponsored architects, born around 1700 and active in the first half of the 18th century.

They inspired a second generation of Palladian architects such as John Wood, Sir Edward Lovett Pearce, James Paine, and John Carr. Wood is perhaps the most significant as the whole of Bath is a testament to his vision of a city of crescents and squares, straight and serpentine streets, and continuous frontages of terraced townhouses designed to create the illusion of great houses, all set in a wooded valley. The two most famous building groups are the Circus (1754–70), suggesting the Roman Coliseum, and the Royal Crescent, with its continuous Ionic columns.

Photo 1.21 Cockington Court, near Torquay, Palladianism and classical revival all rolled into one

33

Unwittingly, Wood left a legacy of stone decay due to the propensity of water-laden air (condensate) to sink in the bowl that is formed by the valley, so encouraging it to mix with the sulphurous gases from the many coal fires of the houses to become a weak sulphuric acid. The Bath stone, an oolitic limestone (calcium carbonate), then converts into calcium sulphate on the surface. The rain-washed surfaces dissolve, whilst the sheltered surfaces break out into miniature volcanoes of erupting calcium sulphate.

Lovett Pearce did for Dublin what John Wood had done for Bath, and his Old Parliament House of 1728 is every bit as grand, as was the addition of the classical façade at Trinity College. A whole generation of English landowners around Dublin built accordingly in the Palladian style, much to the fury of their Irish tenants who saw only subjugation reflected in the resulting sinister and forbidding façades. John Carr was still in vogue in 1780 with his Buxton Crescent in Derbyshire and Basildon Park in Berkshire, all reflecting the Palladian style, as was James Paine at Nostell Priory, Yorkshire and at the garden room at Wardour Castle in Wiltshire. In fact, the Palladian style dominated the much of the 18th century as far as external elevations were concerned, the following influence of the neo-classicists largely concentrating on interior embellishment.

It is interesting to note that the townhouses of the London squares, the crescents in Bath and Buxton, and of course those that proliferated in every Georgian town in England, Ludlow in Shropshire and Blandford in Dorset being prime examples, were in fact cosy havens, easy to heat and maintain away from the bleak and austere country houses of the same period. These emporiums of style were of limited use in a freezing winter, and only served as venues for summer house parties, the focus of which was the concentration of culture in their galleries and their landscaped parks for early morning rides and afternoon tea parties. Georgian townhouses today still provide homes of warmth and elegance, capable of meeting the demands of a sustainable home, especially with the ability of their external walls to act as thermal stores and wick away warm moist air. Their country cousins, the great houses, remain what they have always been: large galleries in which to display the treasures of a nation, with only their service ranges paying any heed to sustainable living.

The neo-classicists

Once again, neo-classicism was a reaction to what had been seen as the overtly repressive style of the Palladian movement and the ability of the sons of gentry to travel abroad even further

Photo 1.22 Buxton
Crescent, Derbyshire

than Italy. The two principal architects were William Chambers (1726–96) and Robert Adam (1728–92). Of the two, the latter was possibly the most influential, dominating the style from 1760 to 1800.

Chambers is most famous for his Somerset House in the Strand, begun in 1776, and the Casino Marino near Dublin, Ireland. Interiors in essence became even more flamboyant than hitherto, with echoes of Baroque (by now termed Rococo), although Adam developed his own singular style of plasterwork that was both light and graceful. In essence there was now even greater attention to responding to the movement of the times, an increasing interest and awareness in archaeology from the 1750s onwards. The Society of the Dilettante had been founded in 1733–34 by a group of rich young noblemen in the process of making their Grand Tour. The aim was to promote Greek taste and Roman spirit, and although the movement started as a purist and archaeological approach, as propounded by James Stuart and Nicholas Revett, it was somewhat opposed by Sir William Chambers.

Robert Adam tended to stay out any such complex arguments, but was by no means a Greek purist and privately resented the restriction on his range of architectural sources. He was an ambitious Scot who, like his forerunner Inigo Jones, soared to fame from seemingly nowhere, although his background was a trifle less humble. His father had been a leading architect in Scotland, and Robert and his brothers took on his practice and travelled to Rome in the company of a son of one of his father's clients. He took care never to use the term 'architect' whilst there, as to have done so would have barred him from engaging socially with the aristocracy. Upon his return in 1758 he set out to create 'architecture of movement, gaiety, and variety', in sharp contrast to the rigid principles of sober Palladianism. In addition, his was architecture on a domestic scale as opposed to that of public buildings for display. It consequently appealed to a nation whose great houses still encompassed the Tudor, Elizabethan, and Jacobean. Adam frequently took such buildings and imposed upon them an eclectic feminine decorative style, inspired by Roman and Greek architecture, as well as the work of Vanburgh and Burlington. Adam thus specialised in interior decoration and structural alteration to existing buildings, creating rooms in contrasting shapes and sizes, made mysterious by columned screens and embellished with a stucco decoration in Etruscan and Pompeian style.

What followed are some of the most fanciful creations in England, notably Syon House in Iselworth, near London (1762–69), Osterley Park, also near London (1763–80), Nostell Priory near Wakefield, West Yorkshire (1765–75), and Saltram House near Plymouth, Devon (1768–69). Adam's external characteristics were domes over the saloon and entrance hall, free-standing columns creating a portico but with a decorative entablature rather than pediment, rusticated basements, and sweeping curvilinear staircases to culminate at entrance level. Delicate swags and a vertical hierarchy of windows with triangular pediments complete the picture as at Kedleston Hall, Derbyshire. Internally the plan would be a juxtaposition of rooms with features such as apses to reflect the shapes found at Roman baths. The plans were, however, quite cohesive with a central entrance hall and saloon surrounded by the normal suite of drawing room, library, ante rooms and even a bedroom. The planning reflects more the realities of family. Kedleston in particular was designed to have four pavilion-style wings, two of which were never built. The private wing to the north-east was domestic in scale, and there was a kitchen/laundry to the north-west. The former was essential, as with many great houses the interior of the main range was not for everyday use. For the greater part of the year it remained closed and shuttered to protect the delicate stucco embellishment on walls, gilding of the Corinthian capitals on the column screens or the gilded figures atop the columns in the Roman ante-room, as at Syon House. The furniture was set against the walls in case covers. This 17th-century style of private apartment, appearing again in the mid 18th century, mirrored that provided at Osterley, with its self-contained drawing room, bedroom, and dressing room, and was thus the nearest that Adam came to creating the sustainable home, easy to heat and maintain.

What followed was a second generation of neo-classicists who were to herald the eclecticism of the Victorian Age, albeit they were still steeped in the classical ideal. Of the most notable were George Dance (1741–85), Henry Holland (1745–1806), James Stuart (1713–88), George Steuart (1738–1806), Robert Mylne (1734–1811), and Thomas Hardwick (1752–1829). Mylne specialised in bridges and country houses; Hardwick in London churches and chapels; James Stuart in the Greek style; and Steuart in such masterpieces as Attingham Park in Atcham, Shropshire, and St Chad's Church in Shrewsbury.

Even more notable and destined to make an enormous architectural impact was James Wyatt, who together with his brother Samuel and near relative Jeffrey Wyattville had an involvement in an eclectic range of country house architecture, ranging from the re-incarnated Gothic style mock castle at Sheffield Park in Sussex to the pure classical style of Dodington Park in Gloucestershire (1798–1813). His lack of stylistic consistency did not stop there, as Fonthill Abbey in Wiltshire can testify. Designed as a country house in the guise of a medieval religious establishment, Fonthill was begun in 1796, only to have the tower collapse in 1825. What was at fault was a lack of understanding of the behaviour of traditional materials, notably brick when rendered with the newly innovative hydraulic lime renders. The latter had a propensity to act as a corset, restricting the tendency of brickwork to expand. Wyatt, however, had an even more notorious claim to fame, being known as 'Wyatt the Destroyer' on account of his prolific restructuring of Salisbury, Hereford and Durham Cathedrals, which was to herald an even more sweeping reform of the structure of numerous rural churches. It is true that most rural churches had received little attention since the last great movement of the Perpendicular, which had faded by the mid 16th century, apart from those very few modified by the Georgians, and that many services were probably being conducted in the rain. What was to follow, however, was destruction on the grand scale in the name of restoration of their medieval integrity.

The end of the 18th century saw a confused picture for classical houses, with an eclectic range of styles from the re-invented Gothic of Strawberry Hill (a former mid 18th-century small country house, extended and gothicised by Horace Walpole), to the re-invention of a form of vernacular – 'cottage ornee' for gate lodges and even 'playhouses' for the very rich, to the extraordinary creation in a mixture of Chinese, Hindu, and Gothic style of the Royal Pavilion, Brighton, in 1815 by John Nash. Equally at home with this or creating Gothic, Italianate or pure Classical style, Nash was the doyen of them all. He firmly believed in indulging his patrons in the delights of all of these styles, termed 'the Picturesque', and in particular merging the rural amenities of the country house park with urban living. The result was Regent's Park (1811–13), with its grand terraces and villas. Regent Street he created for the pleasure of the Prince Regent.

One architect out of all of this interesting assemblage of skills and perspectives stands out amongst the others, that of Sir John Soane. Although very much a designer of the late 18th/early 19th century, Soane, a pupil of Henry Holland, became Surveyor of the Bank of England and had some great notion, possibly inspired by the need to light a building without the excess of insecure sashes, that buildings did indeed need to have natural light from above. By doing so he showed an innovative approach to sustainable lighting, heralding the innovation of the roof light. Top-lighting, often in the spandrels formed by his starfish-like ceiling, became his trademark and he considerably altered his own house at 13 Lincoln's Inn Fields for this purpose, especially so as in fact his alterations were destined to ensure its future as a museum for the study of architecture and allied arts. This use still exists today, with a collection of drawings of his contemporaries, models, paintings, and sculpture, plus miscellaneous antiquities, all scattered in a somewhat random way. The rooms containing this collection are deliberate confections of spatial drama, on different floor levels, with mysterious sources of natural light enhanced by mirrors. In one sense one could perhaps describe him as the forerunner of the naturally lit and therefore sustainable building.

Soane had one other great quirk, that of stunning simplicity, the reduction of classical architecture to the bare essentials of scale, proportion, and symmetry with former overt pillars reduced to gentle pilasters. A bow front and a moulded cornice and frieze with incised Greek key motif were often his only concession to decoration. The Greek key decoration, dividing the storeys in place of a string course is both bold and gentle. His own house, Pitshanger Manor (1800–03) – which became Ealing public library and is now a gallery – was a classic example, as is the ill-fated Pell Wall Hall, near Market Drayton, Shropshire. This wonderful house suffered a tragic fire in the 1980s, revealing the construction of many houses of the picturesque movement: a brick substructure faced with Grinshill sandstone. Sadly it also resulted in the destruction of much of the plasterwork, together with the characteristic ceiling and top lighting. The author had the dubious honour of meeting the former owner preceding the fire, and the greater honour of showing the building to Sir John Summerson during the early attempts to save the building from demolition. It is good to be able to report that this was achieved through the auspices of the later Pell Wall Hall Preservation Trust, although it is largely the shell that remains as a testament to the genius of Soane.

The classical revival

Throughout the 18th century classical architecture was regarded as an expression of an intellectual idea associated with Rome. By the late 18th century and the early 19th century, most architectural connoisseurs were turning away from the complexities of Rome to what they

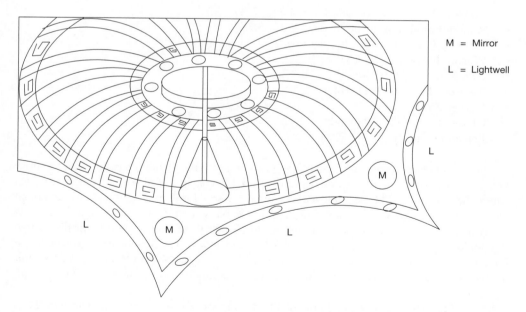

M = Mirror

L = Lightwell

Diagram 1.10 John Soane's starfish ceilings existed at Pell Wall Hall, Market Drayton, Shropshire

Diagram 1.11 A
typical neo-classical
frontage in Robert
Adam style

saw as the simplicity of Greece. In 1762 Stuart and Revett had published *Antiquities of Athens*, which, although it heralded the beginning of Greek Revival, was not to become influential until the early 19th century. In addition, there had been something of a reaction against the muddle (the picturesque) that characterised the end of the 18th century, and in particular what was seen as the extravagance of Nash's designs for Buckingham Palace. It can also be seen that Sir John Soane had also grasped the mantle of Greek design with his simple country houses with their Greek key motif, and in many respects was the herald of this new movement. In particular there was a desire for public monuments to have a more sombre appearance.

The era is reflected in the work of the following architects:

- William Wilkins, the son of a stuccadore, 1751–1815
- Robert Smirke, a pupil of Soane, 1780–1867
- Decimus Burton, son of a builder turned architect, 1800–81
- William Inwood, 1771–1843
- Henry Inwood, 1794–1843
- Thomas Harrison, 1744–1829
- Harvey Elmes, 1814–47
- Charles Cockerell, 1788–1863
- Benjamin Latrobe, 1764–1820

This seemingly inauspicious group of architects designed public buildings, and some country houses, to reflect the mammoth edifices of Greece. The style was also emulated in America where the Greek War of Independence was much admired for its analogies with the American ideal. In addition the period was one of economic expansion and a new system of democratic government inspired grand public buildings for which the Greek style was ideally suited. Benjamin Latrobe in fact immigrated to North America in 1796 where he was involved in the building of Baltimore Cathedral and the remodelling of the Capitol in Washington, DC, but not before he had designed the rather severe looking Hammerwood Lodge near East Grinstead, West Sussex.

Some of the greatest public buildings of all time appeared in this Greek Revival style. They include Smirke's British Museum, the Inwoods' St Pancras Church with its landmark caryatids (a replica of the stone ladies from the Acropolis), Harrison's Chester Castle (still used as an assize court), and the gargantuan St George's Hall, Liverpool, by Harvey Lonsdale Elmes. Wilkins produced the grandest country house of all at Grange Park, Northington, Hampshire. All were characterised by Ionic or Corinthian colonnades of simply breathtaking proportions, none more so than the ill-fated Euston Arch by the rather lesser-known Philip Harrison. Designed as the gateway to the north out of London, for the London and Birmingham and Great Western Railways, of all buildings of the period it was the one that most epitomised the aspirations of the Greek Revival, and yet it was seen fit to demolish it in 1962, in a misguided attempt at embracing the spirit of yet another new age. Its loss remains one of the greatest travesties of the 20th century.

The Greek Revival went on to grow in strength in the north of England and Scotland when its lustre had faded in the south, and became the basis for many a town hall such as that at Leeds, and the Harris Library and Museum at Preston. Glasgow and Edinburgh assumed the role of the 'Athens of the North', especially for public buildings of note.

Needless to say, none of these buildings was designed with sustainability as we know it in mind, but their use of stone on a grand scale and the sheer exuberance of their form has ensured their survival in the face of nature if not in that of mankind. In this sense they have a form of sustainability in the shape of an endurance which we could not hope to emulate.

The Victorians

The Greek or Classical Revival heralded yet another new age, one that was to see some of the greatest engineering feats in construction and some of the most confusing architecture ever to be witnessed. In a way it could be said to be the breakdown of architectural form due to an obsession with the style of an earlier age, notably the medieval Gothic. This is an affliction that seemingly has never left us, although now there is an apparent obsession with recreating Georgian mansions in the countryside. This practice that has little to do with recreating the spirit of the Georgian age or sustainable building construction, and much more to do with creating large and pretentious houses, often out of kilter with their setting.

Victorian architecture is convoluted and complex, demanding not so much an understanding of style as a thorough grounding in the ideals that drove the various progressions. These varied from the writings of eminent novelists such Sir Walter Scott with his Gothic horror stories, through the driving forces of architectural writers and scholars such John Ruskin and A.W.N. Pugin, to the engineered structures that were possible with iron and glass. Interspersed with all of this were revolutions in social and agricultural aspects of life, with the movement away from

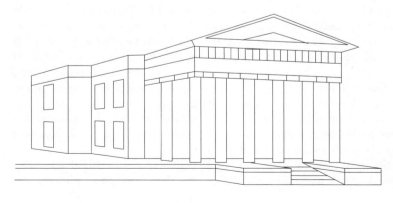

Diagram 1.12 Typical classical revival frontage of Cyclopean proportion

Photo 1.23 Workers' housing in Buxton, Derbyshire

the land and into factories of agricultural labourers, and the transport revolution characterised by the introduction of the railways.

The main periods are detailed below:

- **Early Victorian Gothic:** has its main impetus in the 1830s and is best described as Romantic Gothic, capturing the spirit of the 14th century with its flamboyant 'Decorated' style. Features such as ogee windows captured the spirit of the Middle Ages.
- **Italianate Revival:** 1840s–1860s.
- **High Victorian Gothic:** 1850s–1870s, a harking back to the pure form of 13th-century Gothic, regarded as the correct form for the Middle Ages and the ideal of all Antiquarians.
- **Later Gothic Revival:** 1870s onwards, a mixture of all manner of Gothic styles, often called 'Eclecticism'. At the same time was born a range of other eclectic styles for country houses, referred to as 'Old English Style' for those that utilised vernacular forms, Queen Anne, and neo-Georgian in general.

To expand on these general themes one needs to understand that Early Victorian Gothic was dominated by one Augustus Welby Northmore Pugin. His was a religious zeal fuelled by the belief that Gothic architecture represented 'the unchanging truths of the Catholic Church'. Notwithstanding his fanaticism that almost certainly accounted for his early demise, after completing several mammoth tasks including the Houses of Parliament, his beliefs were to set the seal for first the Arts and Crafts movement and in turn the Modern movement. These beliefs revolved around such concepts as that architecture was true or false, good or bad, and most pertinently honest by revealing its structure and function. He was also the first advocate of using natural materials. He produced some of the most astonishing interiors by embracing the principle that everything in a building should have a single design concept, be it woodwork, metalwork, tiles, wallpaper, or furnishings. In this he mirrored Robert Adam, who also created complete interiors down to the last doorhandle. Another primary aim was to create complete Christian communities of church, vicarage, school, and cottages.

Anthony Salvin (1799–1881) was another well-known contemporary, with an extensive knowledge of medieval domestic and military architecture. He pioneered asymmetry, the functional placing of chimneys and window openings, in the spirit of 'form follows function', and this became a feature of Victorian country houses.

The period generally was characterised by the revival of the Catholic Church in the 1830s and by the Oxford movement, which emphasised the spiritual role of the church. In turn the Cambridge Camden Society (Ecclesiological Society 1845) emphasised the correct forms of medieval building and liturgy.

The Italianate revival

During the 1840s to the 1860s Gothic was competing with the Italian renaissance, to such an extent that by the 1860s there was a veritable 'Battle of the Styles', with once again politics taking sides (shades of the Whig-inspired Palladianism). The Liberal politicians favoured Italiante, and the Conservatives favoured Gothic. Italiante style was also the southern alternative to the northern penchant for Greek Revival. George Gilbert Scott (1811–78) was so affected by this Battle of the Styles that he had been forced to amend his designs for the Foreign Office between 1857 and 1861, to suit the Italianate renaissance taste. The style resulted in some of the finest buildings in London: the Victoria and Albert Museum of 1859, the Albert Hall of 1866–71, and the Natural History Museum of 1871–81. A stylistic feature of the buildings was the use of polychromatic brickwork of various colours to create a pattern, but with an overarching use of red brick.

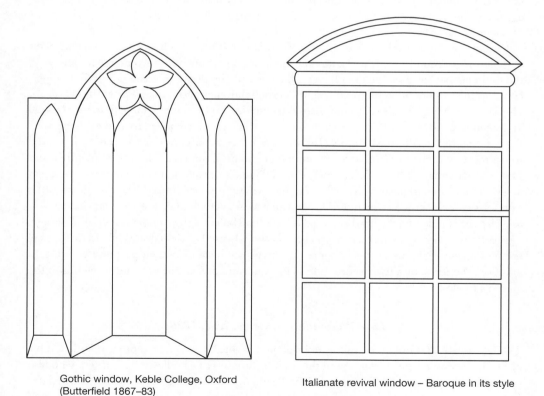

Gothic window, Keble College, Oxford
(Butterfield 1867–83)

Italianate revival window – Baroque in its style

Diagram 1.13 Battle of the Styles: Italianate versus Gothic window features

High Victorian Gothic, 1850s to 1870s

Inspired by a very pure reformed Gothic, that of the 13th-century, High Victorian Gothic represents the height of religious fervour and was characterised by a distinctly masculine and almost violent architecture, where polychromy was frequently used. Needless to say it was churches that were most affected, many being built in the 13th-century style with two centred arches and stiff leaf ornamentation. Architects like William Butterfield (1814–1900), a devout Anglican, specialised in the use of blue or black bricks in red brick walls, as at Keble College, Oxford. The building now stands as a testament to extreme faith as expressed in a strident use of two centre arches framing geometric windows set within a riot of polychromatic brickwork. William Burgess (1827–81), another devout follower of the teachings of John Ruskin (1819–1900), specialised in a rather different approach, taking real medieval castles and transforming them into fairytale confections with fanciful multi-coloured interiors as at Castle Coch near Cardiff, and Cardiff Castle. Ruskin was the driving force, producing a prodigious literary output as exemplified by his *Seven Lamps of Architecture* in 1849 (sacrifice, truth, beauty, life, memory, obedience) as well as *The Stones of Venice* (1851–53). Ruskin was to be a mentor for the Arts and Crafts movement, which marks the close of the Victorian period and the beginning of the early 20th century. Like William Morris, who was to follow his lead, he was deeply moved by the imprint of the human hand upon a worked surface, which remained long after the craftsman ceased to exist. Through it, he believed, he could read the joy and passion of the craftsman in his work. This concept was at the very heart of the natural house of previous ages and would pioneer a return to the use of natural materials modified by the age-old influence of the craftsman.

Ruskin also influenced other landmark Victorian architects, such as George Gilbert Scott and George Edmund Street. Unfortunately Scott was very much his own man with an eye to the main chance, and rather than being influenced by the desire to create natural buildings, concentrated his attention on modifying numerous churches and some 20 English cathedrals. He was driven by a belief that medieval architecture had more claim to fame than the classical architecture of Greece or Rome. Such was his influence that he enjoyed royal patronage, with the creation of the Alfred Memorial (1863–72) in Hyde Park, London, a structure that owed much more to the use of iron as a frame than any natural material, and the Midland Grand Hotel, St Pancras (1868–72). This newly restored building is a testament to the creative use of a very natural material, such as brick, in a somewhat unnatural form. Other architects using iron for the main structure include Alfred Waterhouse, who used rolled steel bars extensively as well as red brick, and the newly created terracotta. Hailed as the wonder material of the age, terracotta was indeed a natural material but was largely used as a decorative cladding for steel frames that were later to be proved somewhat disastrous because of their propensity to rust. In addition, the terracotta itself is prone to acidic gases mixing with rainwater to etch into the glazed surfaces.

Late Victorian Gothic revival, 1870s

The late Victorian Gothic revival saw a brief harking back to a softer, more romantic Gothic of the 14th century, particularly with the architecture of G.F. Bodley (1827–1907), who also pioneered the Queen Anne Revival, J.L. Pearson, and Giles Gilbert Scott's much later Liverpool Anglican Cathedral. Bodley worked closely with C.E. Kempe (1837–1907), a noted church decorator specialising in the design and production of stained glass.

Photo 1.24
Knightshayes Court,
by William Burgess,
one of the finest
Gothic revival houses
in England

Photo 1.25 Tyntesfield
House, near Bristol, an
even more elaborate
Victorian Gothic
house

The architecture of the Arts and Crafts movement

The Arts and Crafts movement was the most significant attempt ever to recreate the true spirit of the Middle Ages and the rebirth of the natural house, and is probably the most tangible of all the architectural periods in its contribution to sustainable architecture. Its architects included the eminent Richard Norman Shaw (1831–1912), who reacted against the Gothic and Italianate for country houses and produced an Old English style using local vernacular styles and materials, as did his contemporary Philip Webb. The result was buildings with tall chimneys, large tile-hung sweeps of roof, and mullioned windows with leaded lights. In the 1880s, Shaw developed a neo-Georgian style and Bryanston near Blandford, Dorset, now a school, illustrates his passion for the Queen Anne Revival. At the heart of these houses was not just the thick external wall of their earlier predecessors, to act as heat stores, but a structure of iron or steel girders, enabling independent planning on every floor, free of the restrictions of structural cross walls. Heating was via a gigantic coal boiler and iron pipes, and electricity was introduced. The natural house was becoming more of a comfort zone, despite the teachings of John Ruskin and William Morris extolling the benefits of a medieval way of life.

Edwin Lutyens (1869–1944) continued the spirit of the age right into the Edwardian era and up to the Second World War. Castle Drogo in Drewsteignton, Devon, shows as succinctly as any building the steady transition from the Victorian Arts and Crafts romanticism through to the severity of the Modern movement. Often described as the last great castle to be built in England, Castle Drogo commenced in 1910 and was finished in 1931. Designed to reflect the origins of the owner, Julius Drewe, whose ancestor had been a Norman baron, it epitomises the spirit of the Arts and Crafts movement and the Victorian passion for everything medieval.

The Arts and Crafts movement as a way of life

By the late 19th century the Industrial Revolution had borne many fruits, notably the use of iron, steel, and glass to create structures, and with it a host of new building types to meet the needs of the transport revolution and industry on a scale never seen before. The social revolution which resulted from this industrial revolution created a deeper division in society in the towns with artisan classes employed on a mass scale in large factories, and a new middle class who rose from a business background. Antiquarianism, once the preserve of the upper classes and the clergy, spread to the newly emerging middle classes. Nonetheless it was the immense power of British Industry that ruled the day, as seen by the Great Exhibition of 1851, set within the vast expanse of the Crystal Palace. The latter was a glass and iron structure on a gargantuan scale, housing everything that was innovative about British mechanisation and symbolic of an imperial power. All of this was far removed from the ideals of the natural house and the hand of the craftsman, and as a result there was a revolt against everything that was machine made, by certain members of society, notably William Morris (1834–96). In 1861 he founded Morris, Marshall, Faulkener and Co, bringing into the fold Dante Gabriel Rossetti, Ford Maddox-Brown, Edward Burne-Jones, Charles Faulkener and P.P. Marshall to manufacture furniture, metalwork, glass, ceramics, wallpaper, textiles, embroidery, and carpets.

The period was one of revived interest in all aspects of British decorative design. The same era was being influenced by the Shaker movement in America, characterised by plain and uncluttered interiors, with chairs hung on pegs on the wall when not in use, accompanying an equally simple and austere way of life, similar in many respects to that of the medieval peasant.

In contrast, Morris's view of the craftsman ranged over a wide spectrum of materials including ironwork and ceramics, but mainly on textiles such as furnishing fabrics and embroidery. His particular interest was historical patterns combined with British floral forms derived from a passionate interest in the British countryside. Thus his designs encompassed the use of natural English plants such as lilies and honeysuckle, whilst the conservatories and hothouses of his patrons were in contrast filled with exotics from far distant lands. British plant forms lent themselves to designs based on net-like frameworks. Ideas from one manufacturing technique such as textiles were used for the design in another, such as wallpaper and tiles.

In 1882, setting up as a rival, John Ruskin inspired the Century Guild, which lasted until 1888. It encompassed some of the most famous men of the era: William de Morgan, the potter whose tiles decorate many a fireplace of the period; Heywood Sumner, a designer; H.P. Horne, a metal worker; and Arthur Mackmurdo, an architect.

The year 1884 saw Richard Norman Shaw gather into his practice some of the most famous architects of the period, such as William Lethaby, Earnest Newton, and E.S. Prior. In a period when true labourers were no longer housed in vernacular buildings but in terraced houses of brick and slate materials transported by the railways, the middle classes embraced the vernacular revival espoused by these architects. Modest country houses for the middle classes were designed with massive sloping roofs and gables, overhangs at eaves, and long bands of mullion windows.

Materials such as a very rough render (harling) vied with stone brick and timber frame, often some if not all of the materials on the same elevation. Sloping buttresses on corners were a common feature, a sort of in-built repair feature, and windows were delicate wrought iron casements, with decorative catches.

Many of these characteristic Arts and Crafts houses were in towns such as Bournemouth, Dorset, and the half-timbered style transferred itself onto the edge of towns by the 1930s, becoming the infamous 'Tudorbethan' that characterised pre-Second World War ribbon development. Design of such houses did, however, come as near to the ideal of the natural house as would ever be seen again. Constructed of sturdy vernacular materials, albeit not always locally to hand, and with good ventilation and heating systems, most of which still revolved around the open fire, they epitomised a reversal in the trend for non-sustainable classical houses of the Georgian era with their leaky sliding sash windows and flamboyant detail such as applied ornament, totally at risk of being destroyed by the elements. Vernacular revival houses in contrast were akin to their medieval predecessors where 'form followed function' and all was practical. The trend for decorative interiors, however, continued with the setting up of the Arts and Crafts Exhibition Society in 1887 encompassing Walter Crane, the book illustrator, William De Morgan, with his tiles and ceramics, and Philip Webb and William Lethaby, the architects. It was set up as a form of revolt against the Royal Academy. Meanwhile in America, Frank Lloyd Wright was experimenting and being very successful with a different style of vernacular

Long low sloping roof, arch braced porch/loggia, elongated
windows emphasised with shutters, tall chimney and gablets

Diagram 1.14 Typical features of an Arts and Crafts house (after Newbold, 1923)

Photo 1.26 The Arts and Crafts at its zenith in this house in Milton Abbas, Dorset

Photo 1.27 An Arts and Crafts modification of an 18th-century house in Henstridge, Somerset – embellishment of an existing Georgian house

building. Again there was a clear emphasis on 'form follows function', with long low buildings with in-built shading for the hot Californian sunshine, but the timber used for their construction was machine cut. Wright did not believe that the hand of the craftsman was essential in the same way as his late Victorian contemporaries in Britain, enshrined in the virtues of 'craftsman made' in their buildings and interiors. Wright maintained that machines could produce objects that were simple and truthful to their respective materials. In Britain, William Morris's 'craftsman made' ideals also became somewhat clouded by the birth of the Aesthetic movement, which whilst it had many aspects associated with the natural way of life such as such as natural garden layouts, reflecting the seasons, and textile or wallpaper designs reflecting natural forms such as birds, animals or flowers, also led to a rather less natural interest in everything Japanese, Indian or Eastern in general.

Photo 1.28 Redcliffe Hotel, Paignton, Devon, reflecting a Victorian desire for Eastern architecture and a lifetime spent in colonial India

This led to huge quantities of such material being imported and marketed through the newly innovated Liberty's department store in London, due to an unprecedented demand by the middle classes. Whilst some of this material would be hand made by natives of the countries in question, some of it may well have been machine made and was thus the antithesis of the Arts and Crafts ideal. Rather than stimulate local home-produced crafts, it did the opposite. In fact, what occurred was an improvement in commercial manufacture of mass-produced goods and the modernisation of established furnishing firms by innovative machinery, for example for the block printing of wallpaper. In addition, of course, we now see that the importation of Eastern objects resulted in sea miles and hence carbon emissions.

Houses continued to reflect the use of natural vernacular materials in that they were built in red brick or brick coated with harling, or timber frames clad with hanging tiles, and in many instances some combination of two or more of these materials and styles. They also had steeply pitched roofs, mostly for architectural effect rather than as that all-essential surface facing into 'wind-driven' rain down a valley, and thus deflecting damp away from the walls. Many an Arts and Crafts house has been constructed with the appreciation of the view in mind, and thus regularly deposits the corner of its long, low roof cladding on the ground in storm conditions. Leaded windows set in large mullions and gardens with sunflowers, the latter being the revered symbol of the Aesthetic movement, completed the effect. Such building formed the basis of idealised communities, emulating Old English rural villages, but peopled by artists, writers, and architects, themselves steeped in the expectations of the new middle class but emulating their peasant forefathers in a bygone age. Such communities, did however, became the inspiration for the garden city movement, and model villages for workers in factories such as Bourneville, near Birmingham, did reap some of the benefits of the thinking, particularly with the provision of enclosed gardens, with staggered plots so that a more rural setting for each house could be achieved, as well as the attendant buildings such as a church, Church of England school and Sunday school. Even small market towns saw this as a way of developing, and a notable example of an Arts and Crafts inspired community exists on the outskirts of Bridgnorth, Shropshire.

More notable examples are the garden cities of Letchworth and Welwyn in Hertfordshire and Hampstead, a suburb of London.

Architects of the later period such as M.H. Baille Scott and C.F.A. Voysey became notable, inspired perhaps by The Red House in Bexleyheath near London, one of the first Arts and Crafts houses for William and Jane Morris, built in 1860 by Philip Webb. In addition, it had been built with the intention of being the centre of a romantic community of artists and craftsmen, although sadly this never really materialised. They were followed by the indomitable Edwin Lutyens, active from the 1890s to the 1930s, who embraced asymmetry, the use of stone interlaced with thin bands of tiles, handmade brick and oak mullion windows. These designs live on, not only in their grand country houses but also in text books such as *House and Cottage Construction* (Newbold, 1923), whereby the spirit of the design was adopted in late 19th-century villas on the outskirts of towns and villages throughout the land. Ponderous but tapering upwards chimney stacks, in long low sloping roofs culminating in porches supported on massive timber and arcaded elements give a feeling of unsurpassed solidity and security. Elongated mullion windows with their more fragile leaded lights set within wrought iron casements with cast iron fretted catches echo a previous age, all redolent of a traditional house built using craftsman excellence. In this way the Arts and Crafts movement sought to redress the balance lost by previous architectural eras, and create a truly natural house.

An architect who excelled in all of this but at the same time can be said to lead architecture almost seamlessly into the Modern movement was Charles Rennie Mackintosh. Active during the early years of the 20th century, his designs reflect all the essence of the Arts and Crafts movement together with a gripping sensationalism in respect of his interiors. In addition, he firmly embraced the principle of Robert Adam and A.W.N. Pugin in that he continued the basis of the design into the furniture, fixtures, and fittings. Thus houses like Windyhill in Kilmacolm, designed by Mackintosh for William Davidson in 1900, Hill House in Helensburgh, for the publisher William Blackie in 1900, House for an Art Lover (his own creation, unfettered by the demands of a client) and the Willow Tea Rooms, both in Glasgow, all embraced extravagant interiors. The tulip motif dominated, be it in paintings on walls, executed in metalwork, incorporated into light fittings, or on overmantles on fireplaces, as well as in the fretwork on his characteristic tall-backed chairs, and in stained glass. Despite this seemingly overt interest in decoration, Mackintosh's interiors did have form and meaning as seen by the L-shaped sitting room at Hill House. Here there is a summer end with a long window onto the garden, furnished with a window seat and book racks, whilst at the winter end there is the fireplace. Such a commonsense approach, born of a vernacular tradition, has been lost in sitting rooms today, which try to be all things to all people and often fail abysmally.

Even Mackintosh's St Matthews Church, with its characteristic Arts and Crafts tapering tower, had the tulip motif on the lectern and a hint of it in the curvilinear tracery of the windows. His Glasgow School of Art building (1897) has the most magnificent wrought iron window supports surmounted by a floral motif. Far from being purely decorative, however, they have a real structural function of supporting the metal transoms, in an exposed location subject to high wind loading. These features underline the 'form follows function' principle that is the basis of the Arts and Crafts movement, and indeed the natural house. The Glasgow School of Art is a unique survival in that it retains it original function, surely a testament to its fitness for purpose in its initial construction, as well as now having the role of housing the collections of Mackintosh furniture and fittings.

The Modern movement – the final demise of the traditional house

Charles Rennie Mackintosh and John Soane have a unique place in history in that they have some hand in the birth of the Modern movement. Viewed simplistically, Mackintosh designs have a strong adherence to 'form follows function', and Soane produced uncluttered elevations with simple classical elegance. In his Glasgow School of Art, Mackintosh showed that even though he was using traditional materials such as brick and stone, the constructional elements were very much to the fore, with the structure of the building been clearly seen and appreciated. This building, above all others, can be seen to be a transitional building, straddling the concepts of the old and the new. Even Edwin Lutyens embraced the Modern movement in the early years of the 20th century at Castle Drogo in Devon (1910–1930). In fact the theories of the Modern movement were considered well before the establishment of any of the well-known architects of the 20th century. Pugin was adamant that there should be no features of a building which are not necessary for convenience, construction, or propriety, a fact reiterated by Nikolaus Pevsner in 1968 in his treatise on *The Sources of Modern Architecture and Design*. In addition, the Crystal Palace had been a massive influence in that it showed the scope of the use of iron and glass, a form of construction that would eventually translate into the use of concrete, steel, and glass. The principle of 'truth to material' was a founding principle of the Arts and Crafts movement, but in fact had its birth in a much earlier architect, Viollet de Luc (1814–79), a French architect and theorist of the 19th-century Gothic revival. William Morris was instrumental in underlying the fact that architecture was not the sole province of the upper classes but could be expressed by the hand of the humble craftsman. This was a move well away from the elevated status of architect enjoyed by those who were privileged to enjoy the patronage of the wealthy in the 18th century.

The early part of the 20th century was not, however, completely swept up in the birth of the modern movement. There was a good deal of adherence to 'traditionalism', particularly for large public buildings, where the Baroque, as inspired by Wren and Vanburgh, and still inherent in French classicism, became a strong influence. Internally, in these public buildings there was some fusion with the ideals of the Arts and Crafts movement with the use of sheets of marble, polished mahogany, bevelled glass, and bronze light fittings and door furniture. There was a curious marriage between a rather chunky and angular classical cladding to the steel and concrete framing that supported it, such as the Liver Buildings at Pier Head, Liverpool (1908–11). Every market town has a bank of this early 20th-century period and style, with overt and dominating classical façades often supported on a rather more prosaic steel frame structure.

The Modern movement 'proper' arrived in England in the late 1920s, and was mainly imported from Germany where it was used in the design of private houses for the professional classes. France and Germany had not experienced traditionalism and the English Domestic/Vernacular Revival as enacted by Richard Norman Shaw and Edwin Lutyens, although France had certainly clung to its Baroque traditions, as any journey along the River Seine in Paris will testify. In Germany there was a strong desire to move away from traditionalism. In contrast to traditionalism, the German-inspired Modern movement buildings utilised flat roofs, steel reinforced concrete frames, and metal windows.

The frame enabled a spatial freedom not enjoyed by vernacular or classical houses of earlier periods, whereby the planform was determined by the structural cross walls. Glass panels filling large areas of the wall were also possible, providing solar gain, although at the time sunshine was considered to feed the cult of the 'body beautiful' rather than to warm the room. External balconies were also incorporated. Adolf Loos was one of the main instigators of the movement, with his Steiner House in Vienna in 1910, which was devoid of ornament. Other primary architects were Walter Gropius, Mies van der Rohe, and Le Corbusier, who all studied under the same master, Peter Behrens. Behrens was active from 1912 to 1925 in finding new forms for new functions, and producing designs where the structure was uncluttered by ornament. His work inspired the Deutscher Werkbund (German Work Federation), which set up a factory for the mass production of furniture and prefabricated houses, the latter being an attempt to improve mass housing. Gropius, influenced by Behrens and also Frank Lloyd Wright, became director of the Bauhaus, a school aimed at teaching arts and crafts to bridge the gap between art and industry. This became an important focus and an international meeting place for the development of the Modern movement. Out of this grew an emphasis on function, with each space in the steel and concrete houses being designed, having a defined purpose to ensure maximum efficiency. The houses exhibited flat roofs, white surfaces, liberated interiors by virtue of their steel reinforced concrete frames, cubicular forms, and long narrow horizontal metal windows. In 1925 the Bauhaus moved to Dessau, and Gropius was given responsibility for the design of the school and living quarters for students and staff. This consisted of ferro-concrete rectangular blocks, with interconnected flat roofs and glass curtain walls. It was Mies van der Rohe, however, who put the Modern movement firmly on the map with his design for a housing scheme of 21 model dwellings at Stuttgart, the Weissenhof Siedlung. The emphasis was on healthy living, hygiene, clarity, and order. Le Corbusier used machine production of mass objects such as glass bottles as a source of inspiration, in the belief that such objects reflected considerable experience in their design, accumulated over many years. In essence the Modern movement catapulted house construction back into a machine age of mass production, quite contrary to William Morris's ethos of craftsman production, and thus ended any possibility of the continuation of a reversal into traditional construction, bringing it to a premature end.

The early 1930s saw the Modern movement being embraced by the very middle classes who had so enthusiastically extolled the virtues of the hand of the craftsman in their buildings, but they were the only sector of society who was able to bear the cost of this new architecture, despite its machine production. Many of the architects who were of the Jewish faith had to flee Germany in the build up to the Second World War, and some fled to England to exert an influence. It was the work of many of the secondary architects of the movement, who followed the lead of Basil Ward, Berthold Lubetkin, Erno Goldfinger, and Maxwell Fry and who designed buildings on the outskirts of towns such as Bournemouth, and even in remote

Photo 1.29 The Modern movement on a domestic scale at Organford, Dorset

countryside, who brought the style to the fore in England. They are many and varied but the most notable are P.D. Hepworth, Oliver Hill, and Edward Maufe.

The 1930s saw the establishment of two firms in England who were to have a major influence in the field: Connell, Ward and Lucus and the Tecton Group. Perhaps the most iconic building of the era, and one which has recently been restored, is the De La Ware Pavilion at Bexhill-on-Sea, East Sussex. This brought this new style of architecture to the masses, who enjoyed its socialist and sociable connotations, its design encompassing ballroom and bars having an echo of the 18th-century Assembly Rooms in market towns. The latter were for the more refined elements of polite society, whilst the Pavilion was for the enjoyment of the masses.

War brought the movement to a standstill, but in its aftermath the desire for a 'brave new world', as well as the German-inspired cult of health and sun, drove the impetus for that which was bright, shiny, and new. Anyone whose parents lived through the war will have childhood memories of the casting out of old furniture and its enthusiastic replacement with tubular steel and glass. Thus it was with buildings, and so began the era of the Crittall window, whose existence is as visible in remote country cottages (although not for much longer as they being constantly removed) as it is in what few Modern movement houses survive unscathed by the cult of the plastic window. Crittall's contribution to the movement began as early as 1919 with a series of semi-detached houses in Braintree, Essex, with their characteristic flat roofs and crisp white concrete surfaces. In 1926, Crittall inspired the design of some of the houses serving Crittall's new factory at Silver End Garden Village, Essex. These were to have a major influence through publicity in the architectural press and served as the inspiration for many similar buildings in the south of England.

One building stands out as model for the post-war era: the Royal Festival Hall (1949–51), designed by Robert Matthew and Sir Leslie Martin and associated with the Festival of Britain in 1951. It came at a time of bleak austerity and was designed to lift the spirits. Coventry Cathedral (1951–62) was similarly an optimistic endeavour, intended to dispel the gloom and carnage of the war.

The 1950s, however, also saw the birth of a harsh architecture, Brutalism, with little or no concession to charm or comfort and with a very heavy emphasis on the new wonder-material, concrete. This is a material which has subsequently been found to have many problems, but at the time, buildings such as the University of Sussex (early 1960s) by Sir Basil Spence, and the Engineering Building at Leicester University (1953–63) by James Stirling, were to set the seal on an era which would only start to show doubts about 'the brave new world' and its modern materials at a much later date. In fact, the steel-framed building with its lightweight rain screen cladding has never looked back, and it was not until the late 20th century that any dissenting voices could be heard, chanting the mantra of the natural and traditional building. It is the early years of the 21st century which has really started to take the manufacture of materials producing high carbon emissions with any degree of seriousness, in addition to the fact that so many modern materials simply have no longevity and are thus eminently non-sustainable.

Conclusion

This excursion through history has taken many twists and turns, but throughout it has illustrated that traditional building forms have eminent sustainability, whilst fanciful architectural forms do not, delightful though they are, and important as they are to preserve for the cultural messages they convey.

One loses count of the number of steel windows dating from the 1930s to the 1960s that are removed on a daily basis, and already UPVC windows that have been installed in the 1980s

Photo 1.30 A pintle-hinged iron and leaded casement can be repaired and given secondary glazing. It does not have to fall victim to plastic window replacement

are due for replacement. Compare this with the sash windows that date from the 18th century which are still in place, and even the 17th-century wrought iron casements on pintle hinges that are still in situ. Sadly they are always at risk of falling prey to the extreme marketing regimes of the plastic window industry, yet they have survived for centuries. They have few champions apart from overworked conservation staff in local authorities, and are always at risk from aggressive marketing of UPVC, but does anyone stop to question how long they have actually been in service, and thus how sustainable they have been, not to mention how this can continue with a little basic repair?

The 1930s reinforced concrete wall, severely diseased by carbonation as a result of the loss of the alkalinity of the concrete surrounding the steel leading to corrosion and expansion of the steel, contrasts unfavourably with the ancient lime mortared stone wall, which has stood for

Photo 1.31 An ancient but durable stone tile roof

centuries with only a minor lime mortar repointing. The flat roof, born of the Modern movement, has a perennial problem with its constantly eroding bitumen felt, in comparison to the ancient stone tile/slate/clay tile roof which has withstood the storms of many ages, needing only a minor replacement of the odd tile or slate.

The list of materials and construction inspired by the Modern movement but rapidly failing is endless, and the comparison with solid traditional construction of an eminently sustainable nature ever prevalent. Despite this there is still a slavish adherence to the use of any new wonder material on the market, with little thought to its sustainability. A prime example is the ever-increasing use of epoxy tape to glue glazing bars of the correct size and moulding onto sealed units of up to 25 mm (1 in) in thickness. This is an attempt to disguise the plastic spacer bars and emulate the 'real thing'. The simulation is, of course, an impossible dream as real glazing bars were delicately thin, especially by the early 19th century, and held glass of no more than 3 or 4 mm (⅛th in). It will be interesting to learn of the eventual demise of these pseudo glazing bars.

In all fairness, however, the Modern movement did engender a love of crisp white and uncluttered interiors, a form of minimalism that is still prized today, and is not that much at odds with the traditional interior of the vernacular house. In an age when personal possessions were few, the interior of a 17th-century yeoman house, with whitewashed walls and ceiling frames, may have appeared minimalist also. In this respect, there is a certain empathy between the traditional and Modern movement house.

Photo 1.32 A brand new Modern movement inspired sustainable house

2

TRADITIONAL BUILDING MATERIALS, THEIR SUSTAINABILITY, AND THE CONTRAST WITH MODERN MATERIALS

The concept of the survival of traditional buildings as a function of location and building materials

The majority of traditional buildings surviving in the rural areas of Britain today are the product of a little recognised social and agricultural revolution generated by the Black Death in 1348/9 and again in 1361. Where previously there had been a lord of the manor with villeins holding land in the form of strips and working the lord's demesne as a form of payment, this was replaced, by necessity, by the lord of the manor leasing off his demesne land to those tenants who survived. This agricultural revolution led to a social revolution as the survivors began to accrue profits that exceeded the income from their previous subsistence living, and ploughed these into substantial domestic dwellings. Whereas the villeins had occupied mud, timber or stone huts, the latter from stones picked up from the fields, the new class of yeoman farmer sought to build for many generations in solid stone or timber frame, or more substantial cob houses. So successful were they that these solid dwellings, or rather their replacements of 16th- or 17th-century date, that they are still with us today and form the bulk of what can be described as traditional houses in the countryside. The extent of their survival is dictated by two important factors: first, when they appeared in the general scheme of development, which varies according to location; and second, perhaps most pertinently, by the nature of their building materials.

The villeins, and their cottager descendants who were not fortunate enough to rise to the rank of yeoman, occupied buildings best described as primitive; they were built of flimsy materials and designed to last no more than a generation, say 25 years, and consisted of a simple timber or mud hut. These buildings simply do not survive, hence the ambiguity concerning their exact form. Today we would recognise such a structure as equating with the many new sustainable buildings built of, say, straw bale on a tyre wall base, although we hope and trust that these modern counterparts will last a good deal longer.

The vernacular buildings that replaced them were in contrast built of good solid local materials, were constructed with only function in mind, the chief of which was to be as a temple to the fire, and by craftsmen in conjunction with the occupier, passing on a way of life known for generations. The date at which there is a clear emergence of this house type in any one area is known as the 'Vernacular Threshold'. In the south-east this might be as early as 1400, in the north as late as 1700. In the former, timber framing was the epitome of what was desirable to a yeoman farmer, with a peg tile roof; in the latter, stone buildings with a stone tile roof were the norm. All were in sharp contrast with polite buildings, designed to be pleasing to the eye rather than simply functional, often using imported materials foreign to an area, and designed to follow a national trend. In a way 12th-century castles are polite in that they satisfy these criteria, and survive because they were built of local stone, although they were over-engineered for their function. Sometimes they are described as supra-vernacular because they straddle both camps. Also supra-vernacular were the Elizabethan or Jacobean country houses, built of timber of stone but to a particular design, often echoing some geometric circular or triangular format. More common polite buildings are the large country houses of the 18th and 19th centuries, with their regular symmetrical façades, built in brick in a stone area or some such contrivance. These also survive in a remarkable way, due again to the use of substantial building materials (see Chapter 1 for illustrations of all these building types), whilst large numbers of flimsy cottages of the same date do not, unless they too were built in the form of model villages in fairly substantial materials. Milton Abbas in Dorset is a prime example, constructed in the 1780s. It is probable that its creator never envisaged that the buildings would last more than say 50 years, but they remain as one of the finest examples in the land, due to the sheer bulk of their cob construction and the continuous refurbishment of their thatched roofs. They are a testament to the old adage 'a good hat, boots, and overcoat' as a reference to the thatched roofs, stone plinth, and lime render.

The key to all this survival is the use of traditional materials, but also the way these materials have been managed over centuries. In a nutshell, such buildings have survived because they have had a lot of ventilation, via what would now be described as leaky windows and even open doors in the midst of winter, plus un-insulated roof cladding (tile or slate) or thatch, and a low gentle heat in one of two locations. This has encouraged damp in walls to stay at bay, and removed warm moist air via the chimney stacks, replacing it with fresh air. These traditional buildings have weathered many a storm, but not the indiscriminate use of non-porous repair materials such as cement in the mid–late 20th century, plus a host of other evils such a concrete tiles and central heating systems, whose on–off heat serves to confuse water permeation through structures and mobilise salts to internal surfaces when in operation at high temperatures.

Photo 2.1 A general view of Milton Abbas, Dorset

Building materials are in addition the key to retaining local distinctiveness. Developed over many centuries, craft construction techniques were entirely geared to the most abundant materials available in any one region, so that the buildings themselves become a catalogue of those materials and the period of their use. Skills and craft techniques were handed down through many generations, often through families, so that craftsman evolved with intimate understanding of how materials behaved, and hence how they should be repaired. The only instance where this is now common is with thatching, a skill that can be passed from father to son. It would be a sweeping statement to say that the First and Second World Wars decimated this knowledge pattern, but one cannot help feeling that this was the case. The last war promoted the use of cement on an unprecedented scale, so that it became the common building material replacing all previous masonry practice. What followed was a complete disregard for traditional buildings and materials, and the birth of tower blocks where once there had been brick terraces, with disastrous social results. This negation of the house as an organic being, integral with the site, the environment, and the life of the inhabitants, resulted in a backlash. Alternative lifestyles in the late 20th century demanded a return to the house as a micro-ecosystem, interacting with the macro-ecosystem of the earth. Natural building materials are so much part of the earth and all its life systems. This entity was termed 'Gaia' in the early 1980s, named after the ancient Greek earth goddess. In a way this was a return to a past understanding of the house being part of a local eco-system, by virtue of being built of local materials, dependent on local energy sources (wood before the days of mined coal, or open cast coal), local food and water, and wastes being recycled locally. This is not how modern cities have developed, being emporiums of steel and glass, and there needs to be a return to traditional materials. In fact there is a huge demand for neo-vernacular buildings in rural areas (and the even less satisfactory neo-Georgian), but in a rather artificial way. Cottages are constructed as replicas of their predecessors, particularly the two-up, two-down version with a central front door, and appear to be in stone; some even have thatch roofs, but there the resemblance ends. The wall construction invariably consists of block work, with a cavity filled with polystyrene insulation. The stone is merely stone wallpaper and the building is not a living breathing entity. Even the thatch can be so under-clad with fireproofing materials that it no longer wicks moisture out of the building. It is for this reason that many people turn towards existing traditional buildings for the health-giving qualities that they possess, and in doing so have stimulated the conservation movement.

The sustainability of natural and traditional building materials

Certain criteria need to be examined when debating the health benefits and sustainability of traditional materials. These are summarised below.

Healthy materials

1. Clean and contain no pollutants or toxins, emit no biologically harmful vapours, dust, particles or odours, either in manufacture or in use. They should be resistant to bacteria, viruses, moulds, and other harmful micro-organisms.
2. They should produce no noise either in production or in use, and have good sound reduction properties.
3. They should be non-radioactive and not emit any harmful levels of radiation.
4. They should be non-electromagnetic, not allow the conduction or build-up of static electricity or emit harmful, electric fields.

Ecological and sustainable materials

1   Renewable and abundant, comes from natural sources, and whose production has a low impact on the environment.
2   Energy efficient and using low energy in production, transport and use, and generally locally produced. Good energy conservers with high insulation values, keeping buildings cool in summer and warm in winter.
3   Durable, long lived and easy to maintain and repair.
4   Produced by socially fair means, that is good working condition, fair wages, and equal opportunities.
5   Low waste and capable of being re-used and recycled, thereby saving energy spent on processing raw material.

(Adapted from Pearson, 1989)

   The discussion of the various materials will reveal that not all traditional materials satisfy all of the above criteria, all of the time, but by virtue of their proven longevity, as seen through the examination of existing traditional buildings, have considerable benefit in the sustainability spectrum.

Stone

Stone is an example of a traditional material which does not satisfy absolutely all health criteria, in that stone for instance produces a great deal of dust during its processing, but is nonetheless regarded as eminently sustainable. It also harbours bacteria on its surface plus a host of other micro-organisms, a factor which would amaze those of us given to hugging a stone pillar out of sheer exuberance and love of the material. Certain types of stone, such as Cornish granite, emit radon gas, and stone can hardly now be described as abundant, or having a low impact on the environment during production. It generates energy costs both in production and transport, the latter only if hauled long distances. This certainly used to be the case in recent times, but the current practice of reopening small quarries designed to serve local building and repair projects has thankfully returned this aspect of the stone industry into the sustainable activity it once was. The downsides all seem very depressing when discussing what at first sight would appear to be the most sustainable material ever produced. After all, most of our churches and castles are constructed of it and have survived for 500 years or more. It is indeed long lived, durable and can be maintained and repaired, although there are caveats on both of some of these. Limestone in particular is subject to calcium sulphate decay, a process whereby calcium carbonate is converted to calcium sulphate by virtue of acid rain. This process has only really become a problem since the late 18th century, at the time of the Industrial Revolution. Prior to that, churches, castles, and stone traditional houses had been the most durable of building structures. With stone decay, accelerated by water-borne salts and acidic gases, comes problems of how best to tackle the question of repair, with sometimes complex methodologies having to be used (see Chapter 7). Generally, however, stone is the most durable of materials, and is widely regarded as the most sustainable because of this. If one is looking for a building that will last many generations, then it is still the preferred choice, as it was for our yeoman ancestors.
   Stone can also be regarded as a form of insulator in that it can absorb heat and then radiate it back into a room, acting as a thermal store. It can also regulate warm moist air by holding it in its pore structure until such time as it can be released into a well-ventilated room or removed to the outside, but only if it's a solid stone wall. Traditional stone buildings are always cool in summer and warm in winter, the former because the stone walls absorb the excessive heat.

Stone produced in this country is nearly always by socially fair means, although this was not always the case. The Purbeck (Dorset) stone industry operated throughout the 19th century in conditions which would make a modern-day quarryman blanche with fear, although silicosis from stone dust remains a hazard. It would be fair to say, however, that in such a period there was little waste, any small units being conserved and used to building random-uncoursed stone houses or walls. Today, sadly, a modern quarry is a wasteful scene of discarded blocks destined to become road rubble, as a result of specifications for ashlar facing stone or stone slips to clad steel-framed or block work buildings. This is why existing stone buildings, however humble, should never be squandered by being repointed, rendered, or internally plastered in cement. These actions reduce the stone to be fit only for road rubble when next required for recycling, and incapable of being re-used to serve many generations. Only careful repair with lime mortars, plasters, and renders will result in a building that can be recycled many times, thus saving energy on processing of the raw material.

Stone has perhaps the longest history of all traditional materials. It provided the shelter for early man in the form of natural rock caves, which illustrate perfectly the principle of the thermal store and the thermal flywheel effect. The temperature below ground lags behind that above the surface. Thus earth warmed by the sun in summer heated the cave in winter, and the earth cooled by cold conditions in winter served to cool the cave in summer. The large thermal mass of the cave flattens out the temperature fluctuations as it absorbs heat when the surroundings are warmer and gives heat back when the surroundings are cooler.

Even in prehistoric times stone was highly prized, being used for megaliths to support chambered tombs in Britain and even grander structures elsewhere, such as the Egyptian pyramids. It is generally accepted that the earliest stone shelters were caves, and many have remained in use for centuries, even if in latter periods as animal shelters only. Dry stone walling may well have been introduced also in very early periods of man's interaction with building materials, long before the supposed accident of burning limestone and putting it into water, ostensibly to heat food, which resulted in a sticky mass of slaked lime. History has long shown that some methods of construction which now survive for humble structures, such as field walls and animal shelter, may well have enjoyed usage by the wealthy in much earlier periods. In

Photo 2.2 Potential quarry waste, possibly destined for road rubble

later periods of British history castles, fortifications, and churches were nearly always stone. The heaviness of the material, the lack of roads and transport other than carts meant that the methodology of putting stone together in a wall was geared to its particular form, and this in turn produced a local distinctiveness, notably flint in the Hampshire–Dorset region, combined with stone or brick lacing courses for the distribution of forces, and also in East Anglia, granite in Cornwall, limestone in the Cotswolds, and bright yellow Hamstone in Somerset.

These different stone types were derived from the basic geological classifications of igneous, sedimentary, and metamorphic.

Igneous rocks such as basalt (solidified larva flow), granite, feldspar, quartz, mica, and flint have a crystalline structure and little or no pore structure. Sedimentary rocks are formed by the weathering and erosion of igneous rocks, deposited in ancient river beds, to form sandstone. Limestone is also formed in this way, but being laid down on the sea bed it contains deposits of fossil shellfish. Such a limestone is Portland stone, which has been famous for its longevity and ability to weather, also so-called Purbeck marble, in reality a polished limestone, used universally by the Victorians to re-create their version of medieval authenticity in re-built east windows and chancel arches of churches.

Metamorphic rocks are igneous or sedimentary rocks, changed by heat and pressure into slate, marble (compressed limestone) and alabaster (gypsum).

In addition, various types of stone were considerably altered by pressure, the action of ice, extremes of temperature causing expansion or contraction, the action of water, particularly sea water which also caused changes within the chemistry of the stone, and erosion by wind-driven moisture or the sea. Even small animals caused fractures with their burrowing activities, as did the root systems of large trees. All of these processes produced breakdown either of the surface or in the form of natural fractures, which enabled stone to be more readily quarried.

Photo 2.3 (above) Panel of knapped flints – note the galletting with smaller flints

Photo 2.4 (right) Yellow Hamstone panel at the base of a church doorway suffering from salt erosion

Stone with a high density due to the nature of its molecular structure is more capable of heating up and acting as a heat store. Stone with a less dense structure and more pores and voids is capable of holding air within these, which will act as an insulator. Unfortunately dense stone has a much reduced latent heat and does need to be heated in order to be comfortable on floors as otherwise it presents a very cold area which will draw heat away from the room. Many traditional buildings had stone floors but gained heat from the fire that was ever-constant in the grate throughout the seasons. The use of on–off central heating systems does not give the chance for the floor to gain heat, thus in existing stone floors it is advisable to administer some form of underfloor heating by lifting and relaying the stone flags, *but* only if this is possible without damage. In new build, the use of stone floors essentially requires underfloor heating and considerable insulation beneath. The type of stone used also needs to be dense to allow for wear, as a lightweight stone with a pore structure might provide insulation but would wear away more rapidly.

Earth

Perhaps the most ancient of all walling materials, earth contains no pollutants as such but can contain dust and micro-organisms. The whole process of producing an earth/cob building by digging out the subsoil of an area produces little or no noise, unless done on a mass scale with mechanical earth movers. It is certainly a material with good sound reduction properties, earth banks being favoured on road construction sites to absorb noise away from neighbouring properties. It is not radioactive or electromagnetic.

Subsoil is always available and thus is a renewable and abundant material. For centuries cob houses have disappeared back into the ground from whence they came, leaving sadly little or no archaeology with which to deduce their form. By the same token they leave no waste. They have always been a symbol of buildings that grow out of the earth, and thus require no transportation costs. What does survive more readily is a cob-like mixture used to daub wattle infill panels in between timbers in a timber frame. Mixed with cow pats to make it more plastic, the material adheres well to the surrounding timber and wattles to provide a waterproof seal.

A great deal of human energy was expended in production by virtue of needing to lift the material on to the ledge formed by the first lift of the material, itself sitting on a masonry or flint plinth (primitive damp course). New construction is now so rare that it is impossible to be clear about modern energy costs either in terms of transport or construction, but new localised build, mostly for non-domestic use, occasionally happens.

Photo 2.5 Cob building returning to earth (photo © Rob Buckley of DCRS)

Photo 2.6 Wattle panel not intended for plastering with a cob daub mixture but used in a farm building for ventilation

Photo 2.7 A rare survival of wattle and daub panels where the cob-like consistency of the daub can be clearly seen

Of all the materials it is a good conserver in the energy stakes, forming the ideal thermal store to keep buildings warm in winter and cool in summer. In addition it regulates internal humidity, acting as a buffer at periods of high humidity, particularly in bathrooms and kitchens, and only releasing the moisture in periods of high ventilation or wicking away it to the outside. It also removes polluted air in the same way. It is highly durable as long as the principle of retaining sufficient moisture for the earth particles to bind together is adhered to. Too much and the material turns to porridge, too little and it turns to powder. Kept just right, with a good hat (roof with an overhang), boots (plinth), and overcoat (render of lime or cob slurry) and it will last indefinitely. It is also easy to repair, any cracks being capable of simple stitching with

Photo 2.8 Cob clearly visible in lifts (layers), which should not be visible, the overcoat of cob slurry having weathered away

the wet material rammed into a ledge formed between the two sides of the crack, or the use of unbaked cob bricks, or simply making up a damaged face. Whole walls can be dismantled, chopped up with water and straw and simply remade in the same location. Unfortunately many a householder takes a lot of convincing about these simple facts, and considerable scepticism is exercised by those more familiar with the ubiquitous block-work construction techniques. What is forgotten, of course, is the fact that blockwork takes a large quantity of energy to produce, whereas subsoil takes but a fraction as it requires no burning.

The material has a proven track record of being durable, adaptable to form almost any shape desired, strong and rot proof (although alas not rat proof), and its longevity in dry climates such as Iraq (former Mesopotamia) is legendary – here the Sumerians built large cities on the plains between the Tigris and the Euphrates, containing houses, temples, and ziggurats (temple towers) using sun-dried earth bricks. Saddam Hussein was so taken with the construction of Babylon that he started a rebuilding programme in the same sun-dried bricks, this time with his name stamped in them. Adobe, as the technique is called, is ideal for arches, domes and vaults, and curved walls. The earth bricks can be various shapes, ranging from cones, spheres, cylinders, and ovoids. The technique was exported from North Africa to Spain in the 8th century, and from thence to the Americas in the 16th century (Oliver, 1997b: 209). The material is dependent on the properties of the sand and aggregate for strength, and the clay for bonding and waterproofing. In desert-like conditions its properties of high thermal mass are ideal for countering the extreme diurnal changes of temperature between night and day.

The nearest British equivalent is the clay bat of East Anglia – a rectangular clay block found in Cambridgeshire, Norfolk, and Suffolk, and the clay lump of Essex. The material was a chalky clay, mixed with chopped straw and water and set into a four-sided, bottomless mould, and turned out almost immediately after which they were dried under open-fronted sheds. Using a plinth of flint or brick, they were then clay mortared in stretcher bond. When used for domestic buildings they were clay or lime rendered, but more humble buildings utilised a simple lime wash or tar coating. In vernacular building terms the use of the material is very late, being c.1800 plus, especially lending itself to the work of philanthropic cottage reformers. The late 19th century brought a realisation that the material was prone, as are many versions of cob

Photo 2.9 Dorset cob house rejected because the gable needed rebuilding

Photo 2.10 Blockwork replacement building for the rejected cob house

used for farm buildings, to rat runs, partially combated by the use of brick outer skins. It staggered on as a construction methodology until the early 20th century.

Devon cob is by far the most well known of British unbaked earth materials, although Dorset cob must run a close second. Somerset also has its share. The material is a mixture of clay and gravel subsoil (small stones, gravel, sand, and silt), with chopped straw and water added, laid in lifts of massive thickness and compressed by laying a thin layer of straw over the top of each lift and stomping up and down along the length of the new wall, or beating with a spade. In some areas of Dorset pulverised chalk was part of the mix, especially where it occurred naturally. This was also true in Buckinghamshire, where the mix was known as 'wychert', and in Devon

Photo 2.11 Panel of adobe bricks used at the Centre for Alternative Technology, Machynlleth, North Wales

Earth Blocks – like the rammed earth walls in the information/shop building these are made from unfired earth, using much less energy than normal bricks!

and Cornwall where shillet, a naturally occurring morass of slate fragments, was used in the mix. The date range for surviving buildings is 15th to early 20th century, but the situation regarding earlier buildings is entirely unknown as they will have disappeared into the earth leaving no trace.

There are two methods by which cob walls were raised, using simple tools such as a spade and hay fork. The quick method used a number of smaller lifts, all laid one on top of the other very rapidly, the evidence for which is a readily discernible slump, particularly if the outer skin is a fired brick. The slow method involved a lift of 300–450 mm (12–18 in) left to cure, before a similar lift was added. A wall thickness of 600 mm (24 in) was not unusual. There is little evidence of shuttering, although wattle hurdles may have occasionally been used to restrain a wall showing signs of lean, but generally it was necessary to leave the sides free to allow them to be pared back to a reasonably even surface, after an initial set.

Finally, one cannot leave this discussion without a mention of the sod houses of Ireland, and those that must have existed in Holland, as one is reconstructed in the Open Air Museum of Arnhem. Formed by laying sods grass-side down, or of peat, with at least a double thickness and staggered joints, and bonding, they relied on root structure for strength and provided lowly homes with good insulating mass. Any items of a cellulose nature, such as bed linen, regrettably develops a black mould, as the Arnhem example illustrates, and life in such an inherently damp structure must have been very tough.

Other cultures in, say, Ghana have experimented with their naturally occurring gravel soils by adding cement to form a product called 'landcrete', or if the soil consists of a fine gravel, a product referred to as 'sandcrete' (Oliver, 1997b: 213).

Mediterranean and North Africa cultures have perfected the technique of rammed earth (pisé) construction, compacting loose dry earth between temporary shutters. The material is naturally stiff, mixed with gravel and shells or lime, and relies on compaction for cohesive strength. This contrasts sharply with the British method of wet earth with chopped straw and water mixed to an appropriate plasticity, which can only be lightly compacted and even more so with the clay lump (blocks) of East Anglia. Some use of the technique is known in Sussex, where there is evidence for the use of boards or hurdles. In Lincolnshire the clay cob was plastered against a framework of uprights, the technique being called 'mud and stud'. A very plastic mix of clay, straw, and dung formed the infill for the panels of timber framed buildings, and no modern method has ever bettered this technique, aided and abetted by a yearly coat of lime wash which built up into a fine plaster in the join between panel and timber.

Nonetheless, the subsoil used varied in its mineral quality and hence its chemistry, and by virtue of this its ability to survive, as a compressed material. Soils are formed by rock erosion and weathering, so that unbaked earth consists of a range of particle sizes for gravels, sands, silts plus silicates, soluble salts, and organic matter, the latter all acting together to form clay particles held together by the surface tension of water coatings, created by the hydrogen bond. Soils are thus a mixture of gravels, sands, silts, and clays, held together by water. The clay is the binder, coating the sand particles and filling the voids. The latter are kept to a minimum in a good cob mix but are in all events compressed out. Clays that are lighter and nearer to kaolin are more useful in a cob mix than those which are nearer to montmorillonite. The clay content determines the shrinkage with a low percentage leading to not enough binding, and a higher percentage causing shrinkage and cracking. A good mix has 2:1 (sand/gravel:clay) mix. In reality it is of course those areas where the mix has these characteristics that have produced successful cob buildings. In addition the presence of ferric oxide, producing ochre or reddish brown soils in colour, is also helpful. The iron, usually a component of the clay, bonds to the water molecules and produces buildings that are both strong and durable. These soils are also easy to shape into blocks and harden faster. Herefordshire has such soils, but the timber framing tradition was so strong that other than clay daub the soils were little used for building.

In areas where the clay content is high and shrinkage is a possibility, such as Dorset, and where chalk or clunch was also readily available in the subsoil, its preferential use avoided shrinkage and cracking. The use of wooden frameworks is also not unknown, and additives such as beeswax or tallow are known to have been used in Cornwall. The sacrificial overcoat of cob slurry was very traditional, but in the 20th century this practice appears to have been curtailed, due no doubt to the constant need to regularly maintain it. Instead a lime render was applied, or if the building is very unlucky a cement render was used, which rapidly led to demise. Cement render cracks, due to the expansion of the cob beneath as it heats up, trapping moisture which pumps through the crack by capillary action, which then washes away the cob. Lime renders were typical in some areas in the south-east of England, from earlier periods, where pargeting, the handmade decorative motifs created whilst the render was still wet, was in vogue. Additives were also used for these renders, such as linseed oil or tallow for waterproofing (which needs tempering with formaldehyde to kill off the mould producing micro-organisms), egg white for extra binding power, and hair or straw reinforcement.

It is interesting to note that where cob leads, hipped roofs follow, as there is less chance of collapse than with a cob constructed gable wall, and better spread of roof loading onto end walls in particular.

Lime renders require regular maintenance, and it is easy to see why, in a culture where maintenance costs can become greater than rental values, and the house-building economy is geared to obtaining a mortgage that building in earth is perceived to have negative factors.

Fired earth – bricks

A particular type of earth subsoil is required for best quality fired clay bricks. Called 'brick earth', it is a mixture of clay and sand found in surface deposits, formerly the beds of ancient lakes and rivers. It also contains a multiplicity of other minerals, such as iron, calcium carbonate, magnesium, and various salts, each of which combine to give the brick its distinctive fired colour, but which also govern the way it behaves during the firing process and in the weathering process.

Traditionally clay was always quarried in the autumn, spread out and allowed to break down in winter frosts, made into bricks in the spring, which were then air dried in openfronted but roofed clamps, and fired in the summer. The technique of rolling and puddling the clay to remove stones and inclusions that would cause the bricks to explode in the firing process unless removed, was done by primitive methods such as kneading the clay until the invention of machinery in the 19th century. Then an Archimedean screw was employed to cut and slice through the clay, enabling the stones to be dropped away through the bottom of the machinery, delivering a batch of pure clay to the bench. Here the time-honoured tradition of throwing it with great velocity into a wooden mould on a stock board would commence. It was only with the invention of more machinery that the process of cutting bricks with a wire as they were extruded through a metal channel, rather like toothpaste, commenced. These factory produced bricks fed the huge surge in terraced housing, itself motivated by the aftermath of the industrial revolution and factory production of all kinds of goods. Factory produced bricks lack the wrinkles on the side face of the brick, resulting from the scooping up of the clay, rather like a butter curl, and the subsequent flattening out of this feature as the clay hits the mould with velocity. They also have rigid sides due to the extrusion, whereas hand-made bricks tend to have uneven sides, due to the difficulty in making clay adhere to every part of the mould.

Firing was again done initially in open clamps, which tended to produce variable firing from total vitrification down to under-fired. This was replaced by more efficient downdraught kilns, and firing today is done in gas-fired kilns. Vitrification was used to create patterns in walls, the so-called diaper pattern being the most familiar, and used in early brick 'great houses' to great effect. By the Victorian period, many estate-owned farm buildings operated the same visual effect, on their rather more perfect bricks, by dint of dipping the end of the brick in metallic sand which produced a dark blue glassy compound.

Brick had a late renaissance in Britain, with Holland being the leading centre of production, but not until the 16th century. Any localised production was very small scale and this continued, apart from the larger 19th-century factories such as Bursledon, Hampshire, until the mid 20th century when all small brick makers were bought out by, or found themselves unable to compete

Photo 2.12 Brick earth being quarried

Photo 2.13 Handmade brick-making in operation

Photo 2.14 Blue brick used as decoration in Blandford, Dorset

with, the larger brick companies. Brick sizes have always been governed by what can be conveniently grasped by an average sized hand, except in the 18th century when a particularly pernicious brick tax on every brick in a building resulted in 'great bricks' being produced, often laid on edge to take up the space of the more numerous smaller bricks. Specials have also been produced since the 18th century when they were needed for the production of rubbed (gauged) brick arches. The 18th and early 19th century is also characterised by the use of red 'raddle' or 'reddle' on brickwork façades. The basis of this is red ochre, an oxide of iron used frequently for the marking of sheep. Mined in its purest form it is a sticky, glossy, unctuous substance, and is ground by hand and pushed through a mesh. Yellow ochre was fired at a high temperature to produce the same colour. It is then used as a pigment in a weak lime wash, and universally coating every brick and joint, it produces a dense even colour that disguised differential firing. It is sadly a treatment which usually remains unrecognised during the refurbishment traditional buildings, as are the special markings of joints, either ruled through the centre or top and bottom of the joint, or the less-frequently used tuck pointing, consisting of a fine strip of lime mortar set within a wide joint.

The brick façade so produced masqueraded as the more expensive gauged brickwork, at least when viewed from a distance. Raddle had more extensive use in buildings; it was used by many Welsh farmers in limewash to protect timber; by medieval man to decorate timber framed

Photo 2.15 Rubbers being sawn against a template with a twisted wire saw

Photo 2.16 Tuck pointing – a white lime putty-based mix is pressed into a line drawn in the red mortar joint

Photo 2.17 A traditional rubbed brick arch

structures; by masons to mark stones; and by roofers to who soaked string to mark out the position of battens for slates. Lime wash on masonry was coloured pink or yellow using red or yellow ochre. The term 'raddle' had all but gone out of use by the mid 20th century.

An important aspect of the process was the drying out of the clay bricks, before firing, done in open-fronted clamps, and leaving the characteristic 'squeeze' of clay on the face, often at an angle to the main plain. Shrinkage of between 8 and 12% is not unusual upon drying, ready for firing, and again after firing, bricks will expand as they again take in moisture. Thus in repair work of new build it is essential to ensure that the expansion process has indeed taken place. The shrinkage and expansion factor was particularly relevant in the production of large terracotta components, designed to be adhered to late-Victorian steel frames. Some terracotta was known in the 16th century but often limited to decorative roundels, but in the Victorian period it was hailed as a wonder material that would last indefinitely. Despite some problems, it is in fact the rusting of the framework that presents the greater challenge.

Bricks have many of the properties that unbaked earth has, but are a great deal more costly to produce in terms of the energy used. Bricks are now dried with warm air, and fired at temperatures around 1,100°C. All this is very costly in expenditure of energy terms, consumption of fuel, and emission of pollutants. The product does, however, have extremely good weather resistance, as well as resistance to acids and alkalis. There is no pollution in use unless, as is often witnessed, holes are drilled for various fitments, when the amount of brick dust is prolific and infects every surface in the vicinity, as well as the lungs of the operative.

Bricks structures can absorb noise in use (but not generally to the satisfaction of the Building Regulations), and even the noise of production cannot be described as being prolific. There is no harmful radiation or electromagnetic fields.

No material is so renewable and abundant in terms of brick earth itself, but once converted into a ceramic the material can only be recycled in its ceramic form. Again it is essential that lime mortar is used, as if bonded by cement, bricks cannot be recycled and become fit only for road rubble. Fortunately those rows upon rows of Victorian terraced houses that are being so haplessly demolished in the name of regeneration, when in reality they should be refurbished, were bonded with lime mortar, so at least the bricks can be recycled.

Whilst a consumer of energy in production, brick walls make ideal thermal stores, as was clearly apparent to our ancestors. Most yeoman houses, constructed in timber or stone, had a brick chimney breast, which acted as a vast storage radiator, particularly if centrally placed. In addition, brick walls can act as a buffer for internal humidity, feeding it to the outside when a brisk wind prevails to wick this moisture away. Buildings with solid brick walls can thus keep cool in summer by absorbing too much heat and keep warm in winter. Misguided attempts at extra internal insulation using closed cell material (urethane foam), which inhibits the movement of moisture and does not allow a brick wall to behave in a natural way, can thus be rather self-defeating. That brick buildings can be long lived and durable is without doubt, with some Elizabethan houses being constructed in the material, as well as the 14th-century gatehouse at Beverley in Yorkshire. Unfortunately the modern use of brick, either as a skin to a timber frame or even brick slips adhered to a block-work background, will never justify the energy lost in their production, by their very limited time in usage.

Modern brickworks usually provide socially fair working conditions and wages, and even those companies who still make hand-made bricks in the old way by throwing the clay into a mould now operate fair systems of remuneration, as well as limiting the time spent at the bench in respect of repetitive strain injury.

Lime and clay as mortars, renders, and plasters

The rediscovery of lime as a binder for mortar, renders, and plasters is hailed as the one single aspect of traditional building repair which has done more than any other material to ensure the longevity of traditional buildings. Since the discovery of cement in the late 19th century and the universal use of the material throughout a period roughly spanning from the 1930s to the 1970s and beyond, irreparable damage has been done to many existing buildings. Cement joints force the masonry to deal with precipitation and the attendant mobilisation of salts or frost crystal formation, instead of the sacrificial mortar joint doing this work. Cement renders act like a corset to expanding masonry, and the resultant cracking encourages acid rain to linger to create calcium sulphate in joints, and consequent map cracking (see Chapter 7). It has taken 30 years at least for the error of these ways to be detected and lime to be hailed as a universally beneficial material, not only in practice but also in terms of ecological aspects. What other material can re-absorb the carbon dioxide emitted from the material when it is burnt? The basis of the material is the lime cycle, whereby limestone (calcium carbonate) is burnt to become calcium oxide, having lost the carbon dioxide. Calcium oxide is highly reactive, and due to its ability to render down any human tissue was used in this 'quicklime' form to dispose of bodies during periods of plague. The newly burnt and much more lightweight stone is then slowly added to water to become lime putty, or has steam fired into it to become hydrated (powdered) lime. This process is called 'slaking' and the result is calcium hydroxide. During this process the heat absorbed during burning is released, producing an exothermic reaction. The resultant material, if the lime is added to water, is a bubbling, hissing, and spitting sticky mass, resembling cream cheese.

It is this that, when added to a well-graded aggregate, forms a mortar which can be used also as a render or plaster, and in essence returns to calcium carbonate (limestone). In the process of taking on a set, carbon dioxide is re-absorbed in a delicate balance with water being evaporated out. Too much water results in a mortar that shrinks, producing fine vertical hairline cracks in a joint or a mass of crazing in a render/plaster, and weak set which will be liable to extreme weathering. The trick is to carefully control the exudation of water, by spraying if exposed to hot sun or drying winds, and covering the work with damp hessian. Alternatively, if it is raining the work must also be protected with dry hessian, leaving an air gap of say 200 mm (8 in) behind to enable carbon dioxide to enter the mix. Lime putty or hydrated lime should never be used in any season of the year liable to frost. Traditionally building work was only done between April and October, and it was the advent of cement, which is fast setting on contact with water, that has resulted in building work all the year round. What is often forgotten is that this frequently results in the introduction of additives to the mix, which whilst entraining air bubbles to create voids to cope with frost crystals, creates a host of other problems. The addition of this saponifying (soap-like) substance, a mixture of a strong alkali and fat, can result in salts of carboxylic acids. The importation of salts in any form is highly deleterious to any mortar, render, or plaster. Salts can be hygroscopic and continue to attract water to mortars, renders, and plasters, creating damp patches that are reparable only by completely removing the material or by extensive poulticing (see Chapter 7).

Lime is thus a wonder material. Burning and re-formation means that a lump of limestone can be converted into a linear strip of limestone in a joint, or a vertical surface of limestone on a wall. In addition, carbon dioxide can be removed from the atmosphere during the setting process, and the danger to the ozone layer averted. Although there will certainly be carbon dioxide emissions during burning, once in use, or even in its slaked form (covered with a film of water to exclude carbon dioxide), the material will last indefinitely, providing it is given care and attention. If it should break down then it can be replaced easily, and in fact lime mortar joints or renders/plasters should be regarded as a sacrificial medium when used in conjunction with a precious masonry material whose replacement is not going to be an easy matter. Use of well-graded sand produces voids which act as inhibitors to salt decay (large pores cope with salt crystal formation). The material is strongly alkali in its early years, and thus will inhibit the formation of moulds and bacteria. Production is generally without undue noise, and the material itself has good sound insulation properties when used as a plaster. There is no danger of radiation or electromagnetic fields. Like all similar materials it is regarded as abundant as it is directly from the earth's crust, although in reality no material is finite. Historically limestone was always burnt and used locally where it occurred.

Where limestone was in short supply so that it was deemed to be expensive material, clay was traditionally used. Clay has many sound properties when used as a mortar, being deemed energy efficient as it did not need to be burnt, and because it was used on the spot, even more energy efficient in terms of production, transport, and use. However, it lacks the durability of lime, and external joints were thus pointed in lime imported some distance. Clay also expands, and this did cause structural problems in some traditional buildings. Having said this, there are innumerable vernacular buildings in Dorset and Devon, and occasionally in the Midlands, where clay is the jointing and plastering compound used with stone. They have existed until now since at least the 17th century, but are falling victim to a regime of complete replacement despite the fact that clay is a highly insulting material.

Both clay and lime can be recycled, the former merely by adding water, the latter by grinding up and using as an aggregate. Sadly this is rarely done, the potential to be totally green being eclipsed by the desire for fast-track, profitable building. In the example above, only the stone will have been recycled as a cladding to a 'barn conversion style' new build.

Hydraulic limes have to a large extent revolutionised the acceptance of lime once again in building construction. Discovered in the late 18th century as a desperate attempt to find a fast-setting binder, they were abandoned once the discovery of that ultimate supposed panacea of Ordinary Portland Cement (OPC) came into universal use, but have re-appeared as a sort of halfway house between lime putty mixes and cement. As early as 1780 the difference between a pale limestone and a grey limestone containing clay (hydraulic) was noted in terms of the

Photo 2.18 Burnt limestone is slaked to form lime putty in a lime pit

Photo 2.19 This smaller version shows the hissing and spitting of the exothermic reaction when burnt limestone is added to water

Photo 2.20 The slaked lime is ready for combining with sand to make a mortar or rendered mix

Photo 2.21 The slake lime putty resembling cream cheese is ready to be combined with sand

Photo 2.22 A range of sands are chosen to add to the lime putty for the correct colour and texture

Photo 2.23 Mixing can be done by cutting and slicing the lime putty into the sand

Photo 2.24 Alternatively, the lime putty and sand can be mixed in a mortar mill or paddle mixer which does the cutting and slicing. A cement mixer with internal fins can be used

Photo 2.25
Clay-mortared
Dorset farmhouse
rejected in favour of a
replacement dwelling

Photo 2.26 The
replacement, doubtless
sustainable but then so
was the original

amount of expulsion of carbon dioxide (described as carbonic gas) upon burning as well as the hydraulic lime producing a stronger set. J. Smeaton, as a result of the construction of the Eddystone lighthouse, produced *A Narrative of the Building of the Eddystone Lighthouse, and a Description of its Construction* in 1791. The primary discovery was that Blue Lias (so-called Bath brown lime), having been burnt, and then slaked by exposure to damp air in open-sided but roofed sheds, could, when combined with an aggregate, produce a render that was fast setting and waterproof. Moreover, it could emulate the golden brown Bath limestone when used as render but lined out to look like ashlar work. Limestone contains a homogeneous blend of clay, sand, and quartz and heterogeneous silicaceous nodules of flint or chert in crevices between the strata. These act as a source of silica and alumina, so that when the limestone is burnt a bond occurs between the calcium and these elements, notably calcium aluminates, and calcium silicates that produce the extra bonding medium. The downside is that argillaceous (clay-like) limestone produce alkali salts also.

The burning of 'clay containing' limestones produced a mortar that was:

- feebly hydraulic (setting time under water was 15–20 days) equivalent to the modern NHL2, or
- moderately hydraulic (setting time under water 6–8 days) equivalent to the modern NHL3.5, or
- eminently hydraulic (setting time 1–4 days under water) equivalent to the modern NHL5.

Non-hydraulic limes (NHL) have obviated the need to add cement to lime mixes in order to achieve a fast set or deal with difficult weather conditions. In fact, during the Smeaton Project to assess 'Factors Affecting the Properties of Lime-Based Mortars' it was deduced that 'the addition of small quantities of cement to lime/sand mortars had a negative effect on strength and durability, and that whilst strong mixes such as 1:2:9 or 1:1:6 (lime:cement:sand) do improve durability, they are often unacceptable in terms of hardness and water vapour permeability' (English Heritage, 1994: 44). This in essence means that they are no better than cements for causing the entrapment of water, the negligible water vapour permeability which causes building blocks to crumble whilst trying to manage salts, and the undesirable 'corset action' of cement renders.

Better setting can be induced by the addition of brick dust (1:3:1 – lime:sand:brick dust), which acts as a pozzolan (again forming calcium aluminates and calcium silicates), and improves strength and durability. The Smeaton Project (English Heritage, 1994) concluded this was a better product than high temperature insulation (HTI) – refractory brick dust. It is interesting to note that historically all manner of pozzolans were used, such as forge scales, burnt coal (frequently seen in Victorian mortars, registering as black specks), and wood ashes. In fact, the latter is particularly relevant as all early limestone would have been burnt using wood, and it had been noted that the burnt limestone at the base of the kiln (called cendré) was much more reactive, producing a stronger set and a more durable product, suitable for floor finishes.

Photo 2.27 Victorian black speckled mortar on a 14th-century building

Gypsum plaster and stucco

It is difficult to think of this modern plastering material as a natural product but in essence it is, being a naturally occurring mineral, resembling limestone in its appearance but with translucence. It occurs in many forms, dependent on its crystal structure, and colour as determined by the mineral content, so that the finest gypsums, such as Satin Spar (fibrous crystals), Selenite (tubular crystals), Desert Rose (plate-like crystals), and Alabaster (fine grained) are used to make ornaments. Gypsum contains a range of minerals, including limestone, clay, silica, and iron compounds, and the impure forms are grey, brown, or reddish brown, these being the basis of gypsum plaster. Its use as a building material in ancient periods was largely confined to those countries which enjoyed a hot dry climate, particularly in the East. The most famous historic source is that mined at Montmartre, near Paris, hence its introduction into England in the mid 13th century as 'plaster of Paris'. Its use in England was limited due to the climate, although similar concerns must have prevailed in Paris in the late 17th century where it was used to cover timber framed houses on a lath base to try to inhibit the spread of fire. It is still regarded as an important fireproof material. In England its universal use was quite late, being also confined to finishing coats on decorative lime plaster ceilings in the 17th century, and finishing coats on lime plaster walls in the 18th and 19th centuries.

The raw material is processed by being ground and crushed and then heated (calcined) to a temperature of 145–160 °C in order to drive off water. This is considerably less than the 900–1,000 °C for lime and the 1,450 °C for OPC. This lends the material credence in the green stakes, as there is less fuel used and less carbon emissions. It does not, however, negate its principal disadvantage that it is to some extent soluble in water and thus unsuitable for use externally in a damp climate.

The chemistry of the material is fundamental to the understanding of its behaviour and the different types of gypsum plaster.

The raw material is calcium sulphate dihydrate $CaSO_4 \cdot 2H_2O$. This means that two molecules of water are chemically combined to every molecule of calcium sulphate. The water is called 'water of crystallisation'.

The calcining drives off 75% of the chemically combined water, leaving only half a molecule of water to create calcium sulphate hemi-hydrate – $CaSO_4 \cdot 1/2H_2O$ (pure plaster of Paris, referred to as Class A). This is the principal component of all modern plaster, although in practice a whole range of hemi-hydrates are incorporated.

It also traditionally contained a whole range of additives, some designed as retarders as the material takes on a very fast set. These included organic hydrogen carbon compounds, such as hoof and horn meal (keratin, an animal protein), combined with lime, citric acid, borax, and sugars combined with lime, such as molasses. These interrupted the set long enough for the plaster to be worked onto a wall, ceiling, or cornice.

If all the water of crystallisation is driven off, then calcium sulphate anhydrate was formed (C_aSO_4) at the higher temperature of 160 °C. This takes longer to chemically bind with the water of crystallisation (see below for the setting process) and was thus also regarded as a retarder.

Accelerators were also used to encourage a faster set. In particular the raw material was ground and used as an additive. The small particles acted as a seed around which the hemi-hydrate or anhydrate would rapidly grow. This is why plastering tools have to be scrupulously cleaned after use as a trowel caked in dihydrate would eventually become unusable, the slurry (described below) being increasingly attracted to it.

The need for accelerators and retarders is more understandable when the following setting process is understood.

The setting process

The material requires no aggregate. The material sets by re-combining chemically with water, in a process which starts with hydration. The calcined dry powdered calcium sulphate is added to water and forms a very liquid slurry, the crystals of gypsum being in suspension. Again, rather like lime putty, an exothermic reaction occurs with heat absorbed during calcination being released, accompanied by expansion. This is why, when poured into a mould some space is left to cater for this expansion, and the wet slurry feels warm to the touch. The process is detailed below.

1 **Hydration:** molecules of water start to combine with the crystals of hemi-hydrate or anhydrate to form crystals of dihydrate suspended in free water, to form a slurry.

2 **Application:** this is applied to a wall or mould for fibrous plasterwork, a fairly rapid process as almost immediately the initial set occurs.

3 **Initial set:** long thin crystals of calcium sulphate grow into a crystal matrix as the molecules of water become chemically bound. Individual crystals become less free to move.

4 **Onset of stiffening:** the process of hydration continues until the interstices of the lattice set begin to fill up with more dihydrate. Crystals of hemi-hydrate $CaSO_4 \cdot 1/2H_2O$ becomes $CaSO_4 \cdot 2H_2O$ plus spare water plus some spare calcium hemi-hydrate yet to bond with water. Any crystals of any anhydrate $CaSO_4$ present start the slower process of first becoming hemi-hydrate then becoming dihydrate, i.e. $CaSO_4$ become first $CaSO_4 \cdot 1/2H_2O$ then become fully hydrated to become $CaSO_4 \cdot 2H_2O$. It is important to note the slower process of conversion of anhydrate to dihydrate, through two stages, as this explains why it is sometimes used as a retarder.

5 **Final set:** all the hemi-hydrate and anhydrate is converted to dihydrate, filling up all the interstices in the lattice so that a solid mass is formed. This still contains an amount of free water which then has to be allowed to evaporate away. The above process indicates quite clearly how the anhydrate takes longer to take on a set and why it can be used as a retarder.

Various types of plasters were traditionally used, depending on the hardness of the set required. These are described below.

Anyhydrate gypsum plaster, previously referred to as Class C or 'Sirapate', was used on ceilings and lower dados on walls where a hard flat finish or a patterned surface which would take longer to form, was required. It contained a small amount of hemi-hydrate to speed up the initial set, without inhibiting the slower full set required to allow for polishing.

Keene's plaster, Class D, often erroneously referred to as 'Keene's cement' (all fast setting mediums were described as cements in earlier periods), was anhydrous plaster in its purest form with no hemi-hydrate, designed to give a slow continuous set, to allow for polishing to obtain a very smooth surface. Again it was used in areas of high exposure, such as dados. Extra retardation to allow for even more continuous working and polishing was achieved by adding alum (aluminium potassium sulphate). At the time of its use, it was chiefly obtained from mining the natural mineral kalinite, and became anhydrous when heated to 200 °C. It is a double salt, and acted, like borax and cream of tarter, as a retarder.

The discovery of this slow-setting but miraculously hard plaster led to the development of Scagliola (mock marble), the material being coloured with mineral dyes, by adding powdered anhydrate to the dye coloured water. A dough-like consistency was then rolled out like pastry and arranged around columns of wood to simulate marble. Realistic nodules of dough were pressed in as well as string soaked in dye. The string was removed just before the final set and the streaks grouted in. The whole was then rubbed down and polished to a high gloss, the shine being reinforced by linseed oil or poppy oil, and finally polished with beeswax. The realism was so good that it is difficult to tell the difference between marble and scaliolia, requiring testing of temperature by hand. The technique was also used to produce imitation marble table tops; a pair of such tables can be found at Uppark House near Petersfield.

Commercial gypsum plaster was referred to as Class B, and was the standard hemi-hydrate plaster plus retarders, including anhydrate.

Plaster of Paris was Class A, called 'casting plaster', and is the pure hemi-hydrate that sets quickly, and aside from being used to create a cradle for broken limbs, its chief use was for fibrous plaster work, the so-called 'stick and rag'. The stick is a reference to the battens used to support the wet plaster in the mould when set, and the rag the hessian used in layers for the same purpose. Modern plasterboard incorporates this material with accelerators.

Ash plasters were used for less sophisticated work, although this was limited as the majority of traditional vernacular buildings continued to use lime plaster until the early 1900s. This was a product of localised burning, and aside from the ash from the fuel used, inevitably contained a whole mixture of unburnt material, the raw dihydrate, hemi-hydrate, and over-burnt material – anhydrate. The latter slowed down the set and produced a harder finish. It can be quite difficult to recognise.

Stucco is a confusing term, often applied to a whole range of materials including lime and gypsum and even cement. In general, it is understood that stucco is reinforced by the addition of powdered marble dust, while lime plaster is reinforced by the addition of chopped hair, be it ox, horse, cow, or goat. The purest stucco was termed 'stucco duro' and was pure carbonate of lime, obtained by burning travertine, the deposition as a result of springs holding lime in solution and then depositing it at the base of a valley. When burnt, mixed with sharp sand and fine marble dust (imported from Italy) it was capable of being highly polished. Its use externally avoided all the pitfalls of using gypsum, although it again employed carbon-based retarders such as milk, blood, and white of eggs. Few but the most wealthy could afford such niceties, so other forms of stucco were invented that were less expensive to produce. All were lime-based, as this was the only medium that could cope with the British climate, although undoubtedly they were meant to emulate their gypsum cousin used in drier climates, whose ability to produce an even polished surface must have been much envied. They were:

- **Common stucco:** hydraulic lime, sand and hair.
- **Rough stucco:** Chalk or high calcium carbonate lime, and sand, more often than not used internally, lined out to imitate stonework, and coloured using various aggregates.
- **Bastard stucco:** A superior mix of fat-high calcium carbonate lime putty and fine washed sand.
- **Trowelled stucco:** a finish coat of bastard stucco, scoured, polished, and painted and water hardened to a finish as smooth as glass.

Sgraffito is not so much a material as a way of using coloured plasters, particularly black and white, to produce a picture not unlike a fresco (produced by painting pigments into wet lime so that the pigment became set within the crystalline set of the lime). The art of sgraffito was devised in the late 18th century, following the discovery of the Baths of Titus in 1770 in Rome. It involved, for example, as in the Arts and Crafts re-invention of the technique shown in Photo 2.28, placing a black fine lime plaster over a white lime plaster and scratching through the surface to delineate a picture, in this case the recumbent form of a female in the best classical tradition. Sadly the example illustrated was removed unceremoniously from a house dating from the early 1900s in Poole by the local water company for the creation of a sand filter bed!

(The above information adapted from Bankart, 1908 and Gwilt, 1891.)

Photo 2.28 Arts and Crafts reclining female figure executed in sgraffito work but removed during demolition

Timber

Timber divides into two distinct species. The softwoods are evergreen conifers and can be grown in very cold conditions, as well as in temperate climates. The former has the distinct advantage of growing very slowly, producing densely packed growth rings with only a short season of growth to produce the lighter summer wood prone to disease. Such timber was imported into Britain in large quantities in the 18th and 19th centuries, from what were termed the Baltic States, and it is often branded with its port marks.

Hardwoods, sometimes termed 'broadleaf varieties', are only grown in temperate zones or in tropical conditions. The former are characterised by the oak, beech, ash, birch, and alder, the latter by mahogany and teak.

Timber is surely one of the cleanest, non-polluting, and thus one of the healthiest materials ever produced. Its naturally formed linear cellular structure formed of cellulose is hygroscopic, the water molecule being attracted to the upside-down glucose molecule that constitutes the cellulose composition of the cell walls. This ability of the material to breathe water in and out renders it able to stabilise internal humidity and act as an air-filtration unit. In a similar way the empty cells, once emptied of moisture, fill with air, acting as an insulator, so that wood is always warm to the touch (heat is not removed from the hand). These cell walls can also absorb sound. Electromagnetic fields cannot pass through wood, and the only health issue is with the terpenes that form the molecular structure of the resins, particularly in softwood, although this is a recent issue and for many centuries appears to have had little effect. Terpenes are thought to disrupt human molecular structure and cause allergies. Turpentine, which has a concentration of terpenes, was the major constituent of the balsam tree, and when distilled from the resin, left a more solid substance, called 'rosin'. The latter, used for violin strings, was also the basis of many varnishes when once again reconstituted with turpentine.

Sustainability is high on the agenda for timber, as few other materials absorb carbon dioxide to make sugars (cellulose) for growth of an infinitely renewable resource. This is regrettably diminished by the destruction of the Amazon rainforests on an unprecedented scale. Not only is this readymade carbon sink being destroyed, but the burning for clearance produces carbon dioxide to add to the greenhouse gas burden. The Forest Stewardship Scheme (FSC) has been initiated to try to reduce the impact of this global catastrophe by encouraging the use of softwood and hardwood from managed sources and not from rainforests. Also being encouraged nationally is the use of timber in the 'round', and a return to the coppicing and pollarding of oak, ash, hazel, chestnut, and willow, also being exploited for fuel. This will in turn increase the removal of carbon dioxide from the atmosphere. One might also like to re-introduce the use of vines such as clematis for tying poles in building construction (a technique used for all early scaffolds), but for many building contractors and building control officers this would be a step too far. Vines grown for the purpose were a feature of traditional forests and sight of them in an ancient forest is now a rare and wonderful experience. Binding and lashing was also achieved using trees and shrubs of the willow family, one species of which, known as 'osiers', were particularly useful for binding, as were the withies, the young willow taken out as thinnings.

Coppicing is thought to be a very ancient tradition, the resultant poles either as thinnings or from pollarding, having their bark removed, but because they are young they contain a great deal of sapwood prone to beetle infestation. They are figuratively used in Saxon hut

Photo 2.29 Pole rafter roofs can sometimes fail but can be replaced if there is a thriving coppicing initiative

Photo 2.30 Pole rafters can be supported with secondary rafters

reconstruction. Their use for pole rafter thatched roofs is indicative of the coppicing tradition but can lead to problems, particularly in areas where whole roofs are done in this tradition. Milton Abbas in Dorset is such an area.

One of the chief reasons why coppicing and pollarding has diminished is the importation of softwood for fence panels, whereas in a previous age thinnings and pollardings would have been in constant use for hurdles, and also for the oak stave and hazel basketwork infill panels of timber frames, required as a base for daub and plaster. This dry construction, preceding the wet trades, rendered some degree of wind and waterproofing, as well as rigidity to a structure during construction. This assisted with the general progress of construction, and the use of timber in the round, and cleavage of round poles to make laths, ensured very little waste from processing. Today acres of timber are wasted in processing, particularly on building sites and developments. Whole skips are filled and sent to landfill, which is supremely distressing.

The recent practice of using draught horses to pull timber from difficult terrain adds to the sustainability of the material, although regrettably material in the 'large round' still requires transit by large haulage vehicles, inviting more thought to be given into how to utilise wood locally. The insulation value of wood cannot be overstressed, and treated or painted on a rigorous maintenance regime, even softwoods can be durable. Maintenance and repair is the key to the longevity of wood, and this itself produces few problems, other than the inability of modern generations to face up to the challenge rather than sit in front of the television, as well as the constant battle to pass carpentry skills on to the next generation. Processing by socially fair means depends entirely upon where in the global economy of wood the production is taking place. Certainly few materials produce waste that is so capable of being re-used (see wood products), and universally through time timber has been decanted from one building to the next. The utilisation of re-used timber in barns, in association with rebuilds for increasing yields and also in yeoman houses, themselves frequently rebuilt from their medieval predecessors, was a common practice. This recycling has led to the much vaunted myth of ships' timbers that permeates every conversation with a new owner unfamiliar with the rites associated with timber frame construction, which also include the incising of magic marks such as W or M, references to the Virgin Mary, and daisy wheels, whose meaning is rather more obscure. These are usually to be found on fire beams, door jambs, and window frames, all timber components, although the occasional barn indicates their use incised in lime plaster.

One of the most obvious methodologies associated with the reduction of waste in times past was the production of shingles for cladding by successively halving and quartering a log, using a cleaving technique until triangular shaped slivers of wood were produced. This avoided the waste associated with the production of wooden tiles using sawing techniques. Cedar shingles are frequently imported from North America, but our own oak, redwood, and larch have sufficient resin to resists decay. Riven/cleaved wood using a froe or beetle, sat astride a contraption called a horse (not unlike a modern workmate bench) and pulling the froe from side to side avoids much of the waste of sawn timber.

Processing whole timber beams traditionally involved the removal of the outer portions of sapwood and bark, using a side axe to make large scoops and an adze to make smaller scoops, turning the log over in line with a chalk-marked string. The finished product was then taken to the saw pit, where the hapless bottom sawyer suffered the indignity of sawdust in his eyes and lungs, whilst the log was sawn into post and beam sections, using the double-handed saw. The saw leaves diagonal saw marks that are characteristic of the angle of the saw. It is possible that trees enjoyed a far deeper symbolic meaning than they now have. Certain traditional buildings show clear evidence of the heartwood centres facing each other across, say, a cross passage in a house, or threshing floor in a barn, as if demarcating the primary space.

Photo 2.31 Shingles as well as boarding on a building in Alaska, North America

Photo 2.32 Rob Buckley of DCRS using a shave horse for cleaving

Timber is time-immemorial material and probably preceded stone. Certainly even in the historic past the Saxons emulated timber construction in their early stone churches, producing turned stone balusters in windows or triangular headed windows, all as if they were in timber. Even their reliquary boxes for carrying about the bones of saints were in stone but resembled timber in their form. This skeuomorphism, as it is termed, is surely a reminder that the makers were actually more comfortable with timber technology but resorted to stone in the interests of longevity. A unique example is illustrated in Photo 2.37 of a stone china cupboard made by an 18th-century quarryman/stone carver for his house.

Photo 2.33 Processing a beam – side axing commences to remove the sapwood

Photo 2.34 The side-axed pieces are removed to reveal the beam

The use of timber reflected the fact that much of Britain was covered in vast tracts of forest, much of which was simply expended in the early Middle Ages to gain agricultural land. Long before this the forests provided the very basics of shelter for hunter gatherers, the simple tent-like structures, linear or round, possibly being the basis of the roof form that was adopted into the British vernacular. Other cultures, for example in North America, developed the log cabin, which reflected the long straight softwood which grew tall to reach the light in a crowded woodland (see discussion below), but in Britain the oak was predominant and although the early oaks grew in close contact with each other, they still had a tendency

Photo 2.35 Adze and side marks on tiebeam in use in the 19th-century granary converted to residential use

Photo 2.36 The side axed beams are then pitsawn into smaller elements or, as in this case being demonstrated at the Wealdd and Downland Museum, the beam is having the sapwood removed

to produce a forked trunk, so ideal for the making of a cruck blade. This created a roof and wall all in one.

The use of long vertical timbers, cut into slabs, manifests in a surviving building at the Saxon church of Greenstead-upon-Ongar. Archaeological traces of the most important Saxon halls show only postholes, with little cognisance about whether the wall cladding was vertical or horizontal boarding. Interestingly the tradition of vertical boarding survives well in Dorset, commencing with surviving late medieval plank and muntin screens and continuing into the 19th century with plain ledged boarding. Log construction may well have preceded timber framing in Britain, being abandoned when straight oaks were no longer available. It was certainly in use for early settlement patterns elsewhere in Europe, and continued in Switzerland where it is not unusual to find it in use in multiple storeys which have survived for up to 400 years (Oliver, 1997b: 249),

The log cabin prevailed in North America for the early settlers in the late 19th century when timber framing was all but fading out for the very poorest social classes and regions of Britain

Photo 2.37 Stone china cupboard, an 18th-century skeuomorphism, such a feature normally being executed in timber

Photo 2.38 Cruck blades in this Herefordshire village house reflecting the use of sturdy oaks grown in ancient forests

in the same period. Prior to this oak timber had enjoyed centuries of use, culminating in the construction of the great houses of the Elizabethan era as well as for vast numbers of yeoman houses in those clay areas where timber was abundant. It was used for roof construction all over the Britain, until the advent of imported Baltic pine in the late 18th century onwards.

No other material can quite illustrate so graphically its one-time abundance or reflect equally well the nature of the material and how it has been grown. By the early 17th century, particularly around 1600, the strain of so much timber use, particularly for charcoal production and the demands of the navy, were beginning to show, and this can be graphically seen by the greater use of small branches, for example, set within the panels of timber frame to form quatrefoils or diagonal bracing.

Log construction corner joints were varied, but of particular interest was the practice of inserting a form of narrow slip tenon between upper and lower logs, a practice that may have some affinity to the tenon used in the 'mortice and tenon' joint which became universal for British timber framing, as did the half lap joint which characterises the interlocking corner joints of log cabins. Also utilised were angled flat cuts to bed the logs together, and the half dovetail.

Again in North America, the vernacular is characterised by clap boarding on timber frames, the building type reflecting not only the predominance of the timber for the frame but also the development of sawing techniques, which by the early 1800s had become very sophisticated due to the use of water-driven machinery. It is also possible to see the use of imported timber in Britain being for the same purpose, especially in Essex.

Sawing has a distinct developmental pattern, starting with medieval and post-medieval pit sawing, with many illustrations showing a trestle being used. This certainly continued into the late 18th/early 19th centuries, although in the 1780s a circular saw driven by a waterwheel was invented. Around 1824 this was supplanted by an inserted tooth circular saw, but at an even earlier date of around 1808 the vertical band saw had been invented.

The material itself has an almost magical quality, when one examines how it is formed. The addition of growth rings of winter and summer wood, the former darker and adding strength, the latter lighter and providing much of the linear cell structure, renders this a material that is warm to the touch, due to the air held in the linear cells. Many modern textbooks label timber a cold bridge when used in roof construction, which is somewhat perplexing because a

Photo 2.39 Quatrefoil and diagonal bracing showing use of smaller pieces of timber by about 1600

Photo 2.40 Log cabin reconstruction of early settlers' housing at Banff, Canadian Rockies

Photo 2.41 Interlocking half lap joint on the corner of a log cabin reconstruction at Banff, Canadian Rockies

Photo 2.42
Clapboarded building,
thankfully unrestored,
in Alaska, North
America

well-seasoned timber with no water in the cellular structure will surely act as an insulator. The linear cells impart a huge strength in tension, when compared to the weight. The resultant material has pleasing natural texture, a variation in colour, proven workability, and a high strength-to-weight ratio, so that it has an age-old appeal. Old timber has a particularly venerable quality, the gnarls, knots, and shakes, and the striated surface formed by decades of rain washing out the lignin producing a pleasing texture worthy of veneration. Ancient timber was always over-sized to overcome any tendency for failure due to inherent defects, although Photo 2.43 shows that sometimes this rule was not always followed. The early 16th-century use of timber components in a plate format for main structural elements, such as side purlins in this instance, had distinct disadvantages.

Photo 2.43 Purlin
failure in a Dorset
medieval house

Timber can be said to be totally sustainable, absorbing carbon compounds from the atmosphere in order to make sugars, and then concentrating that carbon into the ground during the breakdown of the tree, due to the inbuilt ability to develop fungal decay. Wood density, the ratio of dark winter wood to lighter summer wood, is a critical factor in how wood behaves; the higher the density, the stronger the wood. It is the lignin, dark brown in colour, that holds the cells together which also imparts strength, and it is this material that remains when the fungi have attacked the cellulose structure of the cell walls, giving rise to the term 'brown rot'. In time the heart of the tree darkens and becomes the heart wood, the very structural core of the tree, and the sapwood is concentrated in the outer layers, and is very prone to fungal or beetle attack once the tree is felled.

Many sustainability texts encourage the re-use of timber building components, such as old doors and windows, and indeed this is commendable but only if they are dated in relation to their introduction. Older buildings are important archaeological entities, conveying human messages about the past, and it is important not to confuse these messages for later generations.

Timber products

It is rare for timber to be used in its natural state in current daily use. The advent of DIY superstores has promoted the use of many wood-based products, and even the requirements for sustainable insulation have promoted the use of wood fibre boards. Many of these products in all events utilise waste timber, which is fortunate as this renewable resource is being utilised at a faster rate than it can be replaced, despite the recent practice of replacing slower growing hardwoods with faster growing conifers (a recent government-inspired initiative).

The principle demand for wood pulp is for paper, which is why the recycling of paper is essential. It is closely followed by the demand for composite boards. This would be commendable if only waste wood was utilised, but frequently whole trees are in demand for this purpose. Of greater concern is the use of formaldehyde and phenols (see below) for gluing the fibres. This virtually negates the sustainable re-use of waste wood in products because these chemicals can be injurious to human health.

Fibre boards are made by simply compressing the wood pulp, relying on the adhesive powers of the natural resins therein. So-called hardboard and softboard fall into this category,

Photo 2.44 Cross-sections of timber – growth rings in yew showing summer and winter wood clearly demarcated

and even medium density fibre board (MDF), although this sustainable factor is negated by the use of urea formaldehyde (UF) as a surface treatment. This is what renders the cutting of the board so dangerous.

Plywood uses thin layers of wood, derived as a veneer, and laid in different directions. Not only is energy consumed by the creation of the veneer, but the use of phenol or UF is used as the adhesive. Plywood is, however, very important as a sheathing material for modern timber-framed building construction because of the extra membrane strength it affords.

Blockboard is similar to ply, but has a central core of blocks of wood.

Particle boards such as chipboard are pressed into sheets with the aid of large quantities of UF or phenols as is Ordinary Strand Board (OSB).

Boards which have greater environmental credentials include: bagasse, derived from sugar beet pulp; flaxboard, derived from the waste products of the linen industry, flax fibres being the primary component; and hemp board, which is similarly fibre based. Regrettably the sustainable nature of such materials is often negated by the use of petroleum derived resins. A return to treatments against insect attack using borax (derived from the metal boron), soda (sodium carbonate) and potash (potassium hydroxide), and natural wood resins for compressive adhesion, plus surface treatments of linseed oil and beeswax, would be more sustainable. Unfortunately modern industrial techniques geared to the mass use of chemicals are not easily reversed.

The same is true of other natural wood products which have dropped out of favour, such as cork. The outer bark of the cork oak (Quercus suber) rejuvenates itself after stripping. Although not a native product, and thus incurring sea miles, it saw a rebirth of its use in the 1970s as a trendy wall cladding but with little thought to its use as an insulator. Its use on floors is also worthy of a rebirth, as long as the waterproof coating required in bathrooms and kitchens is not petroleum derived. Made from compressed cork strips and granules and formed into sheets, it is resistant to rot and moulds, has excellent thermal and noise insulating properties (recently seen in use as such around steel beams in a 1940s extension to a milk factory, possibly as a waterproofing membrane also), and is durable whilst being flexible. It is utterly without waste.

Cork is one of the main components of **linoleum** utilised in a powdered form, together with wood flour, bound with linseed oil and wood resin, and pressed onto a backing of hessian or jute. Linoleum was superseded by vinyl in the late 20th century, which is regrettable because the latter has no more advantages, and many disadvantages. Vinyl tiles are not warm to the touch as is cork, and require a perfectly dry floor, which is often not available in many traditional buildings. Linoleum will in fact be more tolerant of the slight degree of damp experienced in traditional breathing floors, although in reality the best floor coverings for this situation are jute, sisal, and seagrass. Linoleum is flexible without loss of firmness and there is no toxicity from the adhesives.

Rubber is not generally thought of as wood product because, like many products in modern use, current generations are distanced from its production and its relatively historic post-First World War lineage. Now known as 'latex', it is the white liquid extracted from the rubber tree of Malaysia, although itself an import from South America. It made a fortune for the entrepreneurs who founded the plantations, especially after the birth of the bicycle and motor car which required large quantities of tyres. The process that made this possible was the addition of sulphur, which cross links the long chain of natural polymer to produce a hardened product and also removed much of the odour. Latex is not without its problems for humans, despite being a natural product, and it can produce allergic reactions.

Paper is little thought of a building material except for breather membranes, although tarred paper used in many layers has constituted the odd roof cladding in recent history. It was doubtless a response to abject poverty in early social housing. Newtonite lathe was universally used in the mid 20th century. Consisting of a bitumen-impregnated heavy corrugated paper, it was used up to a line where damp traditionally extends and was plastered so its presence became pretty much invisible. Paper does not have to use wood pulp as its main constituent but can utilise any fibrous material, such as sugar beet pulp (bagasse), all reminiscent of the Egyptian papyrus reed that marked its very inception.

Grasses, palms, leaves, and canes

At first sight these would appear to be unlikely candidates for building construction, except for more exotic cultures, until it is realised that the native British thatch, part of the grass family, is about as basic a material as one can get, and is typical of Northern Europe. Reed in particular has a long history in places like Iraq and Egypt. Africa is still a continent characterised by nomadic tribes who erect temporary huts which utilise palm or banana leaves as their principal cladding. Bamboo-framed houses are characteristic of Eastern cultures.

Set against this background, the thatched roof that is universally recognised as a symbol of British vernacular has particular poignancy. Sadly there may come a time when it is no longer a sustainable methodology unless some fairly drastic action is taken by the Department for Environment, Food and Rural Affairs (DEFRA) to ensure that British farmers can grow wheat straw, which is suitable because it is derived from ancient species and has the required length. The slavish adherence to ensuring that none of the modern strains, which grow short butts topped with big ears of corn, are adulterated by being in proximity to these ancient strains is causing havoc in the thatching spectrum. Thatchers are promoting the use of imported water reed, usually from the Eastern continents, Turkey being a prime producer, and the consequent sea miles are not only increasing carbon emissions, but also making the material very expensive. There is an urgent need to end this unsatisfactory situation. Growing more thatching straw in Britain would diminish the carbon dioxide emissions due to the absorption of these in the growing medium, rather than generate them in sea miles.

Much the same applies to wonderful materials such as jute and seagrass, which are so tactile and potentially renewable and ecologically sound. In essence all such materials are healthy, in

Photo 2.45 Straw in bundles, called 'bottles', ready to be laid on a thatched roof

that they have no concerns with regard to electromagnetic fields, or radioactivity, or harmful chemicals. Natural materials can, however, be dusty, 'farmer's lung' being a disease associated with the use of the threshing machine in times past, although modern combine harvesters isolate the operatives from this problem. Natural plant materials can also be subject to moulds, particularly when wet, and it is for this reason that many of the products described below are chemically treated before they enter the country, somewhat further negating their sustainability.

There is no noise generally in their production or use, and most have good sound and insulation properties. They nearly all demand some renewal at constant intervals; it is inherent in the very nature of natural materials that they have inbuilt degradation mechanisms, which is part of the cycle of renewal. Mankind needs to have more cognisance of the close links between decay and renewal, and the consequent ability of natural materials to be abundant and have a negligible impact on the environment. Many natural materials are castigated as useless simply because their re-use and re-cycling is limited, as by necessity renewal must take place.

Grasses

Straw – wheat, barley, oats and rye – are native British species, all of which have been utilised as thatching materials. Only rye, which is in fact the longest and most robust, does not enjoy a reputation in the thatching field.

Reeds, **sedges**, and **rushes** are also native to British thatching techniques, but reed is limited to those areas where it grew in abundance, such as the Fens, and sedge, a freshwater marsh grass, was nearly always used for the ridge. Native reed can grow as high as 2.7 m (9 ft), although imported reed can be considerably longer. Other grasses, such as rice, elephant grass, broom and marrow, are used on the African continent (Zambia and Swaziland).

Hemp, **flax**, **sisal**, and **jute** enjoy a commonality, but the first two enjoyed great popularity as a cash crop in Britain in the 18th and 19th centuries. This was especially so in any area that enjoyed an abundance of water for the retting (soaking) of the stems to release the fibres from their plant resins. Even modest settlements such as West Felton, near Oswestry, which consisted of only a small number of dwellings and appears to have been waterlogged for much of the winter, had a thriving trade in flax, which was used for making linens. Flax was also used for making broadcloth for sails and ropes so was important when it could be grown and produced even some distance from coastal areas, such as at Bourton, Dorset, on the Somerset border and some distance from the Lyme Regis and Bridport fishing industry. Flax fibre has twice the strength of polyester and is 50–75% stronger than cotton. In addition, it is stronger when wet and is to some extent fireproof, and even if it ignites it merely smoulders (Berge, 2000: 159). Quite why it is not used more and has seemingly fallen out of favour is a mystery. It was probably a victim of the post-war boom in artificial fibres such as nylon. Hemp, a non-narcotic cannabis, is fortunately now enjoying an unprecedented revival as a building material additive to lime to form hempcrete, either made as blocks or as a render, providing a high degree of thermal insulation, as well as preventing shrinkage. In earlier periods its use was largely confined to the making of rope, and it appears to have been so valuable that workhouse inmates were set to unpick rope soaked in tar, an activity known as 'oakum picking'. Cordage and rope were often referred to as 'sennit'. Hemp was also imported from India and Africa, and sisal hemp from Yucatan. Hemp is an extremely course fibre, and was twisted by hand using the knee and hip, coiling in new fibres as required. Three strands are then twisted together in the opposite direction.

Jute, its very name alluding to a braid of hair in Indian culture, was universally used for making a coarse canvas suitable for sack-making as well as rope. It makes a good backing

Photo 2.46 Plaster on reed

material for products such as linoleum, and mats made of jute are ideal for breathing floors, as are seagrass and rush matting.

Reed matting now forms the basis of lime plaster repairs in traditional buildings, but in its inception and use in Dorset in the 19th century the reed was used in bundles, flattened out and battened onto joists, to ensure a key for the lime hair plaster.

Bamboo is technically a grass. It grows very fast in tropical countries, achieving full height and diameter in a matter of weeks. Its high moisture content means that it is left for a number of years, and then cut in the dry season. In those cultures which still use it for building material, it is split into half or quarter canes, and used in lattice work to assist with ventilation, or used whole as posts and beams. Half round split sections can also be used as tiles. Joints are traditionally formed by lashing with rope, or using bamboo pegs. It is a very tactile material which has only traditionally enjoyed a use as garden canes in Britain, and indeed any more intensive use would result in sea miles and carbon dioxide emissions, unless this could be offset by the planting of more bamboo to absorb these emissions. This would be stimulated by demand, and bamboo is, after all, like our own native wood: high in tensile strength but flexible enough to be woven into partitions. Although it is an import and generates sea miles it is eminently renewable, and these two factors need to be weighed against each other.

Cactus is very rarely seen as material in Britain, but the trunk can be dried and then sawn into boards and is remarkably resistant to rot.

Palms normally fall into two categories, coconut palms and date palms. They are characterised by short stems and large leaves, both of which are utilised in various ways.

Coconut husks are very fibrous and thus lend themselves to the making of ropes and cordage. The nuts are cut down in their green state, held under water until they start to rot, when the fibres and fleshy meat inside can be separated by rigorous beating. The fibres are then teased into a coarse rope. Coconut leaves are also used for thatching in areas native to production (Phillipines, Malaysia, and India). Coconut palm leaves are utilised as roof coverings for all manor of domestic and animal buildings in the Dominican Republic, after some processing entailing wetting and pressing. They last about five years. The trunks are again natively used for roof and wall frames.

Date Palms have similar usage. The question arises as to whether any of these products should find their way into usage in Britain. To date only **rattan**, a leaf from a non-self supporting palm, high in the forest canopy, grown in Malay, Borneo, Indonesia, and the Philippines, has enjoyed any popularity in the construction of Victorian furniture, and later garden furniture. In recent years this has been replaced by polypropylene (derived from petroleum distillates and thus not sustainable). Rattan is an invasive plant which clings to trees by a series of barbed hooks, and is pulled to the ground by fearless native workers, stripped of the hooks and leaves, and then dried in the sun.

All of these materials are healthy in their natural state, and although a certain amount of dust will be evident during harvesting it is not biologically harmful. All natural materials are, however, subject to bacteria, moulds and micro-organisms. Certainly they produce little or no noise in harvesting and processing, as much of the work is done manually. This creates a bond between worker and the material that tends to be destroyed by machines. In many respects if demand could be stimulated for these sort of materials, and despite the increase in sea miles that would result, it might lessen the need for the destruction of rain forests if their products could be seen to be more profitable.

The treatment of these materials with toxic compounds can be a problem, and again the use of natural materials such as borax, linseed oil, beeswax, and natural varnishes could be a bonus. They are eminently renewable, embody the energy of the sun, and when used in construction would undoubtedly be of high insulation value, as is straw board, one of the most recent spin-offs. Waste is minimal and natural degradation and constant renewal override its lack of durability. Socially fair means of harvesting is a problem yet to be tackled. Overall they evoke a link with that part of our own past which revelled in the production of natural materials, and thus their use represents continuity with a traditional way of life.

Cotton is another natural fibre which is cellulose based. Cellulose products were in fact the basis of the first synthetic fibres, such as rayon and acetate. Cellulose from trees and plants can be rendered into a fibrous material by using sodium hydroxide to remove the adhesion between the fibres. These contrast with the now more popular polyester, nylon, acrylic, and polypropylene made from petroleum derivatives, many of which first appeared in the 1930s.

Animal products

We are now a long way removed from using animal skins to clad our buildings, but animal products have enjoyed a long history of usage in historic building construction. Beeswax has been used in conjunction with linseed oil or on its own as a timber treatment, whilst egg whites were used as a binder in paints and as a very strong adhesive, as were the yolks in paint – called egg tempera – and egg whites were also used in mortars as an extra binding agent. Cheese was used as glue. Even blood was used in various mediums (mortar, plaster, paint) to enhance or retard setting or as a pigment. Hair – cow, goat, horse, etc. – was used in plasters as a reinforcing medium and to prevent shrinkage. Skin and bone was boiled up to make animal glue, with boneworks situated alongside canals for the purpose of rendering down fat and producing gelatine. Canals were a main arterial transport network in rural areas where such unsavoury activities tended to take place.

Milk casein was made from curdled milk by adding rennet (the inner lining of a calf's stomach). This substance with numerous amino-acids has good binding power when combined with calcium carbonate, forming calcium caseinate. Casein can also be made into a plastic by adding formaldehyde HCHO (methanal formed by oxidising methanol, a methyl alcohol). This was known about in the early 20th century before the birth of the petrochemical industry, as a wood alcohol could be made from the distillation of wood and oxidised to be combined with

Photo 2.47 A long-abandoned bone-works at Maesbury, Shropshire

casein to form an early plastic, frequently used for door handles. Methanal had been known for some time as 'formalin' and was used for preserving biological specimens. Casein is still added to lime renders to encourage a faster and better set.

Other natural plastics include amber (not an animal product but a fossilised tree resin), bone, horn, and tortoiseshell.

Shellac (Seedlac) is another early plastic type of product, now little known apart from in the traditional plaster workshop when it is employed in the coating of plaster moulds so that newly set plaster can be released. It is the exudate of the lac beetle, dissolved in alcohol or methylated spirits, after it has been cleaned to remove much of the red dye. Shellac came in a variety of colours: dark shellac or ebonising stain; dark brown from garnet shellac, made from thin strips broken into flakes so that a build up can be achieved; golden brown, formed by the first stage of making the product (heating and filtering through muslin to make button-shaped discs); and blond de-waxed shellac, used to make a transparent finish and known as 'French polish'.

Wool is by far the best known animal product now in the construction industry. Again we are a long way removed from the felted wool which may well have acted as a floor covering of early tents, and still protects the occupiers of yurts for this purpose and as a wall cladding, but wool has enjoyed a rebirth as an insulation material. Its use stretches back to well before the birth of Christ, as the material is so extraordinary in its properties. The fibres, helical in their form and thus easy to mat or spin, are eminently hygroscopic but as they absorb water they also give off an exothermic (heat) reaction. Wool therefore appears give off heat when it is wet. Thus not only does wool act as a buffer for controlling the humidity and air quality in a room when used in the roof and walls, but it actually makes the building warmer when it does so. In addition the dry fibres literally trap heat and so it is the perfect insulator and soundproof mechanism. Wool comes from a variety of sources as well as sheep, notably goat, yak, camel, llama, and alpaca.

Manufactured materials

Glass has an ancient tradition, although not in Britain. It is a material whose production was familiar to much more sophisticated civilisations, such as the Phoenicians and the Romans, and was in use 4,000 years ago or more for vessels and beads. It is a combination of sand or silica, and soda ash (sodium carbonate) and limestone ($S_iO + Na_2CO_3 + CaCO_3$), heated to around 1,500 °C. Potash (potassium carbonate K_2CO_3) can be used instead of soda ash, and the addition of lead (P_b) was common in the medieval period, because it imparted translucence. Colour was imparted by adding metal oxides, such as gold, silver, iron, etc., in the molten stage, and this was known as 'pot metal'. The metal oxides were later plated onto the surface. All early glass was imported from France or Germany in the early 13th century and used only for the most important buildings such as churches and the most eminent domestic houses.

Its development was aided by the introduction of lead 'cames' to hold the small areas of glass, called quarries, these cames being held in place by being tied by copper wire to horizontal saddle bars. All was set within an iron frame, simply set in rebates in the stone, and marked the beginning of a glass tradition that saw further developments. These were aimed at producing larger areas of glass, achieved by blowing glass into a bubble, then attaching a pin known as a 'punt', and spinning the glass into a circular disc. The outer area was thinner and could make larger panes, whilst the centre was reheated or used for small openings to admit light without vision. Muff or cylinder glass was made on a similar principle but with less wastage, as the cylinder was opened out, to make even larger panes. These methods were used up until about 1840. It is essential to conserve any glass made by these methods, which it is now almost impossible to replace unless by importation, generating sea miles. This early glass is readily recognisable by the bubbles of oxygen trapped in the blowing process, known as beads or reeds according to their shape, and the pattern of blowing which can also be sometimes discerned as circular sweeps in the glass.

It was replaced by plate glass in the late Victorian period, rolled and polished by hand by numerous apprentices wielding a stone wrapped in leather and suspended from an armature. This process is discernible by linear striations in the glass that give it a brilliance that cannot

Photo 2.48 Early glass window at Milton Abbey, Dorset

be matched by glass produced by the machine. The production involved the drawing up vertically of the molten fluid into a set of rollers, and was replaced by the later production of 'float glass', which is glass poured over a bath of floating tin. This was the start of the stark flat sheets that form the basis of all glass production today. In the place of very thin glass, with a character of its own and the hallmark of Georgian sash windows (to be cherished if only for this glass alone), has come clever technological glass that involves the application of a metal film applied as a vapour. This allows the entry of short-wave solar radiation but prevents the loss of long-wave heat radiation. It is this factor that has revolutionised the construction of sustainable buildings. For anyone brought up in a 1930s 'semi' where the walls were filled with single glazing set in galvanised steel frames, the so-called sunshine windows, which were anything but in the middle of winter, this is a development that is very welcome.

The careful husbanding of this resource is long overdue. The reserves of all the raw materials involved in the production of glass, and in particular the metal oxides, are finite. A related problem is the high energy required for production and the pollution which results when tin oxide is applied as a vapour, producing hydrogen chloride and hydrogen fluoride emissions. Glass is not polluting in use, and is in fact one of the easiest materials to keep clean, but in disposal the metal oxides used in its production leach out into the soil. This is why it is essential to recycle not only glass products used as containers but also unwanted construction glass.

Photo 2.49 Georgian sash window due for repair

Iron

The production of iron has a long and convoluted history. Iron ore smelted with charcoal with the intention of removing carbon by forming carbon dioxide as a gas has its origins in Iron Age man, and was later refined by the addition of limestone to attract silica, manganese, phosphorous, sulphur, and other chemical compounds in the iron. The resultant slag has a varied coloured striped appearance. The process was revolutionised in the late 18th century by the use of coke rather charcoal. The resultant product was referred to as 'pig iron', as the smelt was fed out into a series of linear channels from a main channel, resembling the array of piglets either side of a sow. The basic chemical process is $Fe_2O_3 + 3C$ (iron ore plus coke) = $2F_e + 3CO$ (iron metal plus carbon monoxide)

The material then spends the rest of its life absorbing oxygen and water in an attempt to return to its initial state, by forming an electrolytic cell where part of the surface is losing electrons (the anode) and part of the surface is absorbing the electrons (the cathode) and forming a pile of reddish dust, creating the problem which faced mankind for centuries. Wrought iron, which involves re-heating the pig iron and beating it to remove more carbon, has less of a tendency to corrode as there is less carbon to be attracted to oxygen, and most early metalwork on doors and for gates is the result of hours of toil in the process. To see a gate or door furniture developing the characteristic pit corrosion of wrought iron is a source of great dismay, especially when a vigorous rubbing-down and coating with an oil-based paint to exclude oxygen and water will preserve it for years to come.

Wrought iron was so labour intensive that it fell out of mass production by the late 18th century and was superseded by cast iron, which retained more carbon, and later by steel, where the carbon content was strictly controlled. Wrought iron was still used for main beams and ties in buildings, where tensile strength was required but cast iron was used for columns. Needless to say, all such existing buildings incorporating this material, whether industrial in their nature, or those less so as iron frames came to be used for large country houses (to release the room spaces from the constraints of masonry crosswalls), are absolutely vital for retention. They represent a massive amount of embodied energy and a scarce resource.

Photo 2.50 A wrought iron gate c.1820 in urgent need of repainting and repair

Attempts to combat the natural desire of iron to revert to its iron oxide state involve the creation of a zinc coating, either by dipping or by an electrolytic process. The zinc then acts as the anode, and being higher up the table of metals will have less tendency to release electrons, and the iron as the cathode, the zinc providing a sacrificial overcoat. This was a process which was to promote the use of corrugated iron as a building material. First produced in the 1820s in rolling mills, it was then impressed to produce ridges and furrows. This formation had a number of benefits. It made the sheets stronger and more free-draining, so that in fact they could span large areas and form almost self-supporting roofs if curved. The material reached its zenith by the late 19th century, being used for all manner of industrial and railway buildings, as well as public buildings in the colonies, and churches.

Corrugated iron also became a universal material for cladding thatched roofs, sometimes preserving ancient thatch beneath, and all manner of farm buildings, lending credence to the view that it became the ultimate vernacular building material.

It is difficult to be pedantic about the retention of this material as a resource, and although it is possible to recycle it, it cannot be said to be generally attractive to many when the galvanising has failed. Its replacement asbestos corrugated sheeting is an even more unlikely candidate for retention, as this fibrous material has been proved to be extremely toxic to respiratory tracts.

Iron ore itself is still in abundance but is still a diminishing resource, and incurs vast transport costs in terms of energy, but there seems to be little abatement in its usage in steel-framed structures. These rely on concrete for protection, but this is only if the area immediately surrounding the steel remains un-carbonated. The lifespan of such buildings is probably no more than 50 years, yet whole cities have squandered vast resources of iron in this form of construction and continue to do so. This cannot be regarded as a sustainable form of construction.

Cement is a material derived directly from the knowledge gained from hydraulic lime binders. It consists of limestone, which is disintegrated and ground up with quartz, sand, and clay, or just clay. The silica from the sand and the aluminates from the clay are important elements as it is the calcium silicates and calcium aluminates which give the material its setting qualities when burnt and combined with water. The ground material is fired at a very high temperature, 1,500 °C, to form cement clinker. These are then ground again, with gypsum added to control the setting. Ordinary Portland Cement (OPC) was named after the Portland rock it resembled, and indeed some concretes, the result of mixing cement with aggregate, can

Photo 2.51 Corrugated iron church at Maesbury, Shropshire

Photo 2.52 Cracks in cement render caused by expansion of brick substrate

be difficult to differentiate from natural stone. This might be highly desirable for an actual structure, but it is definitely not when it comes to using the material as a mortar or a render. The former forces the building block, be it natural stone or brick, to cope with all precipitation, often resulting in mobilisation of salts, and as a render it acts as a corset and, unable to cope with the expansion of the material below, it simply opens up into cracks which let in water, resulting in yet more salt mobilisation.

The invention of OPC in 1824, and its more consistent use in the late 19th century was the result of the need to produce a fast-setting material to expedite the construction process. Few could have anticipated the havoc which was to result in the mid 20th-century escalation of the use of this material, due to the needs of wartime construction. Aside from its unsatisfactory use to the detriment of traditional buildings, its production requires much higher temperatures than lime.

Traditional paints

All paints consist of a pigment, a binder, and a solvent. The pigment provides obliterating power as well as colour, but also provides an element of the setting agent, being a complex hydrocarbon structure, and at the very least a catalyst for what is called 'drying', but which is in reality cross-linking of hydrocarbons with oxygen. White lead, made by artificially corroding blue lead into a white powder, was highly prized as a white pigment, but in cheaper paints it was bulked out with lime or chalk, particularly for masonry paint. Its universal use throughout the 18th century, not only for paints but as a basis of cosmetics, caused extreme disruption of the brain cells and subsequent demise.

The binder is that which bonds to and carries the pigment, and was traditionally linseed oil, although in finer work and where a very white finish was required poppy seed oil was utilised. Oils needed oxygen for the drying (cross-linking) process, but oxygen continued to be absorbed and in time would result in the removal of electrons in the pigment, thus changing the colour.

100

Linseed oil went more yellow as a result, which was fine if a stone colour or the famous country cream was acceptable, but not if a bright white needed to be maintained.

The solvent is that which makes the paint flow in application but which evaporates to leave a hard surface. Water is the most basic solvent, but turpentine was traditionally used.

Distemper

The simplest paints used on internal plaster were distempers. In its basic form this is crushed chalk as the pigment with water as the solvent, called 'whiting' (not to be confused with limewash). Adherence is improved by the use of size (rabbit skin glue), and this was called a soft distemper. A hard distemper was the addition of linseed oil and casein to the whiting, which went by the trade name Walpamur and was still in universal use in the late 1950s. Modern emulsions are the successor to distemper. The solvent is still water but the binder is a petroleum-derived resin (see above).

Linseed oil based paints, with a binder of linseed oil, and a solvent of turpentine could be exotically coloured but only if the client could afford the exotic mineral pigments, such as zinc white, ultramarine (lapis lazuli), or chrome yellow, themselves often the process of burning or unpleasant chemical solutions. Purer pigments, such as cinnabar (vermillion) and malachite (green), could only be afforded by the very rich.

Earth pigment, such as ochre, umber, and red oxide, were less durable, although lamp black, made from coal tar distillation, was more permanent, and mineral black from powdered coke and bone black from charred bones even more so. A cheap red could be made from red lead, itself the result of burning white lead. This was invaluable on iron work and contained some litharge, a lead monoxide that encouraged the attraction of oxygen into the binder to increase the setting power. Lead based paints have never been bettered for finish, texture, and durability, but their use is banned for all but the most important listed buildings, and requires certification. A number of the major estates are, however, re-introducing the use of linseed oil based paints, as they can be produced in a sustainable way. Linseed can be grown in large quantities and does not rely on the petroleum industry.

Another factor which probably influenced the longevity of traditional paints was surface preparation and the number of coats. Joinery surfaces were primed with a mixture of glue size and whiting (gesso), followed by two coats of oil-based paint which was bulked out with whiting (crushed chalk). The final coats were based on pure white lead and mineral pigments which would resist weathering. Gesso was used extensively on internal joinery, especially where wood carvings were done roughly in softwood, the use of a coarse gesso (grosso) over the top enabling finer carving to take place within the hard white material of the gesso.

Varnish was based on the exudates of the balsam tree, the separated rosin (hard resin) being re-combined with the turpentine derived from the exudate to the desired consistency. Pigments could be added to make a more opaque varnish. This is a treatment given to the interiors of sashes in the 18th century and can sometimes still be found in servant's bedrooms.

Although white lead is not a sustainable pigment, linseed oil is a sustainable binder and all paints which are so based are regarded as considerably more healthy than their petroleum-distillate derived replacements.

Limewash was the universal external coating for masonry and render in a pre-chemical age. Consisting of the lime water, which rises to the top of a tub of lime putty, in a milk-like texture, it can be combined with tallow (best done during the slaking process for the lime putty), linseed oil or casein to form a better bond with the surface and become more waterproof. It needs to be applied in several thin coats and can be pigmented with earth or mineral colours.

Photo 2.53 Linseed oil bath for joinery preservation prior to painting on the Chatsworth estate

Potassium silicate paints react with lime on the surface of a render and forms calcium silicate which binds it to the surface, forming a crystalline layer. The best result is achieved on fresh lime render, which will contain a high proportion of calcium hydroxide.

Modern construction materials

Plastics and other petroleum-derived products

Plastics are largely derived from the petroleum industry and have taken over a large proportion of the role of traditional materials, much to the detriment of both buildings and their occupants.

Crude oil, largely from the Gulf area of the Middle East, arrives as an unctuous dark brown smelly liquid. It contains numerous different chemical compounds ranging from simple methane CH_4 to complicated substances with long chains or rings of carbons, called organic compounds. Oil is divided into the main fractions by distillation. These are gases used for fuel (methane, ethane C_2H_6, propane C_3H_8, and butane C_4H_{10} – alkanes), petrol, naptha, paraffin (kerosine), diesel oil, and bitumen. These are all alkanes, chains of carbon atoms with single covalent bonds between them. They are insoluble in water but can be dissolved by organic solvents, such as benzene and carbon tetrachloride (tetrachloramethane). From the 1930s onwards, demand for petrol meant that unless the landfill sites were to be filled with the surplus heavier fractions, it was essential for scientists to devise ways of using them. This was done by catalytic cracking, which can divide alkanes that have single covalent bonds between carbon atoms, that is, C-C (called saturated compounds) to form new alkanes plus alkenes. Alkenes have double covalent bonds between the carbon compounds, C=C and are called 'unsaturated compounds'. This might at first sight appear to be a rather useless set of information, but in fact is key to understanding the origin of products such as polythene, ethene being one of most common raw materials used. Alkenes are more reactive because of the additional reaction which can take place across the C=C bond. They can form polymers by opening up their double bonds to 'hold hands' in a long chain. Ethene can be manufactured from paraffin oil by catalytic cracking and is a colourless gas with a faintly sweet smell. It burns with a yellow flame, and is insoluble in water. Ethene can be converted into ethane by reacting it with hydrogen. This process of catalytic hydrogenation is what is used to convert vegetable oils (liquids containing alkenes) into margarine (solids containing alkanes).

A product which has been in almost constant use since its invention and which is very indicative of the use of ethenes is polythene.

Polythene (polyethene/polyethylene) is a soft waxy thermoplastic resulting from reacting ethene C_2H_4 (a double bond between the carbon atoms) with a catalyst, plus high temperature and high pressure, as a result of which hundreds of ethene molecules join together to form a giant molecule of polythene, the process being called 'addition polymerisation'.

There are other chemical structures that are introduced, such as an anti-oxidant (phenol) ultraviolet stabiliser (amines). These are discussed below as each can have detrimental effects.

Polythene is tough yet flexible, will not react with water or acids and is a good insulator. It can be simulated using a series of paper clips each joined to the one before and after. The giant molecule is called a 'polymer', the small molecules which add to each other are called 'monomers'. Polyvinyl chloride (PVC) is a similar product, also called 'polychloroethene', and in its rigid form is used in construction for pipes. Neither of these materials are degradable by weather or by bacteria – fine as long as they are in use, not so when their disposal is contemplated.

Ethanol is a widely used solvent, known as 'methylated spirits', and is used in paints and to dissolve resins. It is made by the addition of water to ethene. It is a member of a larger class of compounds called 'alcohols'. The simplest alcohol is methanol CH_3OH. All alcohols contain a hydroxide ions (OH) group attached to a carbon atom.

Polystyrene is the most common building material in this sequence of organic materials. It is manufactured by the polymerisations of phenylethene (styrene), using benzoyl peroxide as an initiator. It is then expanded into foam using pentane to form expanded polystyrene (EPS) and is foamed up with chlorofluorocarbons. Extruded polystyrene (XPS) is also produced, and which resists vapour penetration. Phenyls belong to an organic group C_6H_5, present in benzene, the somewhat lethal hexagonal ring of carbons with hydrogens attached. Phenols are added as an anti-oxidant, amines as a ultra-violet (UV) light stabiliser, and organic bromine as a fire retardant. Emissions during manufacture can include a wide range of chemical compounds, such as benzene, styrene, pentane, and chlorofluorocarbons.

Polypropylene is manufactured from the alkene called 'propene', a colourless gaseous hydrocarbon ($CH_3CH:CH_2$) using a heated phosphoric acid catalyst, and then polymerised. Again phenols are used as antioxidants and amines as UV light stabilisers.

Polyvinyl chloride (polychloroethene) (PVC) is yet another ethane, with the addition of a chlorine atom to form chloroethene or ethylene chloride, and heated using benzoyl peroxide as an initiator. The product is a tough white solid material which then requires plasticisers in the form of phthalates; ultra-violet stabilisers, such as cadmium; lead (when used for gutters, cables and pipes) or tin (when used for window production); and flame retardants. The emissions both during production, especially large quantities of phthalates every working day, and during usage together with aliphatic and aromatic hydrocarbons, phenols, aldehydes, ketanes, and vinyl chloride, renders this product one of the most universally worrying developments of the late 20th century. Its use for windows, conservatories, gutters, and cladding continues apace whilst its health effects remain an ever present concern, not to mention the sheer impossibility of recycling the material or disposing of it in any beneficial way.

At this juncture, now that the process of splitting alkanes to form ethenes used in a wide range of common plastics has been discussed, it is probably useful to introduce other substances that are formed from petroleum fractions. Some of them have already appeared above, so it is important to gain an understanding of their origin. The chemistry of petroleum compounds has so many strands that it makes it difficult to know where to commence the story.

Bitumen is one of the few products whose relationship to the primary substance, oil, is easy to grasp, and is in everyday use as a binder for chippings on road surfaces, known as 'asphalt'.

Photo 2.54 UPVC garage doors do not contribute to sustainability and have an impact upon traditional settings

Photo 2.55 Unfortunate plastic cladding and UPVC windows in a traditional building are not sustainable

Its use mirrors but has replaced the natural asphalt that was imported from Trinidad, in the early days of road construction, which was superseded by the use of coal tar, extracted from coal by condensation. The former had a high percentage of polycyclical (two of more rings of carbon molecules) aromatic hydrocarbons. Both have fallen out of use in favour of bitumen. Its major disadvantage is the leakage of dioxins into soil, extremely toxic hydrocarbons that can cause cancer.

Solvents are lightweight distillates used to break down other petroleum-derived materials or as a chemical monomer for other products. When used as a solvent they evaporate in the same way that turpentine used with linseed oil paint evaporates away to leave the binder plus pigment unchanged. They include the following:

- **Aromatic hydrocarbons** – xylene, toluene, benzene, styrene: Benzene, C_6H_6, is perhaps the most well known, referred to as a 'benzene ring', this being the ring of six carbons to which six hydrogens are attached. It is the basis of creosote, and when mixed with coal tar it made an impregnating coating traditionally used as a wood preservative. Its use has been banned for householders because of the carcinogenic effect of the benzene. Naphtha is one of the main fractions from which these products are derived, and as already discussed they are used in the production of plastics. Xylene, much used for dry-cleaning, is an irritant for mucous membranes, and can affect many of the major organs and the nervous system. Toluene and styrene both affect the mucous membranes, and benzene and styrene can cause mutation. Styrenes can be introduced into the air and breathed in, both during production and usage.
- **Aliphatic hydrocarbons** – paraffin, naphthene: These are non-cyclical (do not contain a ring of carbons) compounds that have no aroma, nor do they have the very deleterious effects of the above rings of carbons, but they can irritate the skin and respiratory tract.
- **Chlorinated hydrocarbons** – dichloroethane, dichloromethane, trichloroethane, trichloromethane (chloroform), trichloroethylene: These are the result of reacting hydrocarbons with hydrochloric acid. The solvent forms are mainly used in paints, varnishes and paint removers, of which dichloroethylene (trade name Nitromors) is well-known product, liberally used for paint removal when in reality it is quite lethal. Dichloroethane is a solvent for synthetic rubber.
- **Polychlorobiphenyls** (PCBs): These are now recognised as so dangerous that they are no longer used as fire retardants in electric cables, and as softeners in mastics. Chloroparaffins are used as flame retardants and softeners in PVC flooring, and in mastics. Unfortunately all these substances are poisonous to major organs in the body, can cause mutations, and are carcinogenic as well as being narcotics. Chlorine is a very reactive atom and can cause havoc to the human body.
- **Alcohols** – ethanol, propanol, methanol, isopropanol, butanol: Alcohols are also used as solvents in varnishes. They can irritate the mucous membranes and damage an unborn baby.
- **Esters** – butyl acetate, ethyl acetate, methyl acetate: These are formed when hydrocarbons react with acetic acid. They are used as glue solvents, and polyvinyl acetate, a related product, is used as binder in water-based glues and paints because it bonds so easily with water molecules. Acrylates are esters of acrylic acid, an oxidation of propylene, and are used as binders in acrylic paints, also for the production of methacrylate resin, used as a glass substitute.
- **Ethers/ether alcohol/ethoxyethane** – methyl glycol, ether glycol: Used as solvents and plasticisers in varnishes.

- **Ketones** – methyl ketone, acetone CH_3COCH_3: These contain the carbonyl group $C=O$ linked by two hydrocarbon groups as shown for acetone, and again are used as solvents in glue. Ketones are weak nerve poisons and can damage an unborn child.

Related chemicals

Chlorofluorocarbons (CFCs) are produced by replacing hydrogen in hydrocarbon chemical structures with chlorine attached, together with fluorine. Both chlorine and fluorine are known as 'halogens' and are highly reactive, so putting both together is very risky. They have been used for some time as foaming agents in plastics. Whilst CFCs are stable in the lower parts of the atmosphere, in the higher parts subject to strong UV radiation, the chlorine atoms break away and react with natural ozone, damaging the ozone layer. They can thus contribute to climate change.

 Aldehydes are organic compounds that contain the group CHO, that is the carbonyl group $C=O$, with a hydrogen bound to the carbon atom, making them very attractive to water. The whole can then be bound to a radical R. In the example of methanal, the simplest aldehyde, and the commonly called 'formaldehyde', the radical R is simply a hydrogen. Formaldehyde is made by the catalytic oxidation of the alcohol 'methanol', and is a white polymer. Further oxidation produces carboxylic acid.

 The simple structure of formaldehyde is HCHO, where there is a double bond between the carbon and oxygen. In water it forms a colourless solution called 'formalin', using methanol as a stabiliser. Formaldehyde is the basis of most modern adhesives, mixed with phenols and urea. Phenols (carbolic acids) are white crystalline solids and have as their basis the benzene ring of carbons, with one of the carbons replaced by OH. The equation is thus C_6H_5OH and it is the basis of phenolic and epoxy resins. Its clear relation to phenyl, the organic group C_6H_5 is a reminder of how fundamentally dangerous this chemical is. Urea, another white crystalline solid $CO(NH_2)_2$, is another dangerous chemical containing nitrogen and is very reactive with oxygen. Natural urea is the major end-product of nitrogen excretion from animals. All three compounds, formaldehyde, phenol, and urea, make a very reactive sticky substance with strong adherence power, but these chemicals are also highly toxic substances. Formaldehyde is carcinogenic, irritates nasal passages, and produces allergies. Phenol produces mutations of DNA (genetic material), and both are poisonous to water organisms.

 Amines are organic compounds produced by replacing one or more of the hydrogen atoms in ammonia (NH_3) by organic groups, for example methylamine CH_3NH_2. The latter NH_2 is the functional group (amino group) which has replaced a hydrogen H. They are commonly used as additives in plastics as a hardener or anti-oxidant. In addition, they are the starting point for the production of isocyanates, the main constituent of closed-cell insulation board, and polyurthenane, also for certain paint pigments, silicone, and epoxy resins. They are an irritant to mucous membranes, cause allergies, and possibly mutation of cells. Isocyanates are very strongly allergenic, and an irritant to skin and mucous membranes.

 Phthalic acid esters, also known as phthalates, are esters that are the result of reacting phthalic acids with alcohols (an acid plus alcohol = ester), for use as a plasticiser in plastics. They can account for up to 50% of the contents of plastics and can again affect mucous membranes, a mutagentic (affect genetic makeup), and are a strong nerve poison.

Building plastics

The formulation of building plastics has to some extent already been discussed, the basis being monomers, which upon polymerisation, be it by addition or condensation, forms a polymer.

Various other chemicals are added, such as the halogens (chlorine and fluorine), hydrochloric acid, nitrogen, and sulphur, as well as phthalates (plasticisers), pigments, and UV light stabilisers. Two types of plastic are the result. First, the mouldable, extrudable, and flexible thermoplastics, such polyvinyl chloride, polyethelene, polystyrene, and polypropylene already discussed. Second, the thermoset plastics which are the result of reactions with hardeners, such as polyurethane, epoxy, and polyester, of which synthetic rubbers are a sub-group. Thermoplastics are capable of being foamed up using halogens (chlorine and fluorine). Plastics have come a long way since the first invention of polyvinyl chloride as early as 1838, although not used commercially until the 1930s. Cellulose nitrate is made by treating cellulose (wood pulp or cotton) with nitric acid. Cotton plus nitric acid produces guncotton, a highly charged explosive, but when combined with camphor produced yet another early plastic called 'celluloid', as early as 1865. Bakelite was a much more commercially successful phenol formaldehyde resin made in the early 1900s and used for door handles in the 1930s. Other thermoplastics which first saw the light of day in the 1930s/1940s, but did not come into common usage until well after the Second World War, have come to be used for drainpipes, impervious sheeting in foundations (polyethylene), sheeting (polypropylene), electrical equipment (PTFE – polytetrafluorethylene), glass substitutes (polycarbonate, polymethyl methacrylates), window frames and gutters (PVC-polyvinyl chloride), insulation (polystyrene), and adhesives (polyvinyl acetate). A number of others are in use for furnishings which are now the byword of modern buildings, such as carpets (polyamide, polyacryl nitrale). Many of these products are also used in paints and adhesives.

The range of thermosetting plastics is just as extensive, the most common being butyl rubber, butadiene styrene rubber, silicone rubber, and polysulphide rubber, used for sealing. Phenol formaldehyde is used for dark-coloured electric fittings, as well as adhesives; urea formaldehyde is used for light-coloured electrical fittings, and extensively for adhesives for plywood and chipboard; and melamine formaldehyde is used for laminates.

The question arises, and to which there would appear to be no concise answer, as to what extent these plastics off-gas when in use, analogous to that which they give off during production (plastics factories emit tons of phthalates per annum) and when disposed of in landfill. Certainly it is thought that phthalates, organic acid anhydrides, aliphatic and aromatic hydrocarbons, phenols, aldehydes, and ketones do so. If this is the case, and taken collectively, then these gases will contain carcinogenic, mutagenic, allergenic, and poisonous chemicals which can affect the respiratory tract, the major organs, and the reproductive system. Add to this the release of unbound monomers and in the case of PVC the seepage of chlorine, hydrogen chloride (if exposed to UV radiation), cadmium (stabiliser), heavy metals (pigments) in landfill, and it is easy to see that whilst seemingly transforming building construction, these plastics have also created vast health problems for mankind as well as the natural world. It is not as if once in place in a building they are there for eternity. Whilst not requiring the same maintenance regime as wood, these materials will deteriorate over quite short periods of time, ranging from 2 years (polyethylene) through to less than 30 years for PVC, which has been known to last as little as 10 years due to the inclusion of styrene, possibly only 8 years. In 2009, due to the success of the government scrappage scheme for cars, plastic window manufacturers followed suit and were offering new for old, thus adding to the burden of losing traditional windows in favour of short-lived plastic replacements.

Silicone has a lifespan of certainly less than 50 years and probably only about 15 years. Deterioration agents include UV radiation, heat, wind, snow, mechanical stress, and even micro-organisms, all depending on location, orientation, and climate. Thus any building incorporating large amounts of plastic in whatever form has but a short life, is going to be deleterious to health, and is creating a mountainous problem of disposal, not the least because of the additives which leach out into water and soil. Whilst recycling plastics may seem to be the key, the practice has yet to develop to its full potential, with few recycling centres accepting anything other than detergent bottles. Very little polystyrene actually makes its way into the indestructible park bench.

Certain plastics definitely exude gases when burnt, and so are problematic in a house fire. PVC is once again the chief perpetrator, with the gases being emitted including carbon monoxide (CO), carbon dioxide, and concentrated hydrochloric acid, dioxins, barium, and cadmium. Polyesters, polyurethane, polystyrene, chloroprene rubber (CR), and butadiene styrene rubber (SBR) emit carbon dioxide and monoxide gases, benzene, styrene, ammonia, acids, as does CR which releases dioxins.

Adhesives are equally a problem, with urea formaldehyde used almost universally, yet it is both carcinogenic and allergenic.

Paints that are chemically manufactured utilise petroleum distillates both for the binder and the solvent, and have been in universal use since the 1950s.

Solvents in modern paints are distilled from oil and can range from white spirit to rather more lethal solvents, such as xylene. Also included are butanol, butyl acetate, toluene, methanol, and methyl glycols. All are injurious to health, and yet the turpentine derived from tree-based resins is considered to be more so and is banned. It is difficult to see why the terpenes in turpentine, a natural product derived from natural tree resin, albeit complex hydrocarbons, are any more injurious to health than the hydrocarbons with radicals attached, in the mineral oil products. Whilst the vapour from natural turpentine can be an irritant to mucous membranes, it is the mineral solvents that cause far more damage to the central nervous system. In addition, these vapours continue to be released for some time after application of a paint system, and can be potentially hazardous for occupants, until all the solvent is vaporised. In addition, natural turpentine can be renewed, as trees can be grown, whilst mineral products cannot.

Synthetic binders mainly fall under the heading of 'alkyd resins', a fraction of petroleum. These produce a thermoset action, and the paint film dries to a hard finish. It is possible to combine alkyd resin with linseed oil, which is done by some of the heritage companies. It is heartening to note that some of the major estates are forsaking alkyd resins altogether, in favour of totally linseed oil based paints, having grown and pressed the linseeds themselves. Thermoplastic paints with an epoxide (isocyanate), acrylate, or urethane base require a hardener to be added. These paints in particular have not only a health problem arising from the use of organic solvent but also by virtue of their constituents. Epoxy based paints are known to be allergenic (particularly for eczema) and carcinogenic, and emit phenols, which are skin irritants, for some time after application. Polyurathane contains the urethane group $NH\cdot CO\cdot O$ (addition of water turns these into urethane foam, also universally used on building sites for plugging gaps), which is basically an isocyanate thought to cause allergic skin reactions and asthma. Again residues can continue to be released for some time after application.

Acrylate paints can emit left-over monomers of butyl methacrylate that can cause allergic reactions. Alkyd oil, the chemical compound formed by combining linseed oil and a polymer such as glycerole or phthalic acid, was one of the first modern paints to be formulated in the 1950s, and due to its heavy texture was considerably diluted with solvents such as toluene and xylene, some of the most disruptive to human cell structure. It is notorious for emitting vapour long after its application, and when used as a varnish also emits alkyl phenols which have an oestrogen-like effect on humans.

Some synthetic resins can utilised with water as the solvent, i.e. the plastic components of the paint, move around in the water in miniature pellet-like form. Unfortunately it often involves the addition of fungicides (formaldehyde) and softeners. One of the earliest of these paints, introduced in the 1930s but not in universal use until the 1960s, utilised polyvinyl acetate as the binder. It has largely been replaced with poly-acrylate because it has better adhesion and durability. Although regarded as much more benign, these water-based products still emit volatile chemicals long after application, containing styrenes, acrylates, formaldehyde, and excess solvent for up to 12 months after application.

Finally in this context it is worth mentioning nitro-cellulose paints made from methyl-cellulose in a polluting process using chlorinated hydro-carbons, with obvious deleterious consequences for humans.

Pigments in modern paints tend to be of the organic or mineral (non-organic type) and rarely earth pigments. White mineral pigments, such as zinc oxide and titanium oxide, are prepared synthetically. Yellows can contain cadmium which is highly poisonous, blues can contain barium manganate and ferrocyanide, also poisonous, as can green containing chromium oxide.

Other additives that may be deleterious to human health include fungicides, which can be based upon pentachlorophenol, formaldehyde, tributyltin to name but a few, which are hazardous to internal environments. Drying agents to shorten drying time can also be added, but these are less deleterious. Softeners and film-forming agents consisting of microscopic plastic particles are added to water-based paints to enable the particles to adhere to each other when the water has evaporated. They are frequently based on phthalate or phosphates and can also be released into the atmosphere, forming a hazard in terms of oestrogen which can affect the unborn child, as well as allergenic and an irritant to nasal passages.

Conclusion

The conclusion of a comparison between natural building products and those that are chemically produced can only be that the former are more preferable because in the main they are renewable, and by and large they do not have a deleterious affect on human health. Unfortunately a vast industry has grown up around the creation of chemically produced products, and has so taken over the world that to suggest dismantling it would result in total dismay and the loss of multiple thousands of jobs. Fortunately oil reserves, which are the basis of most organic chemical products, have only 50 more years to run, so in essence a reversal to renewable naturally produced building materials is inevitable. The sooner mankind starts to reintroduce natural products in building construction the better.

References

The sections on natural materials are adapted from Pearson, D., 1989 *The Natural House Book*, Gaia–Conran Octopus, London and Oliver, P., 1997, *Encyclopedia of Vernacular Architecture of the World*, Cambridge University Press.

All of the information concerning the possible adverse effects of using various chemical-based building materials on human health is adapted from Berge, B., 2000 *The Ecology of Building Materials*, Architectural Press, Oxford (translated from Norwegian by Filip Henley with Henry Liddell). This Chartered Institute of Building (CIOB) award-winning text is one of few where the author, having done extensive research, is prepared to commit to paper the potential effects on humans of our unswerving endeavour as a human race, seemingly to destroy ourselves. Doubtless the claims made could all be contested by the medical profession, but the onus is on them to prove otherwise.

3

TRADITIONAL SUSTAINABLE BUILDING CONSTRUCTION – ANCIENT AND MODERN

Traditional wall construction is typically divided between mass construction and frame construction. Mass walling utilises masonry such as stone, brick, flint, and cobble but also includes the non-masonry material of unbaked earth, referred to as 'cob'. Traditional building construction is characterised regionally in terms of its walling material, although roof claddings do play a part. The cladding of framing, in terms of its infill panels or external cladding, is also both a regional and dating characteristic. Walling material and construction are closely intertwined. Mass walling construction has a dual function. The load of the roof and the floors are carried uniformly by the whole wall, which also acts as a protection against weather. Unfortunately most mass walling construction depends on joints between the building blocks, for both stability and weather protection, and these can be a point of weakness. Only in unbaked earth construction is there a homogeneous mass with no joints, rendering this material very significant for stability and weather protection.

In frame construction, only the timber (or steel) uprights and any cross members are load bearing, the weather protection being the role of infill panels or external cladding. The load of the roof is concentrated firstly on the wall plate, which forms a ring beam around the upper part of the building in conjunction with the tie-beams of the main trusses, and from thence onto components of the frame (uprights and cross rails) to be then distributed via the bottom plate (sole plate) to the foundations. It is, by and large, a myth that traditional buildings do not have foundations, although alarming instances do occur when traditional buildings have been modified. The illustration shows the footings of a 17th-century yeoman house, modified into cottages in the 19th century. The later builders dug out the floors, presumably to gain headroom, and had to resort to pinning the new skirting onto the clay behind. Occasionally this excessive disregard of understanding of building construction is found, but it is by no means the norm for traditional buildings.

Mass walling

It is not difficult to envisage the development of mass walling as a construction technique, although such discussions can only ever be a matter of speculation.

It is possible to speculate that there was little difficulty with stability of low thick walls, but if the building needed to be taller with perhaps an upper floor, then not only was the main beam structure of this floor important for holding boards to stand on, it also performed an important tie function between opposing walls whilst acting as a ceiling frame to the ground floor. A tie-beam at the base of the two principal rafters was also fundamental not only to transmit the roof loads vertically down the walls, but also to counteract outward thrust at the top of the walls. Buildings with both a tie-beam and a floor-beam structure were in general very stable,

110

Photo 3.1 Footings absent in a 17th-century house due to 19th-century excavations to lower the floor

particularly as early timber components were usually over-engineered for the role they had to play. Timber was plentiful in the Middle Ages, as it was being removed from the landscape in order to create arable land, rather similar to the pillaging of the Amazon rainforest to grow soya beans today. Both, in their own periods, led to a glut of the timber being removed and utilised in timber construction, as can be seen from the amount of exotic hardwoods being used instead of oak for modern conservatory building in the last two decades.

Cross walls were also a useful device to prevent outward bulging of long walls. Squat thick walls would never have had this problem, but as soon as the walls became thinner or went higher to accommodate an upper floor, cross walls became fundamental. They are, in all events, achieved by adding another 'unit of living' onto an existing single-cell building, and it would have been readily perceived that this division of the plan created square or rectangular self-buttressing cells. This methodology for creating living spaces is the very basis of early traditional buildings.

Small windows were viewed as not causing any great disturbance in the distribution of load patterns, although to early builders this must have been more of a perception than a scientific fact, and so the size and number of openings were kept to a minimum, which also reduced the possibility of weather ingress. The fashion in the 1970s for picture windows in low-slung buildings not only unbalanced the façade aesthetically, they frequently unbalanced buildings altogether!

A key component of stability was the size of the building blocks, invariably stone in early periods, and the frequency of joints. Large stone blocks, well squared, gave narrow tight joints, with the bulk of the loading travelling through the stone, and resulted in very stable buildings. Small uneven pieces of stone or flint, or rounded cobbles, required large thick joints to create a wall, and the joints were a point of weakness. Thus the poor, who could only afford to pick stones off the field, were always going to be disadvantaged, particularly if they could not afford lime mortar (always expensive due to the need for burning the lime) and instead had to use clay, whose primary weakness was that it was subject to swelling. The use of small fragments demanded thick walls in order to obtain any stability at all. The practice of lacing courses of squared stone or brickwork in wall composed of flint, pebbles or cobbles was a reaction to the problem of using smaller fragments of building material with many thick joints of mortar. It

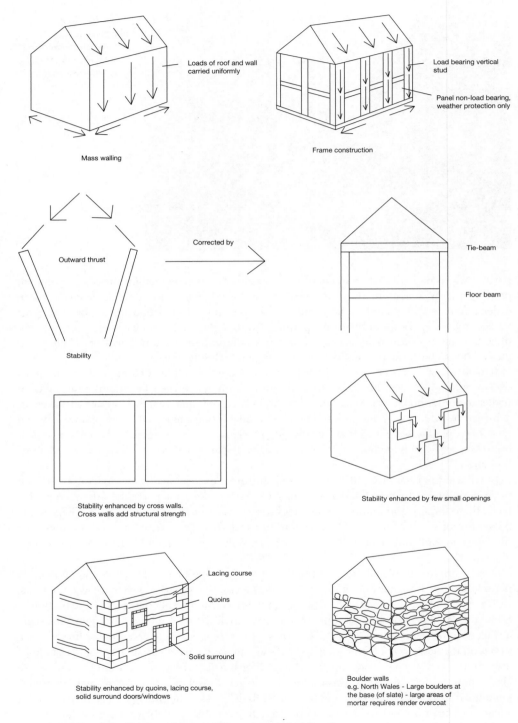

Diagram 3.1 Load distribution of mass walling versus timber framing

Photo 3.2 Glass walls in a traditional building – an unusual rendition of picture windows, now acting as solar windows, in this previously derelict Herefordshire farmhouse

Photo 3.3 Brick lacing courses in flintwork

was not some fairytale technique for creating a pretty façade. For the same reason quoins on corners, plus window and door surrounds, were always surrounded by regular masonry, to ensure that they held upright.

In some parts of Britain, particularly North Wales, the local vernacular involves the use of large boulders at the base of walls in an attempt to achieve some stability where it was most needed, and also because they were readily available. Such large stones demanded very robust wide joints, both for stability and weather cladding. The practice of pushing smaller stones into the face of the joint was called 'galleting' and was intended to cut down the surface area of joints facing the weather, as well as providing a secondary support system for the very outer face. Each small stone took its share of the loading pattern.

Unbaked earth, in contrast, had none of these problems. It is certainly true that the wall have to be very thick, particularly if the building is to have any height, but in general the technique was used for buildings intended to have low walls, and a two- or three-cell structure with small openings for windows and doors. The corners of a building and of any openings were invariably rounded for stability and to accommodate the process of erecting the material, being easier to form than angular corners. In fact, the material is capable of making the most sinuous of structures, providing that there is sufficient thickness to ensure stability.

Of all of these techniques, only timber framing has really survived in any recognisable form and in more than occasional use. New stone or brick buildings abound but the masonry is the equivalent of stone or brick wallpaper, a stone or brick skin to a blockwork backing. Blockwork has taken over the role of the mass walling material. Flint, if used today, is normally set in pre-cast blocks and cement bound. It can be pointed in a lime mortar and even 'galletted' (small flints pressed into mortar joints), but the distribution of flints is normally so limited that in no way can it be confused with a real flint wall. It is generally used as a concession to the local vernacular in terms of aesthetic quality. The use of cobbles or pebbles is virtually unknown today, and the use of unbaked earth is marginal and still awaiting its rebirth. Building regulations for domestic dwellings normally militate against its use because of the degree of damp required to bind the material together.

These traditional constructional forms will now be examined.

Photo 3.4 Large boulders at the base of a wall in a North Wales vernacular building

Stone construction

Ashlar stonework

The ultimate for any high social status knight or lord of the manor was a manor house constructed of regular stonework with thin, tight joints, involving very little mortar. In due course this became the goal of 18th- and 19th-century country house builders of high social status, aristocrats, and even the country squire or wealthy vicar able to afford not just regular stonework but actual ashlar freestone with neat sides and consequently very find joints. In many instances ashlar was bonded onto a rubble stone backing, or bonded over earlier brickwork. Many a country house executed in brick in the early 18th century was not deemed to be fashionable enough by the late 18th century without a cladding of stone ashlar 'stone wallpaper'.

Photo 3.5 Fine ashlar work used on the chimney of a country house

Stone was, however, hard won and very little was wasted. The process involved driving dry stone wedges into cracks, wetting them and waiting until the consequent expansion of the wood produced sufficient movement to allow the stone to be extracted from the seam. The early use of explosive was much more hazardous.

Ashlar is associated with stone beds, which have a natural tendency to delaminate both horizontally and vertically. Thus stone with varying heights and varying lengths was utilised, creating variation in the height of courses. There was little need with ashlar to do anything other than create uniform coursing, although 'snecking', creating a bond between two courses by using a stone which spanned both, was sometimes used. Courses of even height were created where possible, with stones of varying length. There was also little need for quoins (separate corner support), although rusticated quoins were often used with ashlar work.

Quoins were frequently utilised where random, coursed stone was being used for the bulk of the wall, providing a rhythm for the coursing. The classical influence in architecture from the late 17th century saw the development of the rusticated quoin, a deliberate device to create light and shade, sometimes used for the whole base of a building in the late 17th/early 18th century (so called Baroque), to create a mass of light and shade. Internally such areas were frequently basement service areas, but the emphasis externally was upon giving the building a massive cyclopean base, such as Wren or Vanburgh would have been proud of, as a mark of its superiority. Even further embellishment was added to the face of the rusticated stone by vermiculation, a word which conjures up an image of a plate of spaghetti, and indeed the result was reminiscent of this in its effect. The Victorians took the practice one step further with 'rock-cut' faces to echo the very masculine architecture of the period. Quoins were absolutely essential when the wall construction was of rubble stone, usually of the regular type, or where this was given a coat of render, often incised to look like ashlar work. Often referred to as 'poor man's ashlar', it is an indication of the importance of ashlar in making buildings higher social status, and became universal even on quite modest cottages by the early 19th century. Quoins were often faked at the same time, where in fact there was no need for their presence.

Other features that enhanced the social status of a building and which also had some structural function were plinths and string courses. The latter may have helped to expel water run-off from the upper surface of a building, the theory being that water bounced off the edge of the string course away from the lower surface of wall. In practice the upper weathering

Photo 3.6 Quoins are necessary in weaker forms of construction using clunch, a chalk-based building material

surface of the string course could get saturated in storm conditions, resulting in potential dry rot internally. Projecting plinths suffered similar problems.

Irregular stonework

Irregular stonework was much more the province of the poor or building provided by mean landlords. The practice almost certainly began by gathering stones off the fields in order to improve crop yields, but also to make good use of the pile of stones so formed. There are modest buildings in Dorset which have such a variety of quite small stones set in a large expanse of mortar that they cannot even be deemed to be constructed of quarry waste but of field stones. Whilst these building are no earlier than the 16th or 17th century, they must have had similar predecessors.

The construction is also characteristic of stone that is difficult to extract from the quarry and difficult to work into a regular shape, hence its use for random uncoursed work with large mortar joints. Kentish ragstone and Welsh slate are typical examples of this type of stone, whilst millstone grit and granite can only be worked with difficulty and to a degree sufficient to form a semblance of regular stonework. The technique probably also made use of small irregular pieces of stone that would otherwise have been wasted.

The different forms of irregular stonework construction are:

- **Random uncoursed:** stone which is random in shape with no possibility of coursing, and a large quantity of mortar. Galleting was frequently used to cut down the amount of mortar for weathering and to provide a path for forces on the outer face. Walls so formed were often battered, getting thinner as they went upwards. This is also often the hallmark of buildings earlier than the 18th century, and some medieval buildings built of random uncoursed construction. Through stones were used to ensure a tie from front to rear.
- **Random coursed:** stones were random in shape but laid to courses of varying heights – often approximating to the height of the quoins. Bonders or risers (larger stones) which rose up through most of the height of the course as well as most of the depth of the wall were used also as through stones. The technique is not dissimilar to dry stone walling in that there was a careful choice of stone to make up the height of a course, but naturally much more leeway with mortar making up much of the height.

Photo 3.7 Random uncoursed stonework

Photo 3.8 Random coursed stonework in Herefordshire

- **Random squared uncoursed:** this usually means that whilst the stones are random in size, an attempt has been made to square them up, and they are used with a considerable amount of snecking. This interrupts any real semblance of coursing going all the way across a wall. For some reason it is favoured by many modern builders when cladding a blockwork wall behind.
- **Random squared and coursed this** is known more universally as 'regular' and characterised by stones of varying length but the same height, laid to courses. The height of courses varies.

Most of the above required large dressed stones to act as quoins at the corners of buildings, and random uncoursed stonework was frequently designed to be rendered, more to ensure a weatherproof overcoat for all the exposed wide joints than for any desire to embellish, except for the incising of such renders to look like ashlar. Multiple coats of limewash completed the weatherproofing process, building up to a fine plaster often indistinguishable from a thin render. The removal of such overcoats has continued since the Victorian period, to the long-term detriment of such buildings, particularly on churches in line with the much vaunted late-Victorian concept of 'truth to material'. What is even more remarkable is the penchant in recent periods for rustic-looking secular buildings, achieved by stripping away a render coat to expose uncoursed stonework, a fact which would have been anathema to those who constructed and initially utilised these buildings. By so doing the mortar joints are exposed to the weather and erode away. In addition, the stone is exposed to wetting and drying cycles, which lead to the mobilisation of salts and consequent decay.

Uncoursed stonework is very rarely used today, the smaller pieces of stone removed from the larger ashlar blocks or slips being totally wasted, unless one counts crushing for road stone as a viable usage. It is not, and such a shocking waste should be avoided. A whole new role awaits such rejected stone – that of being cast in a lime or cement mortar and used to form mass walling to act as a thermal store. If it can be done with flint, then it can be done with small pieces of stone. Larger pieces of uneven size could be bedded into a mortar face in a dry stone walling way, and perhaps be given a 'watershot' form of construction redolent of walls in the Lake District, whereby the outer stones all tilt slightly downwards so as to throw away

118

Photo 3.9 Regular stonework on this newly revealed chimney, now buried within a house

Photo 3.10 Stonework regular and coursed

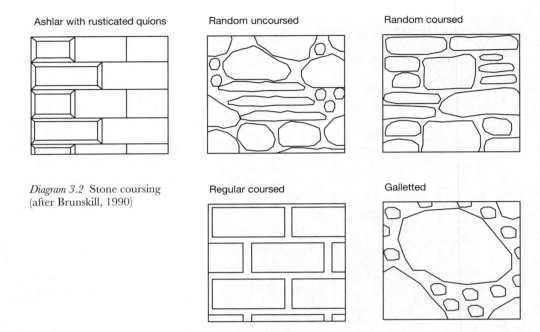

Ashlar with rusticated quions   Random uncoursed   Random coursed

Diagram 3.2 Stone coursing   Regular coursed   Galletted
(after Brunskill, 1990)

precipitation. This traditional technique was devised to cope with the continuous rainfall of this area, but in a situation of climate change whereby the whole country is experiencing longer and more intense spells of rainfall, it could be the ideal answer.

Cobbles and pebbles

These are small rocks worn smooth by the action of ice, rivers, or the sea. As a building material, large boulders and cobbles found service in remote rural areas adjacent to the beach, as in North Wales, and even in fashionable areas such as Brighton, heavily disguised under a coat of render. Generally the material was bedded in clay, this being in line with its very modest social status. Clay mortars are definitely at the lower end of the social spectrum, and rely on a render overcoat for survival. In coastal areas the ready availability of tar for mending boats meant its use as a waterproof coating for a cobble construction.

Pebbles are less than 75 mm (⅜ in) and cobbles up to say 225 mm (9 in) or more. Not always rounded, they can be oval which, like flints, can generate useful tails to bed inwards in the large quantity of bedding material involved, but were also used in decorative diagonal or chevron patterns. Extra support around apertures and on corners was essential, often in brick, as well as lacing courses of the same.

Pebbles and some cobbles could be used to make cast blocks. The material is, however, notoriously cold in use, and prone therefore to condensation. Other than the mortar for the casting, the block would lack the porosity to enable it to act as a breathing wall, although such blocks could be used as an independent external wall as a rain screen cladding.

The wall illustrated below, albeit minus a plaster coating, has massive condensation problems due to the material being always cold. Any warm moist air generated immediately condenses on the wall. The solution would be either an outer lime render overcoat or an inner lime plaster coating. There would be a loss of aesthetic quality of the material either way.

120

Photo 3.11 River cobbles forming the wall of a house – the brown deposit is salts

Uncoursed cobble

Coursed cobble

Knapped flint

Squared and knapped flint

Diagram 3.3 Cobbles, pebbles and flint (after Brunskill, 1990)

Flint

Flint has many similar aspects, requiring large quantities of mortar for bedding, certainly for un-knapped flint (the whole ovoid-shaped flint resting tail-end into the wall) and even for knapped flint (the fatter end removed to create a flat face). Squared and knapped flint requires much less mortar and can be laid to regular courses, but must have been more expensive to achieve, so lowly dwellings used whole flints bedded into large quantities of mortar or clay. As with the technique for uncoursed stone, lacing courses, quoins and the dressing of apertures were done in brick or stone. Early buildings such as parish churches and medieval manor houses in flint areas display a unique chequerboard pattern of stone and flint. The material was so readily available in any chalk area that even the rich made use of it, albeit in a rather superior stone-dressed manner, from the 16th century.

Flints are already available in cast blocks, although largely only used as a decorative medium on the external wall of a cavity. Again the material is very cold, and with the exception of the mortar for the casting is unsuitable to act as a thermal store.

Brick

Brick is not generally thought of as a vernacular material, but in those areas where good brick earth was plentiful, local brick yards occurred from at least the 16th century onwards. The chimney on most yeoman farmhouses, and in many cases the whole chimney breast, was constructed in brick in order to benefit from its propensity to act as a thermal store.

Handmade bricks are characterised by their rather uneven form, and the wrinkles formed on the side of the brick when the scoop of clay from the bench is thrown into the mould. In addition, a linear bulge of clay diagonally across a face is an indication of where bricks have sat in the drying sheds prior to going into the kiln. Variable size and wavy edges meant that large mortar joints were necessary, some of the earliest examples of which are at Hampton Court Palace, London. Variable burning in open clamps meant not only uneven burning but also a range of intensity of colour from pales reds though to dark blues and purples. Builders took advantage of this trait to create all manner of decorative patterns on the surface of walls, including spelling out initials and dates. By the Victorian period, better techniques for brick

Photo 3.12 A rare survival of untouched flint walling

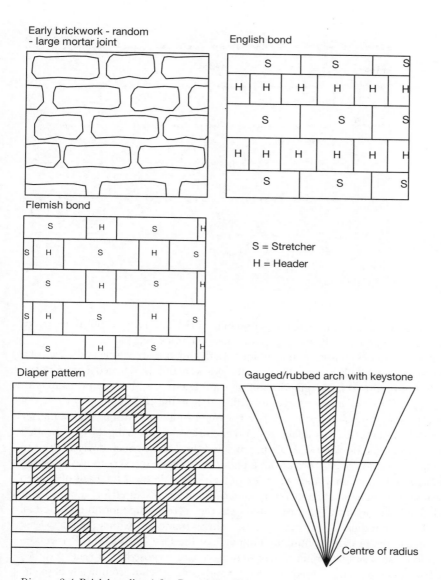

Diagram 3.4 Brick bonding (after Brunskill, 1990)

production, such as extruded bricks, meant less need to use the over-burnt blue ends of bricks, but the desire to use the material as a decorative medium and perhaps the long tradition meant that bricks were dipped in metallic sand and fired so that the headers became a molten purple colour. These were then deliberately used in diaper patterns.

Early bricks will invariably show a variation in size and proportion, as well as pebble inclusions, the latter being difficult to remove by hand. Old bricks are best measured in Imperial, as that is how they were laid, and bricks ranging from 8 to 10 inches (20–25 cm) in length are not unusual, ditto ranging from 3 to 5 inches (7.5–12.5 cm) in width, and 2 to 4 inches (5–10 cm) in height. A good middle road was a brick measuring 9 × 3 × 3 inches (22.5 × 7.5 × 7.5 cm), although

Photo 3.13 Handmade brickwork with blue headers plus traces of lime wash

so-called 'great bricks' were in operation in the period of the brick tax (1784–1850). These were overlarge bricks, often laid on edge in an attempt to dodge the brick tax.

Bonding is what determines construction technique. Early buildings had little in the way of a recognisable bonding, presumably because the brick moulds were unsophisticated and produced bricks of variable size that used large quantities of mortar. There was always a problem with uneven arrises (edge) of bricks, counteracted quite deliberately in the 18th century by the use of ruled joints or even tuck pointing (the insertion of a ribbon of white mortar set within a mortar the same colour as the bricks). The manufacture of precise bricks by squeezing the clay through a linear tube and cutting the brick off with a wire obviated the need for such devices. Prior to this, the only way forward to create a brick with a perfectly straight edge was to use well-fired sandy clay, and rub each brick on a stone to create precise sides. This was the basis of gauged brickwork to be found on high social status Georgian façades. It was also used for 'rubbers' over windows and doors. Each brick was rubbed until it was the correct shape for an arch and all were aligned to the same radius. The bricks were cut using a frame saw with a double twisted wire, using a template, the practice continuing until at least the First World War. Bricks used for external cornices had a similar origin if they were to remain exposed, or if rendered they were often rough cut to shape. The technique of cutting and rubbing is now completely obsolete, except for specialist conservation work, so every rubbed brick arch is a feature to be treasured.

The random early bonds eventually developed into English Bond in which one course of headers was alternated with one row of stretchers, with variations including three or five courses of stretchers. English bond was often used side by side with Flemish Bond in which normally each row had alternating headers and stretchers. Variations on this theme included increasing the number of stretchers between each header.

Timber framing

Simple timber framing traditionally revolved around the creation of cruck frames or post-and-truss frames. No one is brave enough to hazard an opinion as to which came first, the cruck

Photo 3.14 A modern version of a gauged brick arch

being formed from an oak tree with its largest branch split down the centre and with one half rotated to face the other, or post-and-truss construction possibly allied to the lintel-and-post construction of Greek or Roman architecture, or even for that matter to early British structures such as Stonehenge. Certainly cruck buildings were easily formed into open halls, and Britain was heavily forested with oak trees that leant themselves to this form of construction, whereas post and truss requires much more timber processing to obtain the correct components. Mankind has never gone in for work for the sake of it, so cruck construction may be the precursor.

Cruck construction is characterised by what is termed 'reversed assembly'. The cruck members themselves form the main posts (1) to which are then fixed the tie-beams (2) that hold the last component, the wall plate (3).

Post-and-truss construction is termed 'normal assembly'. Here the posts (1) are followed by the wall plate (3), held into the back of jowelled-out corner posts, with this joint then clasped firmly by the addition of the tie-beams (2).

Modern crucks are made from glued and laminated timber with staggered joints and are the basis of some very successful contemporary structures.

Crucks are reared in succession from one end, and the frame is notched into the back of the cruck blade. This means that it is relatively easy to replace the frame should it rot out at the feet, although its propensity for doing so was limited by virtue of this joint not being immediately near ground level. It is not uncommon to find that the framed walls of a cruck building are later than the cruck itself, which may have been re-used and re-sited. Post-and-truss construction, its name suggests, was formed by positioning the sole plates on to an upstanding plinth, then each corner post in turn was angle braced to the sole plate. The normal assembly as described above follows this together with other components of the frame, the uprights and cross rails.

One might immediately dismiss cruck-framed construction for modern build on the grounds that no substantial timber could possibly be found, but in actual fact the construction of curved members in plywood laminate, described above, provides a stunning and innovative component of sustainable new build.

125

Photo 3.15 Traditional cruck roof construction

Photo 3.16 Laminated cruck construction at Bournemouth University

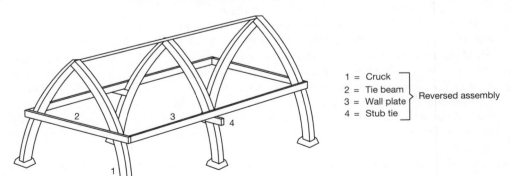

1 = Cruck
2 = Tie beam
3 = Wall plate
4 = Stub tie

} Reversed assembly

Diagram 3.5 Cruck construction (after Brunskill, 1990)

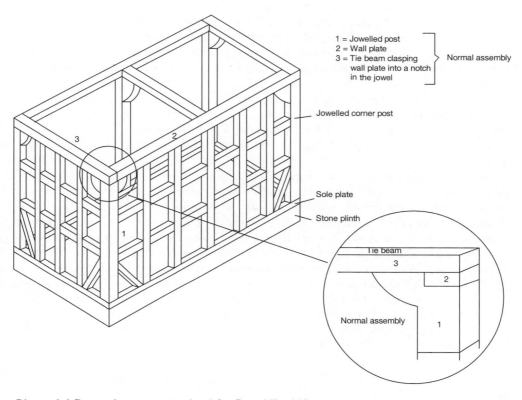

1 = Jowelled post
2 = Wall plate
3 = Tie beam clasping
    wall plate into a notch
    in the jowel
} Normal assembly

Jowelled corner post

Sole plate

Stone plinth

Tie beam
3

2

Normal assembly   1

Diagram 3.6 Post-and-truss construction (after Brunskill, 1990)

Photo 3.17 Traditional post-and-trust framing employed in new build

Traditional post-and-truss buildings are still constructed in counties like Shropshire and Herefordshire where the timber framing tradition is particularly strong (and in some areas where it is not the tradition!). Generally and confusingly the practice of constructing a cheap softwood frame as a support for plywood sheathing and brick cladding is the modern variation, with all its attendant problems of vapour leakage leading to potential rot. The technique is particularly used for social housing.

Traditional wall framing followed various patterns according to timber availability, social status, and period. Large square panels are characteristic of medieval framing, as is the use of wide and shallow uprights. The former may indicate the first grasp of this technique, the latter the methodology of processing very substantial oak trees into slabs as opposed to quarters.

Large curved tension braces spanning from corner post to sole plate provided extra stability in the medieval period, to be replaced by straight braces by the 17th century. As the centuries rolled forward, and particularly by the 17th century when substantial oaks were a diminishing resource, the size of panels became smaller, the scantling of uprights based on the quarter round, and c.1600 the very rich favoured the use of decorative components such as quatrefoils and diagonal bracing. This appears to have been a deliberate attempt to utilise smaller boughs, and may be an indication of how scarce timber had become, the clearance of trees from fields for agriculture being a distant memory, and the mass use of substantial oaks for the navy, and for charcoal burning, taking precedence. Redundant mortises and notches, frequently seen in 17th-century buildings, are redolent of the use of the same timbers in an earlier building, again an indication that by the 17th century timber for construction was becoming scarce.

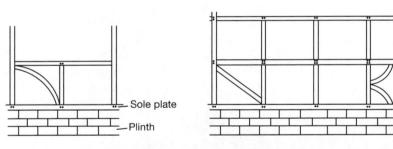

Large square panels and curved brace   Small square panels and straight braces

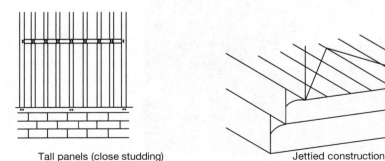

Tall panels (close studding)   Jettied construction

Diagram 3.7 Wall frames in timber frame construction (after Brunskill, 1990)

Small square panels are more typical of the west side of the country, although by the 17th century closed studded framing (tall panels) anywhere was an indication of wealth and high social status, only the rich being able to afford the higher prices of a scarce commodity. This was true of farmers in the south-east of England, who had become wealthy at an earlier period and were building substantial timber framed houses as early as 1400, giving rise to what is termed an 'eastern school' of framing utilising closed studding. A so-called 'northern school' of framing exists where the major posts interrupt the sole plate and sit within the plinth. This may have its origin in earth-fast posts that almost certainly preceded sole plates raised on a plinth, and the considerable distance between the north and south-east, the latter being the major fount of knowledge on framing techniques.

Jettied construction, a form of pre-stressed building construction, was developed by placing one complete framed box on top of another, each successive box overlapping the floor area of the one below, by dint of extending the front- and/or side-facing floor joists. The technique relied on these substantial floor joists, and dragon beams from corners to main floor beams. The latter were supported by corner angle braces to main corner posts. This methodology allowed buildings in towns, where floor space was at a premium and very costly; to grow upwards, and at the same time the upper floors protected the framework and joinery of the lower floors from precipitation.

Photo 3.18 A modern timber frame under construction

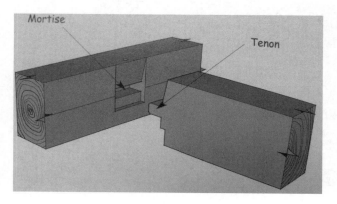

Photo 3.19 A mortice and tenon joint © Rob Buckley

129

Balloon framing

This is the forerunner of modern-day timber framing, and was specifically designed to be clad in wooden boarding. It was often used for lowly cottages, as is evidenced by those at the Weald and Downland Open Air Museum near Chichester, West Sussex. In the early 19th century, the frame was designed to be clad in a lath and plaster (inside and out) or weatherboarding, with little thought to insulation. Now the frame is clad in a plywood sheathing, and insulation, before being brick-clad or weatherboarded.

Timber wall cladding

Timber can be difficult to maintain in exposed positions and as the hardwoods, such as oak, elm, and chestnut, became scarce, softwood framing was developed. From the 18th and into the 19th century, Baltic pine was imported in large quantities. Very slow grown, for long periods, in cold climates where growth was only possible for say three months of the year, the growth rings, heavily impregnated in resin, were packed closely together. This provided an inbuilt fungal inhibitor as well as imparting considerable density and strength. Even so it was no match for English oak, which for centuries had needed no rain screen cladding.

Tile cladding was popular in those areas where oaks had previously grown in good clay soils, as the clay that was being made to make bricks could equally well make tiles. Tiles were hung on laths nailed across the timber uprights or intermediate studs introduced to cut down the spacing, each tile underneath the vertical joint formed when the two tiles are butted together, creating a three-way lap. Handmade tiles were deliberately bent over a saddle-shaped mould, so that they had a two-way camber, from top to bottom, and side to side. The camber is very efficient in removing water from one tile to another down the vertical face or roof slope. Victorian machine-made tiles lack this attribute and not only look boring but are not as efficient, the efficiency being confined to production rather than use. The Victorians did, however, introduce a variety of tile shapes, the most favoured being scalloped – the fishscale appearance lending them the name of 'fishscale tiles'. Coloured tiles were also used, to create variety or

Photo 3.20 Fishscale roof tiles on a clay-mortared Dorset farmhouse; they were also used as wall cladding

spell out names and dates, a very handy tool in later period analysis, and a practice that needs to be revived.

Clay tile regions also utilised mathematical tiles when viewing the frame became unfashionable in the 18th century. These were deliberately shaped as a tile but with the base bulked out into brick shape. When overlapped for weathering the brick shape stood out, surrounded by a gap that could be mortared to resemble brick joints. The tell-tale signs that the façade is mathematical tile and not brick are the wooden fillets that can be discerned on corners and the very narrow depth of window reveals. Rubbed brick arches over windows are often discovered to be incised coloured render on closer inspection.

Slate-hung cladding is similar to tile cladding, and was used in those areas like the Welsh Border and Wales itself, where the wind-driven rain is fierce on west and south-west elevations. Again slates were often shaped at their base to form decorative patterns.

Rendered or pargeted timber frames

Rendering became a common practice in the 18th and 19th centuries. Frames started to rot for lack of maintenance, particularly to infill panels, and lack of suitable timber scantling for repair led to whole frames being covered over. In addition, timber framing had become the symbol of a poor man's building material, and often only survived as a visible asset in lowly agricultural workers cottages, subdivided from a once grander yeoman farmhouse, or in towns in the back streets where artisans rented run-down property. As such it became unfashionable, and lathing over whole frames and coating with a lime render, incised to look like ashlar stonework, became the order of the day. The practice, in all events, had its origins in pargeting, a practice common in the south-east from the 16th century, whereby wealthy yeoman farmers enhanced their high social status by incising a lime plaster cladding to timber frames, in patterns such as basket work, herringbone, fan combed or cable moulding to name but a few. Complicated floral and interweaved trailing patterns of foliage were sometimes created, reminiscent of Saxon zoomorphic predecessors in stone. Features from coats of arms such as the rose or fleur-de-lis gave an indication of political inclinations.

Photo 3.21
Mathematical tile cladding in Blandford, Dorset

131

Photo 3.22 Slate hanging in Guilsfield, near Welshpool

The practice of creating some of these patterns was briefly revived for use on ceilings in the 1980s using Artex, a marble-dust impregnated gypsum plaster. It is now much despised.

Brick nogging

Infilling frames with brick panels, from the 18th century onwards, was possibly a reflection of a backlash against the constant maintenance by lime washing that wattle-and-daub panels required, and the lack of this vital treatment due to sheer laziness. In retrospect it can now be seen that this was often not a good development. No better technique has ever been found for infill panels than wattle and daub, as the flexibility allowed for movement in the frame and limewashing built up to fine plaster in the joint between panel and frame, excluding the weather. Brick panels did not have this advantage, although the use of lime mortar to butter up the joint between panel and frame was more effective than modern mastics. Brick was occasionally laid in a herringbone fashion, and coloured bricks used to create initials and dates.

Insulation problems in timber frames

Methodologies of upgrading panels in existing traditional timber framed buildings for energy conservation is challenging, be they wattle and daub or brick. The removal of either material is not recommended as it is valuable historic fabric, and from a purely practical point of view wattle and daub has never been bettered. Many substitute materials have been tried, such as wood wool slap and urethane foam panels, but the joint formed with the framing remains a problem, however many clever types of mastic are introduced. Insulation of internal walls using a vapour-permeable material, such as sheep's wool and wood fibre board, enables the traditional building form to be read from the outside. Alternatively, cladding the outside with boarding under which insulation is incorporated enables the internal format to be appreciated. It is not possible to retain both the inside and outside form, and this can cause consternation for an enthusiastic owner who wants both good insulation values and the aesthetic quality of the frame to be viewed both externally and internally. Filling the panels with large slaps of insulation material so that they resemble fat pillows is neither aesthetically pleasing nor productive in terms of insulation.

Insulation when frames are externally clad with tile or slate hanging is best achieved internally without disturbing the exterior, but in addition the cladding can be removed, insulation applied across the frame, and re-lathing accomplished onto which the cladding can be re-applied.

Weatherboarding

This was and still is the most popular way of cladding timber frames to exclude weather, being used almost universally for farm buildings in the late 18th and 19th centuries. This practice only ceased with the advent of iron and asbestos sheeting, but has been revived in the build of many portal-framed farm sheds.

Feather-edged boarding with the narrow part facing upward ensured a hefty portion of timber board partially over-lapping the board immediately below. Butt-edged boarding was also used, every board being edged with a bead moulding, as was tongue-and-groove board, with the tongue applied to the lower edge to sit in a channel in the upper edge of the board below. Such techniques were well known for floor boarding, butt-edged boards being regarded in the medieval period as part of the furniture of the building and moved around at will. Some farm buildings had vertical boards with the joints protected with a cover strip, the boards being

Photo 3.23 Rare survival of old weatherboarding, Herefordshire

Photo 3.24 A Dorset farm building with vertical boarding with cover strips

nailed to rails nailed across the frame. This may be a very ancient form of cladding which has wended its way down the social scale from a much earlier high social status domestic use. It has affinity to plank and muntin screens (planks fitted into channels in the sides of much heftier studs) that were medieval screen walls in domestic buildings. In Dorset the practice continued until the 20th century, with the butt boards held in place by a ledge, for domestic partitions.

Boarding was not confined to farm buildings and was used in the 18th century for even quite high social status buildings in towns to disguise the timber frames, by that time regarded as a symbol of poverty, but with features such as jetties still remaining easily discernible, it being impossible to disguise the projecting upper storey. The practice was more popular in the east of England, deal boards being possibly more available due to the proximity of the ports. It can be seen as far west as Ludlow in Shropshire.

Weatherboarding is by far the most effective means of giving an existing vernacular/traditional building an external insulated jacket. The insulation must be vapour permeable, the timber having the same characteristics. It is also a good method for external insulation on masonry buildings, allowing the internal walls to act as a thermal store. The only building material for which it is not suitable, unless in a very exposed position, is cob. Here the material is its own insulation and must be able to freely transport moisture inward and outward.

Roof construction forms

Traditional roof-shape construction is divided between gabled and hipped roof forms.

Gabled roof forms

Gabled roofs are the traditional format, although thatched roofs in the south of England appear to have had a hipped end roof form from quite early periods. Gabled roofs were traditionally very steep and very efficient at removing water. Mono-pitched roofs can be thought of as half

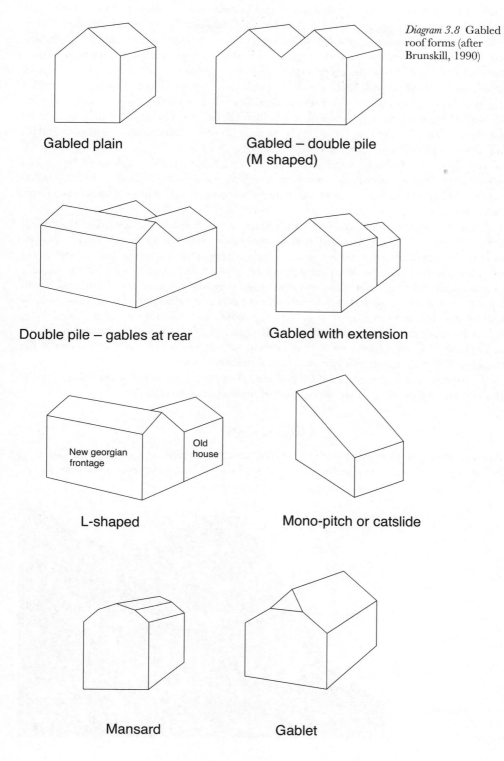

Diagram 3.8 Gabled roof forms (after Brunskill, 1990)

Gabled plain

Gabled – double pile (M shaped)

Double pile – gables at rear

Gabled with extension

New georgian frontage

Old house

L-shaped

Mono-pitch or catslide

Mansard

Gablet

a gable, and are also efficient at water removal. The mansard or gabled mansard is generally a late 17th/early 18th century take on a gabled roof, and possibly reflects an influence from the Netherlands at the same time as brick buildings were becoming popular. Double-pile gabled roofs (M shaped) were utilised for the first country houses, to embrace a square plan format. Dutch influence can also be discerned in the practice of gabled parapets, some in the shape of a Dutch gable, which were devised to protect the ends of battens. Coping stones supported on projecting kneelers also protected verges and ensured that there was no possibility of uplift. Brickwork that tumbled inwards was occasionally used, and must have been challenging in its execution. Other forms of brick decoration reminiscent of the same desire to decorate gables can be found in the late 19th century.

Verges on gabled buildings were also protected with bargeboards, which could be pierced to create a decorative pattern and surmounted with an up-standing finial or a down-standing pendant, although often little attempt was made to protect the ends of projecting rafters, known as 'sprocketed eaves'.

Gabled roof forms lent themselves to a relatively easy method of extending traditional building forms, and still do. Two-unit houses were easily extended by the addition of another unit in the same plane as the roof, or with the ridge dropped down, a practice that still works well with modern extensions. An existing three-unit house could gain a service wing to the rear or a pair of parallel wings that could convert the building into a square plan form, which with the addition of a brick skin and central entrance created a gentrified farmhouse. More usually the old house was fronted at right angles or in parallel, with the addition of a gabled wing, furnished in the latest 18th-century mode with square plan rooms with dado panelling, etc. The old house was consigned to a service function, with the former house-place acting as the farmhouse kitchen, a function it had always traditionally enjoyed.

Gabled extensions onto existing gabled buildings are still very successful, although the parallel format creates valley gutters that are a potential maintenance hazard.

Hipped roofs

Hipped roofs can be thought of as a gabled roof falling backwards at the end of the building. Medieval open halls with thatched roof utilised the form, which made a convenient exit for the

Photo 3.25 A decorated brick gable in Oswestry, Shropshire

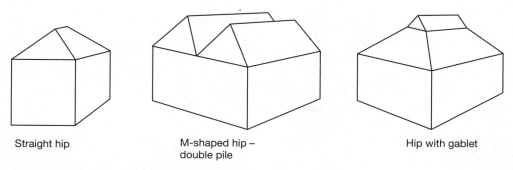

Straight hip   M-shaped hip –   Hip with gablet
                                double pile

Diagram 3.9 Hipped roof forms (after Brunskill, 1990)

smoke from the open fire, through small gablets at the summit of the end hips. Hipped roof forms predominate in the south-west of England, and may be derived from the earlier medieval roof form. Hipped roofed single-cell buildings were conjoined to create double-pile planforms for larger houses, although there was an area of leaded flat roof in the centre of surrounding hipped roofs, thus extending even further the size of the double-pile plan. Mansards, sometimes called a 'hipped gambrel', create a particularly ponderous roof form that can make buildings look top heavy, and apart from a penchant for modern developers in the 1980s to use this form, it has enjoyed little popularity. Generally hipped roof forms are more conspicuous on a building extension than a gabled form, although a mistaken belief that the opposite is true appears to prevail amongst developers.

Roof cladding

Thatched roofs

Currently much angst prevails in perpetuating this ancient form of roof cladding. Many modern thatchers are importers of water reed grown in Eastern European countries, and it is possibly true that water reed has greater longevity, but any organic roofing material is only as good as its lack of nitrogen, as it is the latter that causes rapid breakdown of the fibrous material, being very reactive with oxygen from UV light and acid rain.

Thatching has always responded to the material available. In days before the reaper and binder were available, the long straw (early wheat varieties grew long stems with small ears of wheat) was gathered into yealms or bottles with a miscellany of end butts and wheat ears at both ends. Tied roughly in the middle with a twisted straw, the bundle was tied to the eaves with vine or willow. Later this became tarred rope, and the modern practice of steel ties certainly lacks romance in comparison.

The roof would have been prepared with pole rafters (unadzed larch or ash poles) and cleft oak or chestnut battens. Successive layers would be pegged into place with hazel twisted into a hairpin shape, known as a 'spar'. Evening and winter activity for thatchers involved mass production of these, which required a neat twisting method to enable a bend to be created in the centre. The hairpins or spars were made when the hazel had an element of sap present.

The total cladding of the roof was made by ensuring that each layer overlapped the layer below to a large extent, and a complete roof would have the gables and ridge held down with half hazel rods known as 'liggers' or 'sways', making a decorative criss-cross pattern. This is the

137

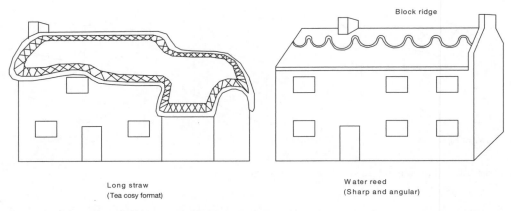

Diagram 3.10 Thatched roofs (after Brunskill, 1990)

Photo 3.26 Steel ties now used for thatching in place of the traditional hazel spars

hallmark of long straw, although rather sadly modern thatchers have started to do the same with combed wheat or water reed roofs, thus confusing the tradition. The unruly ends on a long straw roof were teased into place as the roof cladding was laid, using a legget, a flat board with a handle, in which was bedded iron spikes for the purpose of combing and teasing the material. Large shears were used to trim the eaves, and the finished product resembling a large tea-cosy. The material enabled quite complex roof forms to be swept over, creating a snug continuous roof covering that literally flowed over the building below. Roof pitches were between 45 and 50 degrees and overhangs ensured that the water run-off never inflicted itself on the wall structure. It was thus ideal for cob buildings and boundary walls. In Dorset there is ample evidence that the technique was used for quite grand buildings, such as rectories and up-market gentleman's residences.

Combed wheat roofs arrived with the reaper and binder. The butts could all be ordered to face one way and the ears of wheat the other. The technique of application is similar but the roof adopted a rather neater appearance. Trimming at the end of the job was no longer

Photo 3.27 A recently repaired traditional combed wheat thatch with liggers

required, the bundles being graded up the roof and presenting a more precise format. It has to be said that a newly clad golden-coloured combed wheat roof is a joy to behold, but it sadly soon discolours with a combination of oxygen and UV light altering the bonding electrons in the molecular structure of the straw, and the rapid absorption of carbon particulates via acid rain. The latter also has a deleterious effect, the acid eating into the chemical structure of the straw. The recent three-year spell of saturation has done little for the survival of thatched roofs.

The universal advent of imported water reed, despite the fact that it was only ever used in those areas where it was abundant, such as Norfolk and a very small pocket around Abbotsbury in Dorset, has severely disrupted the ancient tradition of using wheat, barley, or rye straw. In addition, it is surely a non-sustainable material, its importation creating sea miles. It is fair to say that DEFRA and the former Ministry of Agriculture Fisheries and Food (MAFF) have done little to promote the growing of wheat straw for thatching. In fact, they appear to have positively hindered it, discouraging the buying of long-stem seed varieties in case they contaminate the growing of short-stem varieties. The even greater potential of genetically modified (GM) crops contaminating non-GM crops by their proximity, in contrast, appears to have little validity. Some brave farmers are prepared to risk the wrath of the powers that be to supply much needed long-stemmed wheat straw, and long may they do so.

Water reed roofs built from scratch have a mean profile, hence the development of the 'Disneyland' block ridge to try to beef up this vulnerable part of the construction. This is a development that has regrettably permeated the more ancient traditional materials to their detriment.

It is important to maintain ridges, gables, and the areas around chimneys, for all thatching materials as these areas are vulnerable to saturation, resulting in a compost heap, rather than a collection of straw forming mini-gutters to ensure run-off.

Thatching new build tends to look somewhat fraudulent whenever it occurs as a device to make new build fit in with village backdrops. Very rarely is it used for its high insulation quality, to keep buildings cool in summer and warm in winter, although this should be the primary reason for its use.

Photo 3.28 Thatch can become a compost heap where it abuts a flat surface

Photo 3.29 Thatch is a universal traditional roofing material, as seen here in Madeira

Photo 3.30 Moss on a thatched roof means it is no longer functional

Random, stone tiles or stone flags

The use of stone tiles (usually limestone), one of the earliest roofing materials, was popular for very high social status buildings from the Middle Ages to the late 17th century, except in those areas where the material abounds, such as the Cotswolds, where it was used for all manner of building types for many centuries. It was laid in diminishing courses to accommodate the wide variety of forms that came out of the quarry beds, which were removed by dint of wedges. The wedges were then wetted and swelled to create pressure, aided and abetted by frost conditions which naturally laminated the tiles. Stone tiles are laid in the same way as clay tiles, with each stone tile lapped over by the butt joint between the two tiles above it, and at a steep pitch of around 50 degrees. The top of the tiles was dressed to a curved format, and the fixings included oak pegs and sheep bones. At least every fourth course is lime mortared into position to cater for the weight. So strong is the tradition in Dorset, that even when the bulk of the material has long ago been removed, a few course are always kept at eaves level, a practice which has sparked a later vernacular tradition of employing a differential material at eaves than that used on the main roof slopes. Stone tiles lent themselves to the creation of swept valleys or laced valleys, either one being remarkable examples of ingenuity in the placing or gauging of the battens. There is nothing more beautiful than such a valley on a stone tile roof, and when it is replaced by a standard lead valley it is sign of both loss of skill and loss of aesthetic value.

Stone flags are much more robust and considerably larger than stone tiles. They are found in the north of England, and commensurate with grit stone areas such as Lancashire and Yorkshire. They require a more shallow pitched roof, between 30 and 35 degrees, the great weight being totally unsuited to a steep pitch, as they would simply slide down the roof. Again they are shaped at the top, possibly to remove some of the bulk and weight. It is possible that flags were used on eaves, even when the remainder of the roof was another material such as tile or slates, as the flags could hold down the end of the rafters and ameliorate any possibility of up-draught tearing off the roof in exposed locations. The topography of West Yorkshire, for example, means that many traditional buildings were elevated and exposed to such conditions.

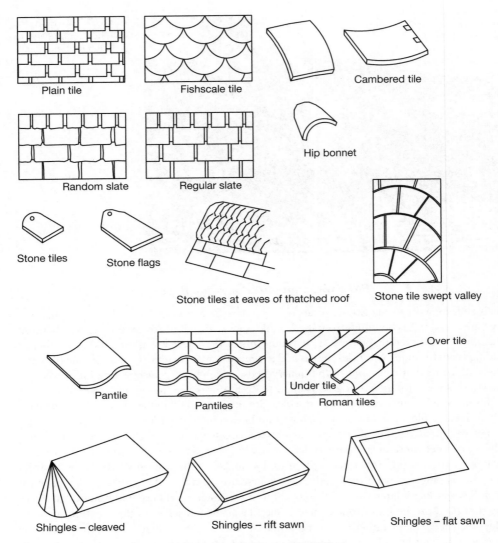

Diagram 3.11 Various types of roof cladding (after Brunskill, 1990)

Random slate

Natural slate started to be used in the canal era (late 18th–early 19th century), when it became possible to transport slate from its source in areas like the Lake District and Wales to much further afield. It required considerable skill to clad such a roof, as the battens needed to be laid in diminishing intervals as they ascended the roof from eaves to ridge, on shallow pitches. This was to accommodate a random selection of thick slate which was then laid with the larger pieces at the eaves, the size diminishing until the ridge was reached. Ridges were traditionally clad in a stone saddle, or with wrestler slates (slates specially shaped so as to have a form of tenon at the top, which rested between the neighbouring tenons on the other side of the slope), the whole

Photo 3.31 A rare survival of a stone ridge saddle

Photo 3.32 A swept valley in stone tiles

Photo 3.33 Random stone tile roof

forming an interlocking profile, or with a lead flashing. Hips were protected with clay bonnets or with a mitred slate meeting over concealed lead soakers, the end hip bonnet being infilled with a mixture of tile slips and mortar and held in place by an iron support, often formed into an attractive sinuous shape. Valleys are swept or interlocking, requiring even more challenging placement of battens to ensure a tight fit.

It has taken long and hard campaigning to ensure that such roof cladding receives the attention it deserves, and there are certainly problems with taking off the material and re-using it. Firstly, because it is thick slate it tends to delaminate naturally in freezing conditions and the smaller pieces suffer more decay. There is thus a shortfall which is difficult to make up in the required sizes. In addition, modern roofing contractors not taught by their fathers or grandfathers but in a technical college environment are much more suited to a straightforward thin, regular slate re-roofing and tend to eschew the difficulties that arise from anything more complicated. It takes a determined owner and roofing contractor to achieve the required product. The demise of grant aid systems for historic buildings has done little to promote the retention of this form of roof cladding.

Regular slate

Regular slate has the appearance of having taken the country by storm in the mid–late 19th century and is as popular today as it has ever been. It owes its initial popularity to the advent of the railway, which ensured that it could be utilised wherever a railway terminal existed. The only difference in the methodology is the use of regular intervals for the battens, ensuring a swift and relatively simple cladding. Later slate roofs were nearly always protected by ceramic ridge and hip tiles.

Other roof claddings, such as cement and even asbestos slates, were later utilised and have proved to be a poor substitute, encouraging the growth of lichen and moss, which encourages moisture retention and thus freezing conditions that encourage delamination.

Clay tiles (plain)

Although clay tiles are regarded as a modern phenomenon, they were used as early as 1400 in the south-east of England, held on by oak pegs and known as 'peg tiles'. The standard rafter and batten format, the latter laid at regular intervals, supports the tiles, each one overlapping two others, with the upper tile protecting the joint between the two lower tiles. Clay bonnets or saddle-shaped tiles were devised for ridges, and for the joins on hips. On some 18th- and 19th-century buildings the hip joints were protected, with a shaped hip tile formed especially for the purpose. Early tiles were not a standard size until 1477, but were always cambered both ways, in order to shed water. Special tile and half tiles were made for closure at eaves. Tiles are almost universally nailed in modern building construction, but traditionally tiles also had nibs from which to hang them onto the battens. These were formed by the pressure of the thumb, leaving its imprint; this is the most tangible evidence of an early tile. As ever, the Victorians capitalised on the ability of clay to be fired into a variety of colours, ideal for the polychromatic phase of around 1860, and shaped into various forms, with fishscale tiles being very popular.

Pantiles

Again suitable for shallow pitches roofs of around 30 to 35 degrees, the material was imported in the first instance from Holland in the late 17th century before being manufactured in the

144

early 18th century in eastern England. Produced in considerable quantities during the 19th century in Bridgwater, Somerset, pantiles infiltrated surrounding counties for all manner of buildings, from the large houses down to farm buildings, it being recognised by farmers that such a roof cladding if not torched (see below) could also offer essential ventilation for animals. Torching was essential for domestic buildings, to prevent the ingress of rain, wind, and snow.

The size was fixed at around $13\frac{1}{2} \times 9\frac{1}{2}$ in (34×24 cm) and the tiles had a considerable side lap and a head lap. Frequently used on hipped buildings, the join was clad with a specially formed half round tile, the ends being heavily mortared, and the same used at the ridge. The problem of exposed verges on gabled buildings was countered by building gabled parapets, a possible source of popularity of this feature in the 18th century. The end of the half round hipped tile, left exposed to the elements, was infilled with a mix of mortar and tile slips.

Roman tiles, Spanish tiles, French tiles

Like their English/Dutch cousin, these tiles are shaped to overlap or interlock.

Roman tiles consist of flat 'under' tiles and half-round or barrel-shaped 'over' tiles. The 'under' tiles are nailed to flat horizontal battens, and the 'over' tiles nailed to vertical battens run between.

Spanish tiles enjoyed a period of popularity following the advent of the package holiday in the 1960s. They are similar to Roman tiles but the 'under' tile is concave and nailed sideways into the vertical battens on either side, and the convex 'over' tile single-nailed to the top of the battens.

French tiles are a machine-made interlocking tile that is hung over the battens, and the interlocking sideways lap helping to keep them in place.

Torching

Torching was an essential component of all traditional roofing construction, and consisted of a triangular fillet of lime-haired mortar inserted on the underside of each roofing component (tile, slate, etc.) the length of the batten. It enabled the roof to breathe, but resist a serious onslaught of wind-driven rain or snow. Constant movement in the roof cladding due to storm conditions (slates tend to joggle, making a noise reminiscent of a clog dance on the roof) tended to loosen this essential component, as did excessive drying out from hot sun on the top side of the slate or tile. Most roof fixings, as opposed to the material itself, tend to have a lifespan of around 100 years, so the torching was naturally replaced, if not done sooner as a form of remedial repair. Sadly it is a tradition that has completely died out as regards common usage. Its replacement in the 1950s by BS 747 roofers felt on top of the rafters and held in place by the battens did little to ensure the longevity of ancient roofs. The material may well have been very effective at preventing wind-blown precipitation but it also had a tendency to sweat, resulting in the demise of rafters due to fungal decay and worm infestation. Its replacement with a whole raft of modern breathable roofing membranes has redressed this problem.

Shingles

Shingles started life as riven oak tiles formed by cutting wood in the round to a required length, then splitting it into successively halves, quarters, eighths, etc. using a tool called a 'froe', a wedge-shaped tool which sits neatly in a crevice and is wiggled from side to side. These are the very best type of shingle, as the riven texture of the surface forms a series of mini-gutters down which water travels, rather than soaking into the surface. The same is true of natural slate. Rift sawn shingles were produced by sawing a quarter log down its sawn face, and had the same tapering point as the riven shingle but without the striations, although the grain of the wood runs across the length of the shingle. Flat sawn shingles, sawn from the hypotenuse of the quarter log, were by necessity diminishing in width, and resulted in considerable waste as the taper of the quarter was reached. This produced a surface that did not have the advantage of the grain running across their thickness as with riven or rift sawn, and as a result they are more prone to cupping in wetting and drying cycles, the chief enemy of all shingles, however produced.

Shingles are normally a minimum of 400 mm (16 in) in length, and taper from about 10 mm ($\frac{1}{2}$ in) at their butt end to about 3 mm ($\frac{1}{8}$ in) at their taper. The taper is course fundamental to achieve the essential overlap, similar to that for feather-edged weatherboarding. Cedar is also the favoured material, as it contains a high residue of resin, essential to inhibit worm and fungal decay, which would otherwise readily attack such slender timber components, although oak and elm were traditionally used.

It is difficult to be precise about the origin of their usage as their lifespan is certainly limited depending on exposure and weather conditions. It is generally that of a thatched roof, around 30 years, so any medieval or post-medieval examples will have long ago been replaced. The advent of steam-powered sawmills in the 19th century saw a renaissance in their use, as did the penchant for vernacular stimulated by the Arts and Crafts movement.

The material is ideal in terms of sustainability, as it is a renewable resource, but as sustainability is also about durability it rather lags behind in these stakes.

Roof construction

Ancient forms of roof construction are many and varied. They are complex in their variety, and their dating and origin is the subject of hot debate amongst academics. The purpose of this text is thus to give an indication only, as it is unlikely that any other than a laminated cruck and basic A-frame would be attempted again for sustainable building construction, although such knowledge is required for restoration of existing ancient roof forms.

Cruck construction

Crucks were formed from the main trunk and one branch of an oak tree, following the line of its growth. Crucks were full, from ground level upwards, although base crucks extended only up to collar or just a little beyond, and were an early medieval form (possibly 13th and 14th century in date), which allowed uninterrupted wide open halls. Raised crucks, resting partly in the wall or on a tie-beam, were a less superior form and continued into the 18th and 19th centuries as curved principles, used in particular in granaries where again an uninterrupted floor space was at a premium. Jointed crucks were used in those areas of the country where clearly there was insufficient timber to form a full cruck, and were very much a poor relation – a situation which dates from the 16th century, particularly in Somerset where they are used in quite high social status houses. The cruck form of construction lent itself to the development of arch braced collar trusses, particularly for base crucks, and this form of construction is characteristic of the medieval open hall.

Other early roof forms

Another early roof form was the **crown post roof**, with a series of posts resting on tie-beams, supporting a linear purlin, in turn supporting a series of collars. These date from the mid 14th

146

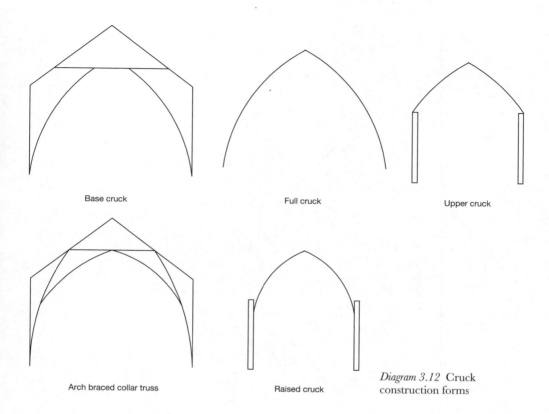

Base cruck   Full cruck   Upper cruck

Arch braced collar truss   Raised cruck

Diagram 3.12 Cruck construction forms

to the mid 15th century, and the crown posts were used as a decorative medium for high social status buildings, being shaped and moulded out of the solid, and braced to both the linear purlin and the collars. They abound in the south-east of England, although this phenomenon is tied to what is generally a higher survival of high social status houses.

King posts are a simpler version of roof construction having their origin also in the early medieval period, but continued in use in the north of England in preference to the queen post, which is more ubiquitous in the Midlands and south of England. King post roofs developed everywhere during the late 18th and 19th century, for farm buildings and buildings associated with the industrial revolution. They initially had pegged mortise and tenon joints at the base of the post where it joined the tie, but this was replaced by an iron bolt by the late 19th century, and even the king post itself becoming an iron bar stretching from tie-beam to the apex of the principal rafters. Wooden king posts were often braced with straight braces to the principal rafters, or had small struts in the spandrel above the collar.

Queen posts were almost universally used for trusses in the Midlands in the 16th and 17th centuries. A true queen post roof has the queen posts supporting the purlins. A rather less satisfactory form is the queen strut roof, with small struts to one side of the purlin, often accompanied by raking struts in the spandrel (the triangle above the collar).

Aisled or arcaded forms of roof construction, so called because the roof swept down over a central nave and side aisles, the post supporting the ends of the tie-beams forming an arcade by virtue of the braces spanning between wall plate and the arcade plate. The arcade plate is in the same position as that of a wall plate in un-aisled buildings. The latter was clasped at the

147

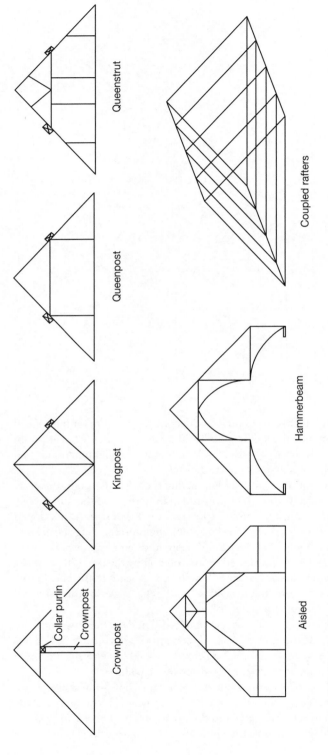

Diagram 3.13 Early roof forms

Photo 3.34 Dorset roof trusses tends to be a simple collar format with no tiebeam, which can often lead to outward roof thrust

back of the jowelled posts by the tie-beam. The whole resembled the roof of a church and indeed must have been the form used for many early church buildings. This form had its origin then, in the early medieval period, but descended down the social scale so that by the 19th century it was typically used for barns, where animals were reared in the aisles and the nave acted as the normal threshing barn.

Hammer-beam roofs were also an early form of roof – 14th and 15th century – used principally for churches. The weight of the roof was transferred down each post to a stub tie-beam clasped between principal rafter and wall plate. The decorative arch bracing completed the ensemble. The whole has the appearance of the central part of the tie-beam being cut away. The hammer beams are the remaining ends of the tie-beam.

Very early roof forms – 13th and 14th century – took the guise of coupled rafters, each pair being cross braced near the apex, so-called St Andrew's bracing.

This brief excursion through the early forms of roof construction indicates that most traditional roofs were over-engineered, but it is a gargantuan leap between this and the modern gang nailed truss, rather reminiscent in its form to the very early coupled rafter roofs mentioned above, but without the traditional mortise and pegged joint. The gang nail is a poor substitute for the traditional jointing and the wood is invariably of poor quality, fast grown with considerable summer wood to attract beetle infestation. Leaving these trusses out in the rain, prior to raising onto a roof, is also a cause for anxiety. All in all, the durability and hence the sustainability of such roof constructions has to be questionable, and the fate of mass housing estates in the long term a point of issue.

Steel frames, where the roof form continues down the walls, are durable in the medium term, but it is questionable whether they are durable in the long term. Steel buried in walls or in roof under-cloaking is going to be prone to moisture penetration from warm moist air generated internally and will attract condensation. Such steel members receive the bare minimum in terms of paint protection, and their long term future must be suspect.

One way of building sustainably is to return to the heavy wood roof truss format as illustrated in the above earlier forms together with a traditional timber wall frame, and although

149

this is undoubtedly costly it does mean that a variety of claddings can be used, together with considerable insulation.

Modern sustainable construction

The Gaia house

The Gaia house, first illustrated in *The Natural House Book* (Pearson, 1989), exhibits the formation of timber framed sunspaces filling all one wall, and the roof could easily be a traditional A-frame construction. The Gaia principle of design for harmony with the planet in terms of siting, orientation, and shelter will be explored later in this text, as will the integration of the house with the biodiversity, such as tree cover and other planting. A fundamental concept of design for harmony with the planet is the use of green materials in an intelligent way that will ensure recycling and re-use if the building reaches the end of its useful life. It is a tribute to many of the traditional constructions discussed above that they are still with us after centuries of use, and in fact the only dismantling that is done is often of a wilful nature, disobeying the conservation ethic of 'add but do not subtract'.

The Gaia principle of design for the health of the body and peace for the spirit revolves very much around creating an indoor environment, and is very much concerned with materials and construction. Buildings are a combination of colour, texture, and the solidity that emanates from substantial chunks of timber and solid stone walls. Not only does this form of construction allow a house to deal with warm moisture, 'breathing' it outwards to the external environment where it can be whisked away by the breeze, but such structures create a feeling of safety and security. Enhanced by natural light and safe from pollutants, the occupants experience an uplift in spirits that cannot be matched by any amount of retail therapy.

In the Gaia house, the street side of this urban building is protected from traffic pollution and noise by a grass bank, trees, and shrubs that filter harsh sunlight in summer and buffer the house from cold winds in winter. On the other side of the building, a two-storey sunspace attracts solar gain in the winter, and is sheltered from intense sunlight by deciduous trees in the height of the summer. The structure is mass walling so that it acts as a thermal store, but the walls are porous so as to encourage the exit of warm moist air. The mass walling also acts as a heat sink in summer so that the house never overheats. Together with the composting of organic

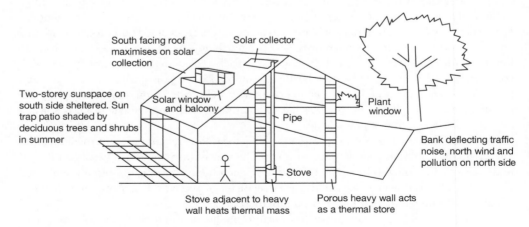

Diagram 3.14 The Gaia principle of construction (after Pearson, 1989)

waste, the recycling of grey water for toilets and garden, and the intelligent use of solar gain coupled with a wood-fired stove in the winter, the house was the ideal of all who aspire to a more sustainable lifestyle, and yet 30 years on the Gaia message is only just beginning to permeate.

The earth-covered house

Another form of building that has struggled for recognition is the earth-covered home. There is probably something vaguely reminiscent of a troglodyte existence about such buildings, and furthermore the average British homeowner is conditioned to an urban culture of rows of more or less identical houses in the street, each with a symmetrical layout of windows and a front door. Ask a child of even a very young age to draw a house and this is what is produced. It implies that somehow we have as a culture been so conditioned that this perception may even be formed in the womb.

Compare and contrast this to the most ancient form of dwellings: the cave. Ancient cave dwellers would have perceived, although would not have been able to explain, the 'thermal flywheel effect'. The earth is a natural moderator of temperature, below that point to which frost penetrates. As the surface is approached there will be more chance of a variation in temperature in response to air temperature, but the earth will still act as a moderator of this. Earth is thus efficient at slowing the passage of heat gained or lost so that *the heat gained in summer permeates into the house in early winter, and the cooling of the soil in winter will not reach the house until early summer*. In effect the earth around the house will cool it in summer and warm it in winter, with a time lag for soil temperature ranging from 4 to 6 months. Any shortfall in heat required in the winter can be made up using a mix of passive solar gain and a renewable heat source, such as wood for a wood-burning stove. Any secondary heat source would require significantly less fuel. Pearson (1989: 74) quotes an earth-covered house in Australia experienced only a change of 12 °C in a period when the outside air temperature varied between −2 °C and 42 °C.

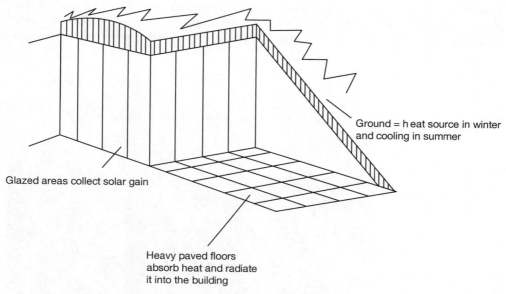

Ground = heat source in winter and cooling in summer

Glazed areas collect solar gain

Heavy paved floors absorb heat and radiate it into the building

Diagram 3.15 The earth-covered house (adapted from Lengen, 2008)

Earth-covered houses can be surprisingly light by virtue of the use of fully glazed front elevations on one or two sides, and a roof lantern (double or triple glazed) incorporated into the essential bulk of earth on the roof.

One of the less sustainable aspects of such constructions is the necessity of concrete walls structures, using cement fired to a high temperature, and membranes derived from petroleum fractions. These essential components are needed to keep the interior dry and free from damp, but the concrete is durable and suitably protected from the membrane and should last indefinitely, and act as a thermal store. This ability for dense wall and floor construction to absorb heat and then reflect it out is an essential component of the construction.

The green roof

Closely allied to the earth-covered house, the green roof is fast gaining a degree of popularity. The principle is simple: a membrane is placed on a flat or very shallow pitched roof, and well supported around the edges; the trough so formed is then filled with a bulk of earth to support a grass and wildflower or sedum-based vegetation. The secret is then to be patient, as it will take some months in a prime growing season for grass to take a firm hold, and sedum will take considerably longer. Wildflowers are notoriously fickle so no great expectation should be entered into, and every flower regarded as a bonus. It is clear that the whole of this construction is highly dependent on the quality of the membrane, so no expense should be spared on this aspect. Another aspect which needs careful attention is the ability of the roof construction to support the weight of a human being and a mower, as from time to time grass-covered roofs will need to be mown, albeit even if only at the cessation of the flowering to allow them to self-seed. In all events, grass roofs start to look very untidy if dead grass heads are allowed to manifest. Sedum roofs do not require this attention, but are very much slower to develop.

The sunspace

This is a device which has gained in popularity as a living space and a haven for semi-tropical plants, not because of its ability for essential passive solar gain, although most homeowners have a perception that conservatories extend the seasons. By excluding wind-chill factor alone,

Photo 3.35 A mature green roof at Craven Arms, Shropshire

Photo 3.36 A less mature green roof at a Somerset education centre

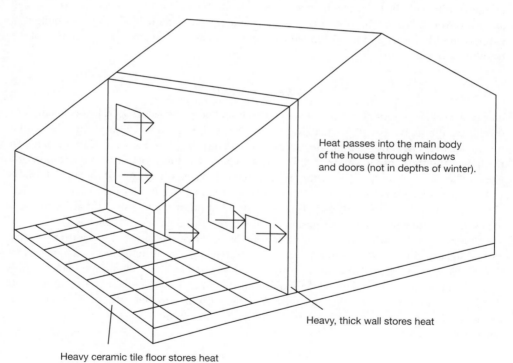

Heat passes into the main body of the house through windows and doors (not in depths of winter).

Heavy, thick wall stores heat

Heavy ceramic tile floor stores heat

Diagram 3.16 The sunspace

quite aside from short-wave solar radiation, the ambient temperature is raised. What house-owners must grasp is that the best location for a sunspace is against a wall which is adjacent to other main living spaces and which can act as a thermal store, together with a heavy-weight floor, preferably clad in a dark ceramic tile, both of which will absorb heat gain and reflect it back at night when the temperature drops. In periods of excessive heat, cooling devices must be used (blinds, roof vents, and shading from deciduous trees), but in the cooler seasons of the year this solar heat gain is fundamental. In addition, although it is essential that double-glazed doors and windows must be part of this wall in order to satisfy building regulations, it should also be realised that during the day the air warmed by the sun needs to be allowed to flow into the house by natural convection. Fans can assist this process but are ideally driven by photovoltaic panels. This warm air flow will only occur, however, in the warmer periods of the year, notably spring and autumn. It will be useless to expect solar gain to flow into the rest of the house in the depths of winter.

The addition of sunspaces to existing masonry building construction is usually a viable option, especially if the design and materials are complementary to the building. A heavy wood construction is nearly always suitable for stone buildings or brick buildings. The stone or brick interface wall should ideally be a solid wall construction, as opposed to a cavity wall, and will thus readily act as a thermal store.

Two-storey sunspaces should also be considered on taller buildings, as they have more of a design affinity with the height of the structure, and will serve to heat upper rooms also. Gabled and lean-to designs generally work better with traditional one and a half or two-storey buildings. Complicated P shapes and multi-hipped roofs create a jarring tone, alien to the nature of the traditional construction.

The solar wall

The solar wall, an essential component of the back wall of a sunspace, naturally occurs in most dwellings, although a cavity, particularly if it is insulated, is going to diminish the ability of a house wall to act as a solar wall. In new build it should be possible to construct the wall between the house and the sunspace in mass walling using concrete, stone, or brick. The addition of a glass wall in front of the solar wall, with the latter painted black, will create the so-called Trombe wall, named after Dr Felix Trombe. The very brave might also consider a series of waterfilled glass columns as heat stores instead of the masonry, but unfortunately sunlight also generates algae, and cleaning out such structures would be time consuming. Water, which is a weak acid, has a habit of eroding whatever container confines it, so the propensity for leaks is ever present.

The solar wall is usually designed to contain a series of vents at high level and low level so that air from the house is fed in at the lower level, warmed by the wall and fed back into the house at higher level. It is more efficient if this air is behind a sheet of glass so that there is no danger of it dissipating into the sunspace. Another option is to clad the solar wall in thick terracotta tiles to increase its ability to retain heat, which will prove less of a construction challenge.

Solar windows

Solar windows are closely allied to the technology of sunspaces in that in essence the solar space is one giant solar window or solar collector of short-wave radiation, which if suitably treated with a thin metal film (low-emissivity glass), incorporated on the inside of the external wall of glass forming the double or triple glazing, will also inhibit the escape of heat gained. This heat

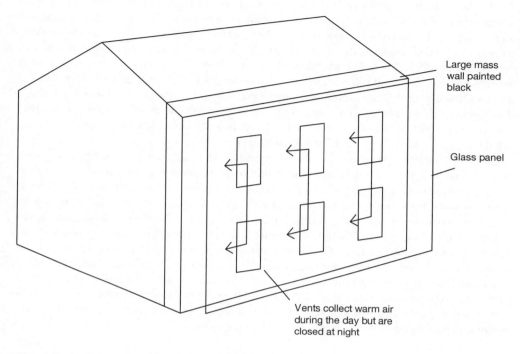

Large mass
wall painted
black

Glass panel

Vents collect warm air
during the day but are
closed at night

Diagram 3.17 The solar or Trombe wall (adapted from Lengen, 2008)

travels into the interior and is stored in the various materials contained therein. In order to ensure that these materials are capable of storing this heat, attention should be paid to coating the surfaces of lightweight construction and floors with ceramic tiles.

The vexed question of existing traditional single-glazed windows will always arise. What needs to be balanced here is the loss of sustainability by scrapping an existing perfectly good window against any possible gain in sustainability by preventing the loss of long-wave radiation from the fabric and air contained in the building. The answer must surely lie with secondary glazing.

An ideal new build construction would contain solar windows at high level as well as low level, so that the high level windows could obtain solar gain in the spring when the sun is beginning to gain height, and the low-level windows in winter when the sun is at its lowest level. Closing vents in high level windows and the use of heavy curtains is essential on a winter night to retain heat. High level window vents have a useful part to play cooling sunspaces in the height of summer, as do blinds or heavy curtains.

Solar cooling systems

Ancient civilisations in all manner of hot and arid climates have had this completely sewn up by the creation of tall courtyard houses with heavy weight walls. The lower rooms around the courtyard were occupied by cattle and humans in the heat of the summer as they were very cool, being shaded by the high walls and any heat being absorbed by the mass walled construction. Upper rooms, fronted by a gallery, were occupied by humans in cooler seasons (and the cattle sent to the fields), as these areas benefited from solar gain, with heat absorbed by the thick courtyard wall structure to be reflected back at night. Heavy beam and slab roof constructions also retained the heat of the sun in cooler seasons, preventing its egress to the rest of the building, but stored it and released it in the cooler nights associated with clear night skies, and also provided shade to the upper rooms. Walls were deliberately painted white with limewash in order reflect heat rather than absorb it. Windows were kept deliberately small or even nonexistent to avoid solar gain. The Roman houses long buried under volcanic ash at Pompeii display this device, the monotony being relieved by brightly coloured paintings. Devices to catch any breeze were situated on the roof, scooping it up and feeding it down into the courtyard, and the use of a central pool or fountain in the courtyard, or even terracotta jars filled with water, served to cool the air down further.

Many of these principles can be employed as cooling devices in the heat of summer, although many would be superfluous in what can only be a brief season in Britain. Vents are one of the essential components in a sunspace, as they can used throughout the night to cool down the 'thermal store' wall conjoining the house. By the next morning, the thermal store will be sufficiently cooled so as to be able to absorb the heat of the day. Shading from blinds or heavy curtains inhibits solar gain. Photovoltaic panels can be used to drive fans or even air conditioning (reverse refrigeration). Cross-ventilation can be made use of by ensuring that the sunspace is on a south or south-west elevation, and windows on a north or north-east elevation. Cool air drawn through from the north side by the natural attraction of built up heat will be drawn through into the sunspace on the south side and out through upper vents. Cross-ventilation, even in rooms without sunspaces, can be one of the most efficient means of cooling rooms and only requires that windows be sited on opposing walls.

Mechanical systems

Although not part of construction, being add-ons to existing buildings or incorporated into new build, it is worth mentioning that heat exchangers, or 'heat recovery ventilators' as they are now more commonly known, especially if powered by a photovoltaic system, can be very useful in introducing much needed ventilation for the summer, as well as providing a boost to winter

Photo 3.37 A plethora of solar and photovoltaic devices on a roof may well be the future for all buildings

heat. Heat from warm stale air, leaving the building, is transferred to the cold fresh air coming in. The airflows are separated.

Central heating by virtue of oil or gas fired is commonplace in most buildings. Existing traditional mass walling does not benefit from having systems that operate for a few hours, morning and evening. The best method is a low gentle heat obtained by turning the radiators down to the bare minimum, according to the occupation level of the room (ensure that rooms are divided into those which require to be warm and those which can be cooler), and running the system continuously in the very coldest periods. This gives the walls and floors a chance to act as thermal stores, and avoids the necessity for mass walling to be heated from scratch. Radiant heat sources are designed to warm air, and so should not be utilised other than for short bursts, unless the radiant heat source is a stove, which will hold the heat of the fire and reflect it out over a long period of time. Some open fireplaces are made of brick or ceramic tile and were designed to hold the heat of the fire long after it had died away. There appears to be something of a backlash against retaining the mid 20th century brickette fireplace on aesthetic grounds, but it can be a boon in this respect.

Photo 3.38 A discrete location for solar panels on a medieval building

Photo 3.39 Ground source heat pumps require external installation for the pump mechanism

Most mechanical add-ons revolve around the addition of solar water heating panels, solar photovoltaic, and wind turbines. The photographs illustrate that however desirable these are, they do have an impact on the building, and in existing traditional buildings it is advisable to seek locations where they are not so prominent. Even the highly desirable ground source heat pump requires some externally placed equipment. Ground source heating has revolutionised the gaining of free heat for underfloor heating systems, but the pump has to be driven by electricity. This leads to a desire for a wind turbine to power the same, but unless a rural site is available these can be an intrusion into a traditional street scene. It is perhaps best to leave such electricity generation to the larger installations.

Photo 3.40 Wind turbines producing electricity on a mass scale, and are best divorced from buildings

Photo 3.41 Canal Central at Maesbury Marsh, Oswestry, a new sustainable construction which is adequately cooled in summer by cross-ventilation, avoiding the energy costs of mechanical devices

4

THERMAL PERFORMANCE
AND THE BREATHING WALL

Modern insulation methods

The acceptable modern methods of insulation, whilst efficient in controlling internal temperatures, vary in their impact on the ability of the structure to deal with trapped moisture, and on the health of the occupants. This dilemma is dealt with in the following discussion. First it is necessary to examine the general forms of modern insulation.

Insulation of suspended floors

Existing timber floor structures at ground floor level will inevitably be a suspended floor with an intended zone of ventilation beneath. This is because the joists will be resting on sleeper walls forming a honeycomb beneath the floor, and vented to the outside by terracotta or cast iron grilles. This was the great invention of the Victorian Age. It resulted in rugs and carpets that levitated on windy nights and very cold feet, but at least the joists survived. Inevitably the outside grilles were subsequently blocked by any means to hand. The modern practice of lifting such floors and giving the gap between each joist a mesh (plastic or metal) box to hold a bat of insulating material can only be welcomed, although lifting floorboards can result in their damage. Modern suspended floors should have their joists suspended from joist hangers attached to the masonry inner leaf so that there is no danger of contact with moisture-laden masonry. This has been the plague of traditional building construction since the earliest inception of timber floors. The bearing of oversized ceiling beams will sometimes stand up to the rigours of being embedded in masonry but floor joists hardly ever do, and much important information about earlier floor structures has been lost in this way. Fortunately many a ceiling frame had its joists resting on a bressummer pegged into a timber framed wall, or resting in a slot in a masonry wall. This was then the first line of defence. The current reason for avoiding contact between masonry and bearing ends is that this will also create an air infiltration path. The zone of ventilation below the insulation must be maintained assiduously.

Insulation of solid ground floors

Many existing buildings have been given a solid ground floor in place of a former suspended timber floor or in place of stone flags or brick paviors bedded on clay or earth and sand. Where such brick of stone flag floors still exist, very special care should be taken to lift the existing floor covering and use the material re-bedded onto a limecrete floor (see Chapter 7 for a detailed description).

Limecrete floors are also far superior in new build in rural situations where they can be accompanied by a French drain, rather than the standard practice of a concrete slab. There

159

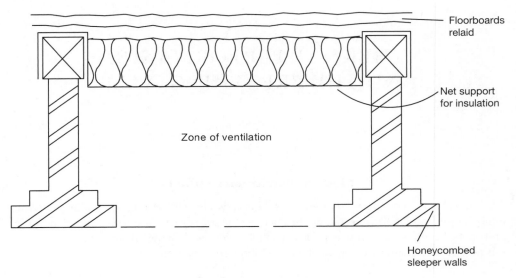

Diagram 4.1 Insulation of suspended floors – can be retro-fitted

has to be a concern for all solid floors that the heat loss from the new floor will be greatest along the perimeter, adjoining the outside walls, with consequent heat loss to the outside. Certainly in a limecrete floor because of its greater porosity, and also for a concrete slab, the centre of the floor will start to heat up after a time, supplemented by solar gain. The U value (a measure of heat loss, fully discussed in Chapter 5) of the floor as whole will depend on the ratio of the perimeter (P) measured in metres, to the area of the floor (A) measured in square metres: A, ratio, P/A. The higher the value of the square metre area the lower the ratio will be. Thus large surface areas will retain more heat than the perimeter will lose. It follows that small surface areas will need more insulation (Nicholls and Hall, 2006).

With solid concrete much of this insulation will take the form of solid bats of polyisocyanurate or extruded polystyrene rigid board. These boards are much more readily available and sadly

Photo 4.1 Special care is needed when lifting stone flags to rebed on an insulated floor

Photo 4.2 A limecrete floor being laid

(they are manufactured from chemicals) do work out much more cheaply than the use of, say cork board, as well as being damp resistant, so they will often be the builders first choice. If a limecrete floor is used, the clay balls (leica) or foamed glass used to resist rising damp will provide the necessary insulation and are a much better option. Limecrete floors tend to be more expensive, but this is all a problem of supply and demand. If demand increases then supply increases and the price automatically comes down. The whole problem with sustainable building construction methodologies is that they are still suffering from too low a demand, and as a consequence too high a price.

Insulation of walls

Traditional cavity and solid wall insulation

Wall construction methods have embraced a form of insulation technique since the cavity was invented in the late 19th century. By stopping rain penetration from the outside travelling inwards (although not interstitial condensation which as a vapour can travel across the cavity from inside to outside), there is an automatic resistance to cooling down caused by water. Wall ties between the inner and outer leaf are twisted to ensure that the clever water molecule cannot track across them, although some early wall ties are actually simply strips of wrought iron. The air in the cavity was also an insulator.

Initially cavities were not completely filled because it was thought that water would once again track across from outside to inside, and yet again cause damp internal walls and consequent mould growth. It should be pointed out at this juncture that this very rarely happens with solid traditional walls with no cavity because they are often very thick and thus only the outer portion of the wall wetted up. The universal use of lime for the mortar with its high porosity ensured that the joints did most of the water absorption and readily breathed this moisture out again. Therefore very few traditional solid wall constructions actually suffered from water penetration, although continuous saturation from intensive storm conditions could

Photo 4.3 Poor cement mortar repointing, preceded by even earlier poor repointing adjacent

result in the inner surface cooling down, and consequent condensation problems. Weak lime mortar is more prone to saturation, but this is rarely the problem. More pertinently the use of cement mortar for re-pointing, forcing the masonry to do the work of breathing moisture in and out, has given such building constructions a bad name. Building blocks were never meant to do the work of a joint, and consequently do not perform as evaporation units as well as joints when dealing with water absorption.

Drylining has long been popular with traditional solid wall construction (but usually without insulation often due to the mistaken belief that cold walls with a film of water on their surface were suffering with penetrating damp. In reality they were simply attracting warm moist air which then 'rained out' over the wall surface. It involved the use of wooden battens fixed to walls with plasterboard over. The battens should always have been isolated from the masonry wall by a linear damp proofing membrane but were frequently not, so such a system that has been in place for some time should be thoroughly checked for any possible dry rot infestations. A much lower U value (ideally U values should be as low as possible) can be achieved by the use of insulation between the battens, but there is a danger that the value of the insulation will be diminished by damp ingress filling up the voids previously filled with insulating air. A similar system used plasterboard that was backed with polystyrene fixed to the wall with plaster dabs. These systems all cause problems with the depth of cills and the finishing of openings generally.

None of these systems should have been used where there was any danger of obscuring architectural detail such as cornices, but sadly they have, particularly in nursing home conversions of former gentry houses. Such systems are also at odds with the capacity of solid walls to act as breathing walls and thermal stores, as discussed below.

Insulation of modern cavity wall construction

Cavity wall construction has readily embraced further insulation being provided by adhering insulation boards to the outside face of the inner leaf of the cavity. A standard cavity around 100 mm (8 in) was initially filled with a maximum of 60 mm (2½ in) of insulation, retained by special clips to stop it slipping out of place. This technology has now been replaced with a full cavity system that inhibits moisture from tracking across and producing cold bridges, although building inspectors generally prefer such buildings to be fully rendered. The system used is frequently tongue and grooved expanded polystyrene (EPS) with an intermittent projection on its outer face to maintain some semblance of a cavity. The joints are angled downwards away from the inner leaf to avoid any possibility of moisture ingress. The EPS boarding is waterproof in its own right, so the system is fairly failsafe, but it is not a particularly green material.

Retro fit systems are also available and useful for modern construction prior to the age of cavity wall insulation (1930s–late 1970s), although there are again concerns about some of the materials used and their potential to off-gas (see Chapter 2). This is certainly preferable to external cladding, as the latter produces problems in terms of the edges of existing door and window openings, re-positioning of rainwater goods and external details such as cornices. It is a totally no-go area for any building of architectural merit externally, but can be useful for the most exposed wall of, say a simple brick cottage, where insulation and cladding in a horizontal timber boarding will not be architecturally out of place.

Photo 4.4 External boarding and insulation may not be out of place on a simple brick cottage

Three materials are commonly used for cavity wall insulation: EPS in bead form; chopped mineral wool (a misleading term for a quite unpleasant silica-based fibre); and the even more worrying urea formaldehyde foam. The fibre is not likely to penetrate internal walls, although when laid in lofts it can make breathing difficult, but the foam may certainly off-gas.

Internal insulation for timber frame structures

Traditional timber framing

Traditional timber framing gave no credence to insulation save through the unique quality of the clay daub wrapped around the oak staves and hazel wattle basket weave. In addition, a well maintained timber frame had the gap sealed between the edge of the wattle and daub infill panel and the miniature timber frame in which it was snugly housed, by virtue of a yearly coat of lime wash. This built up into a dense coat of plaster sitting in the gap, which has never been bettered no matter how many clever devices like compriband (traditionally a bitumen-soaked sponge on a roll, now a polyurethane foam tape) or mastics have been invented.

Newly restored timber frames tend to leak as well as infiltrate air, because these new materials cannot hope to emulate that which the passage of time has created. The message is loud and clear: interfere with tried and tested traditional systems at your peril.

Many attempts have been made at trying to retrofit insulation into existing timber frame buildings. The end result can resemble a series of pillows that have been stuffed into the panels, to even more detrimental systems such as blockwork inner skins. Examples exist of misguided attempts to wall the entire frame up on both sides in order to stop damp ingress. Dry lining is probably the most successful option, but with the new breed of wood fibre boards (discussed

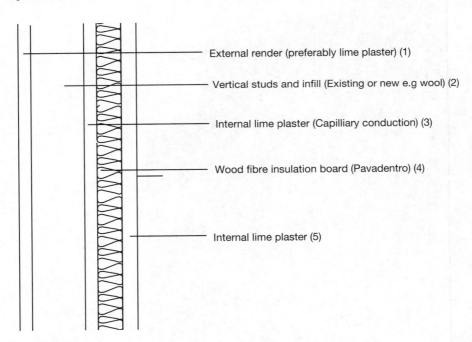

External render (preferably lime plaster) (1)

Vertical studs and infill (Existing or new e.g wool) (2)

Internal lime plaster (Capilliary conduction) (3)

Wood fibre insulation board (Pavadentro) (4)

Internal lime plaster (5)

Diagram 4.2 Using wood fibre board for insulation of traditional timber framed buildings (after Natural Building Technologies, 2010)

Photo 4.5 This traditional timber framed house has survived from c.1600 by virtue of a yearly coat of limewash on the panels

later), combined with hemp cotton or hemp batts. Many owners will mourn the loss of visibility of the timber frame as viewed from the inside.

Insulation of modern timber frames

Modern timber framing is invariably not seen at all, as the studs are never designed to be seen. They are in position to hold a sandwich of insulation, usually of the rigid board variety and invariably Styrofoam (polyisocyanurate) or EPS (polystyrene). The warm side is given a polythene vapour barrier, and then plasterboarded. The external side is against a cavity, shielded from it by a breathable but water-resistant barrier and the whole is then given an external sheath of masonry, usually a brick skin held away from the frame on brackets. When construction of large buildings is contemplated, the same technique is used but the timber is replaced by steel frames clad with metal sheathing, which itself encapsulates an insulating material.

Vapour barriers – their role in insulation

The major disadvantage with most modern insulation methods is the use of vapour barriers. Their role is to totally prevent internal warm moist air from penetrating the structure. This means that it must be ventilated out of the building by other means. The only usable materials are plastic sheets or metal foil. This is the equivalent of putting human occupation in a plastic bag and forcing all of the building structure to breathe moisture from external climatic sources outwards again. This must concentrate moisture within the outer layers of any traditional building materials (stone, brick) where it can mobilise salts more rapidly, causing faster decay of the pore structure and hence the material itself. In addition, holes and tears in the vapour barrier will almost certainly occur, and some warm moist air will penetrate the insulation, causing condensation. Alternatively, vapour checks can be used. They do not cut off the moisture completely, but limit its impact. Many building contractors favour their use in roofs

165

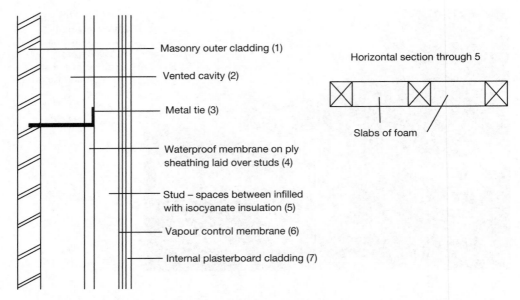

Masonry outer cladding (1)

Vented cavity (2)

Metal tie (3)

Waterproof membrane on ply
sheathing laid over studs (4)

Stud – spaces between infilled
with isocyanate insulation (5)

Vapour control membrane (6)

Internal plasterboard cladding (7)

Horizontal section through 5

Slabs of foam

Diagram 4.3 Modern timber frame construction using isocyanate (after Natural Building Technologies, 2010)

rather than the standard BS 757 roofing felt, but are reluctant to use them on walls. Sustainable builders use wood fibre boards as vapour checks. A rule of thumb is that the resistance to vapour diffusion on the inside must be five to ten times higher than the wind-proofing layer on the outside to give the vapour direction. In essence what is being created is a modern version of a breathing wall, as discussed below.

The concept of the breathing wall versus insulation

The discussion of whether to use insulation in the accepted sense of modern materials (such as polystyrene, polyurethane, polyisocyanurate, and phenolic foam) versus more breathable materials (such as sheep's wool or cellulose-based fibre) is probably vital to mankind's survival. The core of this discussion is the concept of the breathable wall. This is not to say breathing in the accepted sense of breathing air, but rather the interaction between water as a gas or a liquid and the building materials of the external envelope. At the basis of this is the concept of hygrospicity, the ability of materials to absorb and release water vapour. Buildings are like people in that they need to be in good health to function adequately. Humans create a very great deal of vapour of the warm, moist variety, via the constant boiling of kettles for that all essential cup of tea, boiling pans of vegetables, the drying of clothes (bad cases of bright blue mould can be seen in student bedrooms due to the drying of wet clothes on radiators), and the current obsession with showers and extreme cleanliness. Compare this to the lifestyle of our ancestors in stone, cob, or timber houses or even the upmarket Victorian villa. Heating was via one source: an open fire provided with a large chimney. A great deal of the moisture from cooking and clothes drying, plus boiling of kettles and pans over the open fire, sent water vapour up the central stack (the ideal passive stack ventilation), and in all events the stack enabled at least four air changes an hour whilst the fire was in place. Personal cleanliness would take place once a week in a tin bath in front of the same passive stack ventilation. By the time the last occupant of the bath took their turn, inevitably water vapour would be at a minimum – that is, the water would be cold. Astonishingly, we expect these traditional houses to continue to perform in the same way as just described, despite the fact that the fire probably no longer even exists, except perhaps for Christmas day, or the stack may well be sealed up. Against this is pitted the gargantuan quantities of water vapour being produced on a daily basis.

There are three key concepts to understanding how buildings need to cope with this problem:

- vapour permeability
- hygrospicity
- capillarity.

Key to all of these is an understanding of the water molecule. Each molecule consists of two positive hydrogen atoms and a negative oxygen atom. Water is inherently clever in that each molecule, be it in liquid water or in water vapour, is attracted to its neighbour by virtue of the hydrogen bond, the immediate attraction of the positive hydrogen of one molecule to the negative oxygen of a its neighbour. In addition, the inside face of each pore (void) in a building material is oxygen rich, as are all external surfaces. These negative oxygen atoms rapidly attract positive hydrogen atoms from water by virtue of the hydrogen bond, the water running rapidly across the inside surface of each pore, and then filling the small pores completely by capillary action. This latter action can easily be tested using one thin test tube and a wider test tube in a bowl of water. The thin test tube will rapidly suck up the water because the negative pull of the oxygen on the internal face is greater than the pull of the oxygen in the air of the test tube. This is a very explicit illustration of the power of capillarity.

The situation is no different when the water comes as vapour out of the spout of the kettle. It has to leave the building by some means and will head for the nearest permeable material, or failing this it will simply condense or rain out on the nearest cold surface. The temperature at which this happens is called 'the dewpoint'.

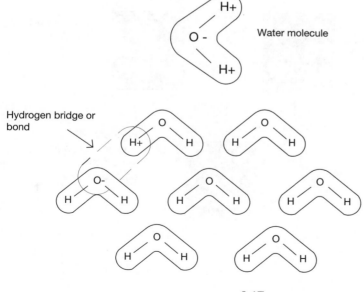

Diagram 4.4 Water molecules and the power of the hydrogen bond (Torraca, 1988)

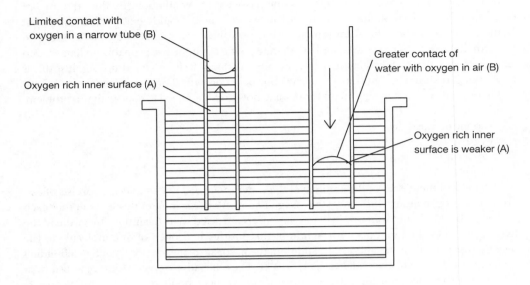

Limited contact with oxygen in a narrow tube (B)

Oxygen rich inner surface (A)

Greater contact of water with oxygen in air (B)

Oxygen rich inner surface is weaker (A)

Capilliarity is forcing the liquid up the left tube. Force A > Force B
In the right tube the inner surface pull is weaker. Force B > Force A

Diagram 4.5 Using test tubes to demonstrate capillary action

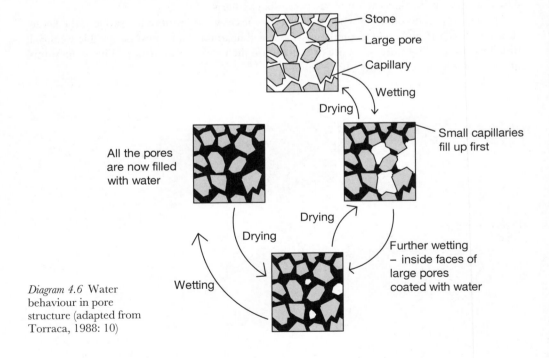

Stone

Large pore

Capillary

Wetting

Drying

Small capillaries fill up first

All the pores are now filled with water

Drying

Drying

Wetting

Drying

Further wetting – inside faces of large pores coated with water

Diagram 4.6 Water behaviour in pore structure (adapted from Torraca, 1988: 10)

168

The rate of passage of the water vapour (very excited water molecules forming a gas) through any porous material it may encounter will vary according to pore size and density, the latter depending on the thickness of the material. A thick stone wall will have a higher density of pore structure than, say a thin brick wall. Pore size is crucial as a range of pore sizes with a preponderance of large pores over small pores will have a capacity to hold a greater amount of vapour. A material with many small pores and few large pores will rapidly fill up. This would certainly be the case with the thin brick wall, as generally brick has many small pores. The rate of passage of water vapour is governed by vapour pressure, the movement through the pore structure, as governed by pore size and density. Vapour is said to move from a zone of high vapour pressure to a zone of low vapour pressure, according to these factors. This is because the negative oxygen lining the pores of the dry area of structure attracts the positive hydrogen atoms in the volatile vapour. The rate of transmission will depend on the pressure difference. A large number of water molecules held as a gas constitutes a high relative humidity (RH). Such conditions herald grave problems for human beings, through the development of moulds and bacteria. To these can possibly be attributed the many problems of young children developing asthma, as well as respiratory diseases amongst the elderly.

Structures also are affected. High moisture levels can diminish thermal performance insomuch as the more a pore structure is in filled with moisture, the less able it is to hold air, which reduces the ability of the structure to trap warmth within it. The result is cold rooms and clammy walls. In many older structures penetrating damp is immediately blamed, whereas in reality the problem can often be traced to the use of appliances which release large amounts of water vapour. Portable gas heaters, for instance, release 1.5 litres of water vapour for every l litres of gas used. An example comes to mind whereby a large farmhouse had black mould all the way up the rear wall of the staircase enclosure. The occupants had instructed a builder to re-point the outside of this wall with a cement mortar, but were ignoring the fact that they were trying to heat this large house in a very cold winter with numerous portable gas heaters. The tell-tale signs are large areas of black mould on the internal face of external walls, behind furniture, and in corners of rooms. It is sometimes possible to trace the joist pattern in a plaster ceiling in this way, with the gaps between the joists holding more black mould than the joists

Photo 4.6 A bad case of black mould due to condensation on the cold areas between the wooden joists

themselves, the joists being hygroscopic wood, capable of wicking water vapour through the cellular structure of the wood, to a limited degree.

The science of vapour permeability

Permeability is measured as the resistance to moisture movement. This resistivity (r) is measured as GNs/kgm (giga Newton seconds per kilogram metre *or* MNs/gm (mega Newtons seconds per gram metre).

Vapour resistance G in, say the structure of a wall is measured as:

G = r × thickness of the wall in metres.

Comparing G values in Table 4.1, we see that clay plasters are by far the most efficient at transmitting water vapour outwards from the interior of a room. Lime plaster is also efficient at wicking moisture away but surprisingly not as efficient as gypsum plaster. Unfortunately, when the latter becomes saturated it loses all ability to bond and simply erupts from the wall either in sheets or in volcano-like eruptions. Thus it is far better to use lime plaster.

Comparing G values in Table 4.2, it can be seen that clay boards, mineral wool/sheep's wool/flax are the most efficient transmitters of water vapour, closely followed by wood fibre boards, whereas high insulation boarding, based on cyanates, is guaranteed to hold water vapour within the building.

Paints traditionally consist of a binder, a pigment, and a solvent. In the case of most modern emulsions the solvent is water but the binder is based on alkyd resins (derived from petroleum). These reduce permeability. Binders such as epoxy and chlorinated rubber will reduce permeability even further. It is thus not surprising that indoor emulsion, limewash and silicate paints have a low G value, as shown in Table 4.3, allowing the maximum penetration of water

Table 4.1 Typical values for wall coatings (May, 2005)

Material	Typical r value (MNs/gm)	Typical thickness (mm)	G value at typical resistivity (MNs/g)
Cement plaster	100	20	2.0
Lime plaster	75	20	1.5
Clay plaster	40	20	0.8
Gypsum plaster	50	20	1.0

Table 4.2 Typical values for wall linings (May, 2005)

Material	Typical r value (MNs/gm)	Typical thickness (mm)	G value (MNs/g)
Brick	50	100	5
Gypsum boards (plain)	60	12.5	0.75
Gypsum boards (foil backed)	4800	12.5	60
Clay boards	90	20	1.8
Expanded polystyrene	150	50	7.5
Extruded polystyrene	1,000	50	50
Polyisocynanate with foil	43,000	50	2,150
Mineral wool, sheep's wool, flax	6	100	0.6
Wood fibre insulation boards	25	100	2.5

Table 4.3 Paint coatings (May, 2005)

Material	Typical r value (MNs/gm)	Typical thickness (microns)	G value (MNs/g)
Indoor emulsion paint	1,500	100	0.15
Masonry paints (outdoor emulsions)	1,500	120	1.8
Silicate paints	300	100	0.03
Limewash, usually 5 coatings	250	100	0.025
Solvent-based gloss paint	20,000	120	2.4
Alkyd resin varnish	80,000	120	9.6
Epoxy resin-based coating	200,000	120	24
Chlorinated rubber-based coating	350,000	120	42

vapour. External emulsions, often advertised as being capable of vapour transmission, are less vapour permeable.

Hygroscopicity

This is defined as the capacity of a material to absorb water from the air as a gas (water vapour) and then and release it back into the air. Hygroscopicity is dependent upon the pore density and volume, and the size of the pore structure itself. Materials with many small pores will be more hygroscopic, due to capillary action, than materials with a range of pore sizes, where the large pores will hold more water vapour than the small pores. The latter will have a greater hygroscopic capacity but will take longer to release the water vapour to the air on the outside of a building. Some schools of thought regard a hygroscopic capacity as being more important than the ability to release the water vapour through the wall or roof cladding, and refer to this as 'buffering'. It is a vital concept to grasp.

It can be seen from the above tables that many materials of a traditional nature, such as lime, clay, and untreated wood panelling, have a capacity to absorb and release water as a gas (water vapour), depending on the relative humidity of the air (RH). Provided the materials are resistant to rot, they can act as a store/sink/cushion for moisture, which is after a while released back into the room (minus any which have permeated to the outside through a permeable walling material) but must then be vented outwards by purge ventilation. This is absolutely ideal for bathrooms, many of which suffer dreadful damp and mould problems because they are lined with vapour barriers. Instances of using highly porous bricks with a lime mortar, to balance humidity in a bathroom, have been shown to be highly successful. In many ways this is an energy saving measure. No mechanical ventilation is required. It is also the basic tenet of how all existing traditional stone, brick, or cob buildings operated over a long period of time. The walls acted as a moisture sink, provided that the amount of condensed moisture in the wall is low compared with the material's capacity to hold it. Likewise, the water absorbed from precipitation during a damp period can evaporate in drier times of the year. The methodology is only brought into disrepute if the amount of precipitation in the wetter times of the year is excessive, and this may have happened in the two wet summers of 2007 and 2008. Nonetheless, the capacity of breathing walls to buffer moisture and release it internally where it can be removed by purge ventilation, or externally where it can be released in optimum conditions such as a drying wind, is fundamental to understanding how buildings perform.

Therefore hygroscopic materials can be beneficial in creating a stable internal environment by controlling air humidity. By doing this, surface condensation is considerably reduced. The water vapour transmits through the internal wall covering and from thence through the wall structure (or roof structure). Materials with a high hygroscopic capacity will tend to hold water vapour by virtue of having a pore structure which is well connected, and in all probability a preponderance of small pores into which vapour can be readily attracted via the draw of the negative oxygen-rich surface of the pores for the positive hydrogen atoms in the water vapour. Materials with many small pores will not hold the water vapour but wick it away at a faster rate. This can be seen by comparing clay bricks, which have many small pores and readily wick moisture outwards via the capillary action of these pores, versus calcium silicate bricks, which have a range of pore sizes. In the latter there is greater hygroscopic capacity as water will sit in the large pores, having travelled rapidly through the small pores by capillary action, but will take longer to be wicked away to the outside.

Wood is a material with a good hygroscopic capacity if it well seasoned (moisture content generally is below 20%). This means that only the cell wall of the lumens will be capable of attracting water vapour, which they will readily do because the cellulose molecule is very attractive to water molecules, as well as the internal face of the pores being oxygen rich. This is what causes wood to swell and why external wooden doors are often more temperamental in the spring when RH is higher. Swelling is also consistent with saturation, but here the water is actually being held within the cell lumens, which is what happens to the same door in a wet winter. The key to avoiding this problem is to choose a wood which is close ringed, that is, slow grown so that the dense winter wood is in more preponderante than the more sap-filled spring and summer wood. Such wood is characteristic of 18th- and early 19th-century joinery, whereas modern fast-grown softwood replacements perform badly in relation to water vapour or saturation.

'Woodfibre' insulation boards are also very adept at wicking away water vapour, but if the pore structure of the wood fibre becomes saturated this will negate the ability of this material to act in a hygroscopic way, as the water will then be held as a liquid in the cell walls. A diagram of wood structure can be found in Chapter 7, but a simplified version is set out below.

Hygroscopicity is also dependent upon equilibrium moisture content (EMC). A hygroscopic material will reach this state at a fixed temperature and a fixed humidity of the surrounding air. If the humidity exceeds 95% most materials will cease to be hygroscopic as the water vapour will become liquid and this travels through the pore structure at a much slower rate than water vapour, particularly if the number of large pores exceeds the number of small pores.

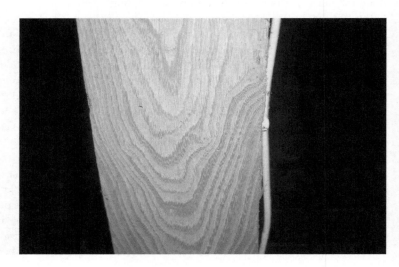

Photo 4.7 Imported softwood used in 19th-century construction with a preponderance of dark winter wood

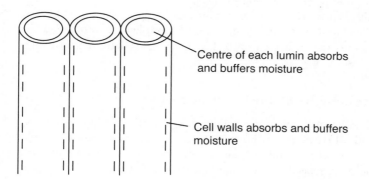

Diagram 4.7 Simplified diagram of wood structure

Centre of each lumin absorbs and buffers moisture

Cell walls absorbs and buffers moisture

Hygroscopicity will obviously be affected by the mass and thickness of the material. Thick stone walls, which in all events have a very variable pore structure, will wick away water vapour at a much slower rate than a thin brick wall.

When assessing the ability of a wall to act as a hygroscopic mechanism in order to control the internal environment, it is vital to consider not only the EMC but also the material itself and the thickness of the wall.

Hygroscopic materials

Hygroscopic materials can be divided into:

- **Highly hygroscopic materials** – timber, cork, fibreboards, woodwool slab (magnesite bound), unfired clay/earth, lime plaster and mortar, gypsum plasterboard and plaster, coconut and other natural fibres, and cellulose insulation. These materials must be finished with vapour permeable beeswax, low-density binder emulsion paints, or limewash.
- **Medium hygroscopic materials** – wood particle board, plywood, hardboard, cement mortar, fibre cement sheets, limestone, and linoleum.
- **Low hygroscopic materials** – bricks, pumice concrete blocks.
- **Zero hygroscopic materials** – mineral fibres, concrete, ceramic tiles (if well fired), glass, plastic foams and finishes.

(Adapted from Harris and Borer, 2005)

Interstitial condensation

Water vapour can be deposited as interstitial condensation if the temperature of the wall itself is low, causing the water vapour to rain out. The higher the wall temperature the further outwards the water vapour will travel. It is fair to say that the warmer the room, the more water vapour will be held in the air (warm air holds more moisture than cold air), and the more chance it then has to permeate outwards through a wall or ceiling. In addition, the warmer the room, the more warmth will be held in the wall, and this in turn will encourage water vapour to avoid condensing within the wall structure. This unfortunately creates an immediate conflict with modern building regulations, which demand that heat should be kept in rooms with maximum insulation. In essence, there is a conflict between modern insulates which discourage permeability but retain heat, and the need to transmit water vapour in order to control internal

173

environments. Proponents of non-porous wall linings and insulates maintain that interstitial condensation is an unpleasant eventuality, and in essence are expressing a preference for warm but moisture-ridden internal environments rather than naturally controlled environments. This will mean some interstitial condensation, which surely has to be better than moisture-laden internal environments.

The benefits of the breathing wall

The rate by which materials adjust to changes in humidity is a powerful tool in the buffering of peak humidity in households where mornings are the time for multiple showers, for example. Baths and showers produce high relative humidity and consequently large volumes of water vapour held in warm air. The greater speed at which materials such as end-grain timber, unfired clay, and natural fibres such as sheep's wool can absorb water vapour and wick it away to the external atmosphere, the less chance there will be of silver fish and mould growths forming on cold surfaces such as tiles in kitchens and bathrooms.

Minke, in his text on earth construction (2000: 16, cited by May, 2005) using a procedure for testing materials at 5% and 80% humidity at a constant temperature of 21 °C, showed that in order, clay plasters, lime plasters, and gypsum plasters absorbed water at a faster rate. Timber has a much greater capacity to hold water over a long period, due no doubt to its ability to hold water within the cell walls, and only when these have reached maximum capacity, hold water in the cell lumens themselves. For this reason timber has a greater holding capacity than many other materials. It is this holding capacity that determines how useful a material is in buffering humidity, and this factor is inevitably dependent upon density. Loose natural fibres like sheep's wool or hemp absorb water vapour quickly but have little capacity to hold it, leaving the masonry to do this job. Hygroscopic capacity is dependent on weight, not volume. It is important also to ensure that there is a sufficient depth of the hygroscopic material. Clay cob walls behave in a particular way in this respect. According to Minke, the first 20 mm of the wall starts to lose its rapid rate of absorption within the first day and reaches its capacity by the end of three days; only then can the next 20 mm start to have a significant buffering effect (Minke, 2000, cited by May, 2005).

Photo 4.8 A new cob construction at a Somerset sustainability centre indicating the potential for new build in this material

Buffering can only be perfect in internal walls. In external walls the inner part of the wall will act as a buffer to moisture but then feed it outwards.

May (2005) concludes that, as a very rough guide, the materials listed below have a speed of hygroscopic take-up of water vapour as indicated. This is useful in the context of traditional building construction.

- Cement render: slow/medium
- Hydraulic lime render: slow/medium
- Concrete: slow
- Fired clay bricks: medium
- Unfired clay bricks: medium
- Softwood, end grain: fast
- Softwood, transverse: slow
- Mineral wood insulation: medium
- Wood fibre board insulation: fast
- Cellulose insulation: fast
- Flax/hemp/sheep's wool/hemp insulation: fast

It should be noted that air conditioning systems will seriously affect the ability of materials to behave in a hygroscopic manner. It is fair to say that the use of such systems in buildings with hygroscopic materials is a contradiction in terms.

Capillarity

Capillarity has already been mentioned, in that small pores which are usually narrow-walled voids attract water more rapidly than large pores (see discussion above regarding the narrow and wide test tubes in a dish of water). Capillarity is thus the action of water in its liquid form. Hydrophobic agents such as silicones coatings are said to block only the larger pores, leaving the small pores to be vapour permeable, but in reality unless all these small pores are linked they are not going to be effectively capable of transmitting water vapour in and out of the structure. There is also no way of knowing if hydrophobic agents are actually blocking all modes of water molecule transfer.

Capillarity in individual building blocks can be tested by placing a standard cube of the material in water, with all sides sealed bar the bottom. The material is then weighed over time spans to test for an increase in weight.

Solid bricks have a very high rate of absorption, as is evidenced by a simple water spill on a brick floor, and this should be borne in mind when indenting bricks in a wall needing repair. The brick should always be well saturated in a bucket of water to avoid the moisture being sucked out of the mortar before it has a chance to take on a set. Plunging a dry normal brick into water will explicitly show the oxygen in the pore structure being replaced with water, via the tiny bubbles which emit from the face of the brick.

Wood in comparison is slower to take up water by capillarity, but is faster than clay. Clay takes up water by virtue of the oxygen-rich mullite layers being attractive to water and parting slightly to accommodate it. This is why clay swells when in contact with water (see Chapter 7).

It should be noted that absorption is faster than desorption. This is because the oxygen in the air is slower to absorb the water molecules as they exit than the oxygen-rich pore structure than the latter was to absorb the water molecules in the first instance. The fact that wetting is faster than drying is a key point in building material behaviour. It accounts for north elevations, particularly on large masses of masonry on for example churches, remaining saturated for long periods and taking on a green algae coating.

The capillarity of various materials can roughly be summarised as follows:

- **Plasters:** Gypsum plaster has a more efficient take-up of water than clay plaster, and surprisingly both are more efficient than lime plaster which equates with cement plaster. Gypsum cannot be trusted to do this job, however, as it is water soluble and will simply disintegrate.
- **Building blocks:** Fired clay bricks, with many small pores, has a much more rapid take-up than either concrete or unfired clay bricks.
- **Woods:** End grain has a much faster take up than transverse sections of wood. Plywood has by far the slowest take-up.
- **Insulates:** Cellulose, due to the cellulose molecule which is so attractive to water, has by far the fastest take-up of water by capillarity, and far outstrips flax/hemp/sheep's wool, followed by wood fibre. Expanded polystyrene has some limited capillarity, whilst the closed-cell polyisocyanates have none.

Paints and capillarity

Pure limewash, casein-enriched limewash (casein being a milk protein found in sour milk that readily bonds to calcium forming calcium caseinate, a secondary binding mechanism), and silicate paints exceed masonry paints in their ability to absorb water by capillarity.

Pottasium silicate paints are formulated with potassium silicate as the binder, otherwise known as 'waterglass' and sometimes called a 'mineral paint'. They become an integral part of the substrate, rather like a stain. They can be used on masonry, concrete, lime renders and plasters, and earth plasters.

Other alkyd resin/epoxy/chlorinated rubber paints have no capillarity capacity.

The chemistry of potassium silicate paints is

$$2KOH + S_i(OH)_4 = K_2S_iO_3 + 3H_2O$$

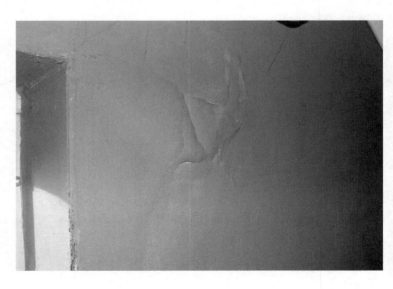

Photo 4.9 Internal gypsum plaster exploding due to water ingress from an exposed external gable elevation facing across a valley

that is, potassium plus silica acid gel gives potassium silicate plus water which evaporates away when the paint is applied.

Conclusions on vapour transmission

Materials are very variable in their response to vapour permeability via hygroscopicity and capillarity. A typical example is mineral wool. It is indeed vapour, open in comparison to say a closed-cell insulate, but although there are voids there is no pore structure as such which is lined with oxygen-rich molecules. Thus it has no hygroscopic capacity and limited capillarity. It is more prone to getting wet by total saturation of the voids, and if this happens drying out is very slow, and mainly downwards via gravity action. In a situation where the mineral wool is laid over ceiling joists, this can be disastrous as it is a recipe for dry rot in the joists. Haplessly this is precisely the situation which could occur, as the majority of warm moist air from baths, showers, and kitchens rises upwards to infest insulation.

A new product on the market under the trade name of Earth Wool is in fact recycled plastic bottles. This is very laudable, but its properties are no different from glass wool or rock wool (mineral wool) in that moisture once trapped will descend downwards. It is, however, kinder to humans insomuch as it can be manually handled and does not have fibres that ostensibly could enter the respiratory passages.

Insulates such as natural fibres are a very different matter, having molecular structures that are in themselves attractive to water molecules. This includes flax, hemp, and sheep's wool, and unadulterated wood fibre. They are thus eminently hygroscopic and also have good capillarity capabilities that enable them to dry quickly. They are very successful in not only insulating, but also enabling the transfer of moisture vapour outwards to the external atmosphere, in a wall and upwards through the vents of a ventilated roof.

Most plasters are efficient at vapour transmission, although May (2005) states that lime plaster is not the most efficient breathable product. Most lime plasters have low hygroscopicity in comparison to gypsum or clay plasters, unless they are of the hair fat lime type, but their variable pore structure does give them a capacity to hold moisture as a sink (buffering), and then expel

Photo 4.10 Loose sheep's wool insulation being laid in a 17th-century roof (photo © Rob Buckley)

it into the masonry and then to the outside or expel it back into the room where it can be cleared by purge ventilation. In addition, they are not prone to disintegration as is gypsum plaster. Hydraulic limes are probably less vapour open than weak cement, according to May (2005). Considerably more research is needed, particularly when it is known that lime plaster has a variable pore structure in size, and the oxygen-rich pore structure will attract gaseous water molecules.

Paints behave much as expected. Those with a weak binder, such as limewash, in its purest state without casein or linseed oil/tallow binders added, and distempers, consisting only of crushed chalk and water, are capable of good vapour transmission. Chemical pigments, because they have complex molecular structures, can also act as a binder and need to be carefully watched in this respect. Earth pigments, as opposed to chemical pigments, have no limiting effect on capillarity. Such finishes can be seen to be actively working, especially when they meet water as a vapour or as a liquid they physically darken, only becoming light again when they dry out. Some paints, such as the silicate range, have the capillarity physically removed by the addition of a hydrophobic agent, but are said to still remain vapour open.

May (2005), basing his evidence on Minke, argues that hygroscopicity is not really relevant to paints as they are so thin, but the possibility that a paint will affect the hygroscopicity of the wall it is covering needs to be considered. On balance, Minke (cited by May, 2005) found that, with the exception of boiled linseed oil or latex used as a binder, most paints had little effect on the ability of the clay cob to perform hygroscopically, but there are many who would say otherwise. Limewash and silicate paints certainly have less resistance to vapour permeability (the G factor) and thus must be a better option for cob than any other form of paint covering. Certainly limewash was traditionally used.

Manufactured chemical-based versus natural insulation materials and the effect on human health

Central to any discussion on vapour permeability and the desirability of removing moist air for the protection human health must also be an examination of the materials themselves and their capacity to endanger human health or alternatively preserve it.

Nearly all insulation materials are manufactured in some way so that during the process some modification of even the most natural material is necessary, and this has to be taken into consideration when calculating the potential for damage to human health and the impact upon the environment. In this section all the major insulation materials will be considered in respect of their effect on human health, using as a primary data source the excellent *Green Building Handbook* (Woolley et al., 1997, 2002) in conjunction with Bjorn Berge's equally informative text on *The Ecology of Building Materials* (2000), translated from Norwegian, in order to make an assessment of the impact on human health. These pioneering texts have been adapted by the author in order to highlight the primary concerns for human health.

CFCs and HFCs

The main concern with manufactured materials based on petrochemical products is the use of Chlorofluorocarbons (CFCs) for the blowing of foam. These are complex chlorine compounds where the hydrogen atom has been replaced by the halogen fluorine. Whilst this as a gas is stable in the lower parts of the Earth's atmosphere, when it reaches strong sunlight the whole chemical structure breaks down, releasing the chlorine atoms which react with the natural ozone and break down the ozone layer. Even the addition of hydrogen as HFC still has a significant effect; hydrofluorocarbons were thought to be more acceptable but in reality are not, even though they are minus chlorine. Alternatives are carbon dioxide and pentene, and blown materials should be carefully checked to ensure that these are used instead of CFCs, etc. Of concern also is the high dependency of manufactured insulation on the provision of oil, with all its attendant pollution problems as well as the use of vast quantities of energy. Of even greater concern is the potential for emissions of aliphatic, aromatics, and chlorinated hydrocarbons, leading to internal pollution for which even greater ventilation is required, at a time when ventilation is in reality being reduced from that which is traditional. In addition, the greenhouse gases (GHGs) carbon dioxide and sulphur dioxide are released during the refining process of oil into its various fractions.

It should be noted that:

- **Aliphatic hydrocarbons** are paraffins, naphthenes, and hexenes.
- **Aromatic hydrocarbons** are xylene, toluene, benzene (carbon rings to which hydrogens are attached), and styrene, manifest usually as solvents. Benzene is added to coal tar to make creosote.
- **Chlorinated hydrocarbons** are formed when hydrocarbons react with hydrochloric acid to form solvents, such as trichloroethene, trichlorethane, dichloroethane, and dichloromethane. The latter in particular are used in paint strippers, and are known to penetrate skin and travel to vital organs.

The efficacy of thermal materials is based on their thermal conductivity or k value, measured in W/mK, the lowest k being the most desirable. Regrettably the materials which involve the most use of chemicals give the lowest k value, notably polyurethane foam, polystyrene foam, phenolic foam and urea (formaldehyde foam), and they are considered below in ascending order, whilst natural materials tend to give a less-desirable higher k value (Woolley et al., 2006: 42).

Of the more natural materials, and using the same informative source, wool ranks the highest followed by cellulose fibres, corkboard, vermiculite, foamed glass, softboard/wood fibre boards, wood-wool slabs (the modern version of which is Heraklith), and compressed straw slabs. The now more popular wood fibre boards are very worthy of consideration.

Manufactured chemical-based materials – a consideration of human health and environmental impact

Polyurethane foam and other isocyanurate foams

These extremely popular products, particularly as they are usually available in sheet form, are derived from distillates from petrochemicals (oil and gas), using processes that consume vast amounts of energy. These particular foams use volatile polyisocyanates, which is an extreme irritant to skin and mucous membranes. Certainly installers may well be at risk from volatile compounds, and the question must arise as to the vulnerability of occupants in this respect. During a fire the material will give of hydrogen cyanide gas. The material cannot be recycled as it is a thermoset resin.

Nothwithstanding the poor prognosis for the supply of oil, like the other plastic foams discussed below the initial derivatives involve the emissions of particulates (view any oil refinery from a distance and it will always be through a grey cloud of particulates), phenols, and heavy metals, plus hydrocarbons together with the effluent of chemical compounds used in the

Photo 4.11
Isocyanuarate panels
being used as
insulation in a modern
building

processing. Rain becomes saturated with this cloud of emissions and forms the gases, nitrous oxides and sulphurous oxides, which then react with rainwater to produce nitric acid and sulphuric acid. Both are highly detrimental to the longevity of limestone, the latter in particular changing the limestone (calcium carbonate) into calcium sulphate, which is water soluble and washes away.

Thus from the outset the environmental disadvantages of the product are numerous. Add to this the fact that the foam needs to be produced using a gas that contains carbons, fluorons, and chlorines. These not only affect the ozone layer but chlorinated hydrocarbons are thought to be poisonous to human organs, including the liver and kidneys, and are almost certainly carcinogenic (Berge, 2005: 147). There is also risk of damage to mucus membranes, particularly during installation, from volatile polyisocyanates. Of greater concern is the danger of fire, when the material will release hydrogen cyanide gas. Formaldehyde, a chemical produced from the oxidation of alcohol, to which phenol and urea are added, is also present. Even in small quantities this can cause irritation to the eyes, itching in the nose, a dry throat, and sleep problems. In the longer term serious problems in the respiratory zones are likely, and the substance is registered as carcinogenic (Berge, 2006: 395; Woolley et al., 2006).

Rock wool, slag wool, and glass wool

Rock wool, slag wool, and glass wool are made in a similar way from quartz sand, soda, dolomite, and lime. Glass wool has around 30% recycled glass. Rock wool has as its basis basalt and olivine. The constituents are melted and drawn out into thin fibres – and therein lies the problem. These can cause extreme irritation to the skin, the eyes, and the respiratory passages. Sustained exposure can lead to bronchial problems. The primary school of thought is that mineral wools do not behave in the same way as asbestos in that the fibres do not linger long enough in the lung, but like many aspects only time will tell. It is safer to use these materials encapsulated in a quilt, some of which are now produced with a foil backing to reduce emissivity. In addition, the material can off-gas aliphates due to the addition of aliphatic mineral oils, plus amines and ketones, if it gets damp. It is also thought that the nitrogen content attracts mould

180

in large quantities. The production process produces large amounts of waste, and involves the addition of phenol-formaldehyde resins, silanes, and siloxanes (Berge, 2000; Woolley et al., 2006).

Polystyrene foam

Two types of polystyrene foam are available: expanded polystyrene (EPS), foamed with pentene; and extruded polystyrene (XPS), foamed with HCFCs, a moisture resisting material not open to vapour. Made from a styrene to create a polymer, the material incorporates phenols and bromine compounds, and the beads of polystyrene expand and fuse together. Emissions during manufacture include many of these compounds. This off-gassing can continue during installation, with particular respect to the blowing agents and the unstable compounds of styrene. During a fire it gives of both carbon monoxide and carbon dioxide, as well as smoke. Disposal can incur leakage of some of the additives (Woolley et al., 2006).

Phenolic foam

Phenols are known carcinogenic aromatic organic compounds (carbons and hydrogen). They are the major constituent of this fire-resisting material (Woolley et al., 2006).

Urea-formaldehyde foam

Even small doses of formaldehyde can irritate eyes and mucous membranes. It is also registered as carcinogenic and can cause allergies. Although not as problematic as phenol, as it oxidises to formic acid and then carbonic acid, it is nonetheless not a pleasant substance and has been banned in furniture manufacture (Berge, 2000).

Natural alternatives

Examining more natural alternatives in order of efficacy as insulators, there is a sharp contrast to the above, although it is virtually impossible to find a material that can be used to enhance thermal performance that does not itself require energy to produce. In addition, many naturally derived products contain undesirable additives. The financial incentives to recycle such materials once their building lifespan has finished are also sadly lacking.

Wool

Probably the most healthy and renewable option on the face of the planet, the universal use of wool is hindered by its production and transport costs. It has a unique ability to absorb and release moisture, and is thus the most efficacious component of a breathing wall. Even its ignition is not a problem as it is mineral treated against fire, and tends to melt as it moves away from the ignition source and then self-extinguishes. Organophosphates used in sheep dips might be the only problem if they are not rigorously removed in the manufacturing process. The material is usually treated with borax to achieve insect and mould resistance. It is not resistant to rust, alkalis, oil, and fat, but it is capable of absorbing the high volume of formaldehyde that may be emitted in the rest of the building.

Sheep's wool batts

'Thermafleece' is the most widely known of this product, which largely consists of wool but contains polyester reinforcement to create the batts. They are further treated against pests, fungus, and fire, this being essential as wool will degrade if subjected to saturation for a long period of time. The key feature with the wool batts is that they can absorb and release moisture without loss of thermal resistivity, and are less problematic with regard to interstitial condensation. They are also effective in reducing transmission of airborne sound.

Hemp batts

Hemp is the acclaimed insulating material of the 21st century, which has been held back by an unjustified connotation with it sister drug-related form. Hemp is very fast growing, maturing to a height of 4 m (13 ft) in around 100 days. This means that weeds have no chance to grow, so no herbicides or pesticides need be used. Admittedly the hemp batts are bound with a thermoplastic binder (polyolefin fibres) and treated with ammonium phosphate salts to provide fire and pest resistance, but despite this they have to be a better ecological option than a totally polymer-based product. More pertinently they have none of the irritant factors associated with handling mineral fibres, and are hygroscopic providing a capability for buffering moisture. Hemp is also available on a roll. Hemp batts can be used for lofts, floors (though not ground floors), walls, and ceilings, but not anywhere that is subject to prolonged exposure to water.

Cellulose

Cellulose fibres are made from recycled paper and treated with borax for fire and insect resistance. The processing results in a fluffy product that can be sprayed into voids. There may be a danger if the particles are so small that they can be inhaled. Certainly the product has huge advantages in terms of its use of an unwanted material, but it cannot cope with excessive moisture content, as it will not breathe out the moisture as fast it takes it in. The cellulose molecule is very hydrophilic (water loving). Known by its trade name Warmcell 100, the cellulose fibre is derived from waste newspapers and treated with fire retardant and biocides.

Photo 4.12 A hemp batt used as a base for lime plaster

Photo 4.13 Cellulose insulation in a mock-up of a modern timber framed wall

It is blown into lofts and floor voids. Whilst undoubtedly effective for vapour permeability, it lacks the convenience of being in batt form.

Corkboard

Another naturally grown material, corkboard is produced from the bark of the evergreen oak, Quercus suber. It has to be imported from Portugal, Spain and North Africa, so there is a sea-miles component to the equation. Although its production requires energy for the boiling of the material to encourage the granules to bond themselves together with their own resins, it is a renewable resource provided the trees are well managed in the 10–12 years they take to re-grow their bark. Cork board can be used for flat roofs, in batt form, or as a loose-fill material. Very popular in the 1970s as a decorative wall material, corkboard needs to enjoy a rebirth as a wall and floor insulation. It is naturally resistant to fungus and water penetration.

Foamed glass

If made from scratch, using limestone and sand, foamed glass is very energy intensive, requiring the manufacture of the glass, the grinding of the same and the re-melting to ensure the incorporation of CO_2 bubbles of gas. There is an emissions problem with glass making with both sulphur and nitrogen causing extra loading for acid rain. The continuous extraction of sand and limestone has ecological and environmental implications, as does all primary quarrying activity.

Clearly there is huge potential for the re-use of glass, which in the building construction industry in particular is regularly wasted in landfill (replacement of single glazed units with sealed units). It is obvious that the material has no vapour permeability and is thus not suitable for the breathing wall, unless ground up and used in a lime plaster (termed 'glaster'), but it does provide good floor insulation in modern build. It is suitable for those existing traditional

Photo 4.14 Foamed glass for use in the sub-base of a limecrete floor

buildings where the floor needs to breathe, by virtue of using a crushed glass product in the sub-base of a limecrete floor to resist rising damp.

Softboard

This is manufactured from ground wood pulp rendered into fibres. The bonding into sheets is achieved without the use of additives, using the fibres as an interlocking device together with the natural adhesive resins in the wood. Ideally it should be made from waste from other wood processes. In practice there is no reason why the material could not be shredded and recycled, but there appears to be little financial incentive to do so. The material is ideal for use in conjunction with breathing walls, as a lining, as it has vapour permeability.

Some softboards are bitumen-impregnated and are thus not suitable for breathing-wall applications. They are designed to be used where moisture has to be resisted. There is a danger that they can be used in the wrong locations.

Vermiculite

The mineral vermiculite, which on the face of it seems innocuous, requires energy to heat it up so that steam within the mineral causes it to produce light concertina-type granules, which contain asbestos fibre. Sources from South Africa do not. It is used either as loose-fill between ceiling joists or as an aggregate in plasters (renovating plaster), composite boards, or lightweight concretes.

Wood wool

The most popular wood wool is Heraklith, made from wood shavings bonded not with cement but with magnesite. Traditional wood wool boards are strands of wood shavings bonded with cement, and were standard fare for the repair of panels in traditional timber framed buildings in the late 20th century. The wood shavings are waste, but the composite board is not capable

Photo 4.15 A wood wool slab

of being recycled once it falls out of use. They are principally used to eliminate thermal bridges, providing acoustic insulation for floor voids, and insulation of flat and sloping roofs.

Another product, Calsitherm Climate Board, is a calcium silicate board described as multi-porous with a high capillary action and insulating properties. It can be applied directly to brick or blockwork, thus avoiding the use of studs.

Compressed strawboard / flaxboard

Wheat, hemp, flax, or barley straw is heated and put under pressure to create a solid slab, bound at the edges, or coated with paper to give them rigidity. This requires the use of an adhesive which, if formaldehyde based, can reduce their environmental credibility. Some strawboard does use a polyurethane adhesive in the compressing process which has health implications. Flax needs to be boiled before it can be processed. The material has huge potential as an insulator and can cope with vapour permeability (ideal for lining of breathing walls) and as thermal insulation for flat roofs, but not with any excessive moisture. In this respect it is very similar to wood fibre in that it is a cellulose product.

Wood fibre boards

Known by the trade name of Pavatex, there is a whole raft of different types of board for different locations, and they can generally be said to have revolutionised the building of breathing walls as well as the insulation of existing buildings. They are made from wood pulp, mixed with water and then heated to activate the lignin to act as a glue to bind the fibres together. They have low thermal conductivity, high vapour permeability, and are interlocking to reduce airtightness. They can also be used to achieve acoustic performance. Used as external wall insulation they give all the benefits of an external render in creating thermal mass, and because they store heat, condensation is avoided. Their best quality is that they allow moisture to pass from inside rooms to the outside, thus avoiding all the problems associated with vapour barriers. In addition, this same thermal mass provides for cooling in summer by delaying the peak of external daytime surface temperature permeating to the inside. A reduction of internal temperature of around 4°C is possible, compared to structures with insulation to the same u-value, but with other forms of manufactured insulation. The systems have LANTEC (Local Authority National Type Approval Confederation) and BBA (British Board of Agrément) Certification.

A system for roofing provides water resistant interlocking wood fibre boards (called Isolair sarking board) over the rafters without thermal bridges, and without the necessity for external membranes or internal vapour barriers. The whole purpose of wood fibre boards is to provide an excellent system of vapour buffering, and thus a breathing roof, whilst utilising up to 95% wood waste, and when the material is no longer required it can be composted or recycled. Another similar system (Pavaboard) provides for insulation under floors that have to carry loads, again helping to control internal moisture levels. It has a k value of 0.042 W/mK.

Pavatex Dithutherm is perhaps the best known product and is available for externally cladding steel frames, and for timber frames externally and internally, to achieve a vapour control layer as well as providing airtightness. Externally, weatherproofing in a proprietary mineral based thin render is necessary, and it is also necessary to provide good overhangs, flashings, and set windows/door back into a reveal behind the rear face of the board. Insulation in the void is necessary, using hemp cotton or sheep's wool to continue the vapour control system whilst providing thermal insulation. Timber constructions are best boarded, but a

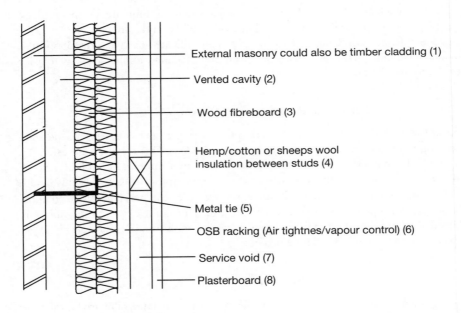

External masonry could also be timber cladding (1)

Vented cavity (2)

Wood fibreboard (3)

Hemp/cotton or sheeps wool insulation between studs (4)

Metal tie (5)

OSB racking (Air tightnes/vapour control) (6)

Service void (7)

Plasterboard (8)

Horizontal section through 4

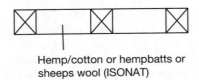

Hemp/cotton or hempbatts or sheeps wool (ISONAT)

Diagram 4.8 Wood fibre board used as an alternative to closed-cell materials for modern timber frames (after Natural Building Technologies, 2010)

specific form of Pavatex (called Pavaclad utilising Pavatherm-Plus) can be provided for external cladding with facing bricks, rendered blockwork, or tile/slate hanging.

This enables the masonry to continue its role as a thermal store and vapour buffering layer, so that the masonry plus wood fibre boarded sheathing provides a complete breathing wall system.

The Pavadentro Renovation System, is specifically designed to be used internally in existing traditional buildings, be they masonry, timber, or brick. Its role is to reduce interstitial condensation by confining it to the insulation layer where it can be released back into the room to be removed by purge ventilation when conditions are propitious. It is not, however, suitable for areas where there is constant high humidity, such as bathrooms, where it would be better to use a clay board. Similarly, it is not suitable for walls constantly subjected to storm conditions. Here a lime plaster externally and internally will provide a more durable moisture buffering regime (adapted from Natural Building Technologies, 2010).

Calsitherm climate board

Calsitherm climate board is an alternative for internal insulation and can be applied directly to an internal plaster without the need for studs. It is made from calcium silicate and is microporous, with a high capillary action to enable it to buffer the moisture content of a room. It is mould resistant because of the nature of the material.

Reed boards/mats

Another internal cladding for existing traditional buildings which also provides vapour permeability is reed board made from lime, clay, or gypsum on a reed base. Reed mats are also available on a roll, emulating the traditional reed base used in the south-west of England, especially during the 19th century. Plastered in lime, they once again provide vapour permeability. This time-honoured technique utilises a renewable material that can be composted when no longer required, and the lime plaster ground up to provide aggregate for a new mix.

NOTE: Care must be taken when introducing any cellulose-based product into a building which has had an outbreak of dry rot. The fungus feeds on the cellulose molecule and will romp away, causing total devastation.

Photo 4.16 Traditional lime plaster on reed was used throughout Dorset in the 19th century

Clay board

Clay board manufactured using clay, reed, and hessian is reputed to have outstanding thermal and vapour diffusion properties, thus making it the ideal material for buffering moisture in bathrooms, for example. In addition, it absorbs odours and acts as a sound insulator. It is one of the most ecologically sound materials, using reeds from managed reed beds.

Concretes as thermal insulators and breathing structures

Concrete has a bad name as a sustainable building material, as the cement generates high energy costs during production, together with carbon dioxide emissions. The aggregate similarly has high energy costs in terms of transport and quarrying.

The three forms of concrete listed below are useful as insulating materials, and the longevity of the material, provided that it is not wilfully removed in the interests of high fashion or the latest trend, may off-set the initial energy costs.

*   **Foamed concrete:** Foamed concrete has better insulating properties than normal concrete. Made from Portland cement and fine sand in equal proportions, it requires tensides (acid esters) similar to those used in the production of detergents (anionic or cationic) to create small voids.
*   **Aerated concrete:** The constituents are finely powdered quartz, lime, gypsum, and cement plus aluminium powder, the latter reacting with the former to release hydrogen and create a porous structure. The material is hardened in an autoclave. The material is ideal for breathing walls, but demands a (hydraulic) lime render on the outside and a (hydraulic) lime mortar in the joints of the blocks or prefabricated units in order for it to breathe out moisture.
*   **Light aggregate concrete:** Here the aggregate is of a mineral-type, light expanded clay, pumice, vermiculite (mica derivative), perlite (a natural glass of volcanic origin), or slag.

Fired clay / clay products

Clay can be expanded into pellets for use as loose fill or for incorporation into concrete or even lime renders used internally, where they will contribute to thermal mass. They are particularly beneficial when used under a limecrete floor slab as a capillary break. Certain brick types are very micro-porous, such as low-fired brick or brick with a high proportion of lime.

'Ziegel blocks' are specifically manufactured to produce an aerated structure to trap air and hence increase insulation properties. Manufactured in Germany (the word 'ziegel' means 'block'), they are used extensively on the Continent for internal blockwork and by doing so create thermal mass. They are particularly effective in hot climates for buffering heat, and in cold climates will store heat produced internally.

The concept of the wall as a thermal store

In addition to considering the use of natural materials as insulators to prevent heat escaping should be the concept of the wall as a thermal store. Thermal mass is also called 'thermal capacitance'. It is the mass of the material multiplied by the specific heat capacity (or the number of moles of molecules, n, times the specific heat capacity c). The more molecules present, the more chance there is of heat being passed from molecule to molecule. Warm air also travels

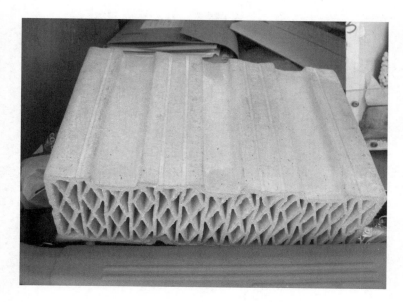

Photo 4.17 A ziegel block

through the pore structure (provided it is dry) by convection. Warm air rises and warms a cooler surface, including the internal surface of the pore structure of dense materials.

Modern practice and indeed building regulations have simply skipped over this scientific fact in the desire to create thermally efficient and energy conserving buildings, to the point where even existing thick stone walls are required to be insulated. This negates any possibility of the thick masonry wall to act as a thermal store, as well as negating its ability to act as a breathing wall, as invariably a vapour barrier will be deemed desirable.

The molecular structure is the key to the ability of certain materials to hold heat, to retain it during the day from solar gain, and then slowly release it during the late evening. It has long been known that both brick and stone have a high capacity for absorbing heat from solar gain. The Roman hypocaust system is based on the principle of the ceramic floors absorbing the convected heat from underneath, generated by a solid fuel fire (see also Chapter 5). Many cultures have utilised this concept, particularly in very cold climates, with the fireplace occupying a central core of the building, and in particular tile-covered concrete or iron stoves in Germany and Holland still remain popular. They stay warm for up to 14 hours after the fire has gone out.

Water also has an incredible ability to absorb heat from solar gain but will release this heat much faster, and as such has rarely found favour in the same way as porous building materials.

The traditional British house, typical of that built from the medieval period through the 16th, 17th and 18th centuries, with massive stone walls and a large brick central stack, also acted as an enormous thermal heat store, absorbing the heat from the fire that was kept in the ingle day and night (banked down), summer (for cooking) and winter. The bulk of the masonry holding the heat released it slowly during the night, to be recharged the following day. The idea of a fire on a hot summer's day is not appealing, and there is some evidence in Dorset ingles of a smaller flue with narrow iron bars designed to hold a small cooking pot or boil a kettle during warmer weather.

Such houses never in fact overheated in the summer, the walls again absorbing the heat from the sun and then sending it skywards via night sky radiation. Traditional construction is thus

Photo 4.18 A Dorset ingle with summer fireplace in rear left corner

Photo 4.19 Detail of the summer fireplace

absolutely ideal for human comfort, storing heat in winter to release it when the central heat source cools down, and buffering excess heat in the summer, but has had a bad name due to the following misunderstandings.

Such solid wall traditional construction takes on average about two to three days to heat up, but once this is achieved they will hold this heat at a low continuous temperature. This negates entirely the use of on–off central heating systems that come on for, say two hours in the morning and three hours in the evening. It is little wonder that the occupants of such buildings complain bitterly that they are always cold. It also negates the use of such buildings as weekend cottages. By the time the thermal heat stores consisting of walls and chimney breast have reached a point where they will hold the heat, this is likely to coincide with the occupant's departure date pending a return the next weekend. Installers of wet systems (radiators) that have tended to supersede dry systems, such as storage radiators (miniature heat stores), have failed to understand the needs of a traditional building. The storage heaters have far more to offer as they provide a low constant gentle heat, and can be just as economic, particularly if the electricity supply can be generated from renewable resources. Modern storage heaters are in all events more efficient than their 1980s predecessors.

Traditional ancient timber framed houses, which are in themselves inherently lightweight in their construction, had a massive brick central stack. This too heated up with the fire, which was an ever present feature, and then gently exuded this heat to the surrounding rooms. The traditional place for the master bed was with its head to the stack, shrouded in a ceiling tester and curtains to cradle the free heat from the central stack. Similarly, the stack absorbed solar gain on a hot summer day, which was then directed up the stack by convection, the stack then acting as passive ventilation.

The thermal capacity of earth buildings (chalk or clay cob) is also notable in this respect. There are whole villages in Dorset and Devon constructed of this amazing material, used from the late medieval period onwards right into the 19th century, which acts as the perfect heat store. It is by far the most environmentally sound of all the heat store materials, and yet has long been castigated. It can behave badly if subjected to water saturation, turning to 'porridge', the most common cause being cement renders that trap water-soaked cob. Conversely, it can

Photo 4.20 A half round chimney breast, typical of Dorset and Somerset, acted as a storage heater

turn to 'powder' if foolishly the wall has a damp course inserted and the material dries out. The bonding within the material is highly dependent upon the water molecule.

Sadly concrete blocks, with low, medium or high density insulation properties, have taken the place of cob in most modern construction. The high density block has the lowest CO_2 emissions by weight and volume in the manufacturing process, and is only slightly higher than a low energy rammed earth house, the difference being marginal in a life-cycle study (Roaf et al., 2003: 61). Despite this promising statistic the earth house appears to offer greater human thermal comfort, being always warm in winter, provided that a low level of continuous heat is supplied, and cool in summer. The cob acts as a vast store for the conducted and radiated heat outside the building. The thickness of the wall is an important factor. Cob walls are naturally thick to ensure stability of the material during the construction process, be it in the 'slow process' of having deeper lifts or the 'fast process' of laying several lifts at once, and risking a slump. The conductivity of the material appears to be low. Roaf et al. (2003) indicate that the optimal depth of mass is 100 mm (4 in) for each exposed surface, and 150–200 mm (6–9 in) for rooms backing on to each other, but cob walls are often around 500mm (20 in) in thickness.

It therefore follows that other masonry-like materials with a pore structure, such as concrete and limecrete used on floors, can be particularly useful in this respect in new traditional building construction. The conservatory is a typical example of how a thermal heat store can be put to good use. What is required is a ceramic tile floor surface in a dark colour, which attracts heat faster than a light colour, laid on a bed of concrete, or preferably limecrete, the bed being at least 60 mm (2½ in) thick, and preferably as much as 120 mm (5 in). On a summer's day the solar gain will be in excess of what is required to heat the conservatory but will be absorbed by the concrete bed and released in the cooler evening (excessive solar gain may have to be shaded out). What is in effect happening is a miniature thermal flywheel, with the thermal mass absorbing heat when the surroundings are hotter than the mass and giving heat back when the surroundings cool down.

The same principle applies to the masonry wall between the conservatory and the building. If this is a heavy mass structure it will behave in the same way, giving a valuable heat gain to the internal space behind it. It would obviously be beneficial if it was a dark colour (often not desirable aesthetically), or clad in natural terracotta tiles whose rich earthy red/orange colour performs well as a thermal store in addition to being aesthetically pleasing. Various options have been tried to enhance the capability of both the floor and the wall contiguous with the house. The former includes the use of rock stores beneath the floor, water storage within the conservatory (a minefield as light attracts algae and large-scale water containment is very problematic), and even reversible walls of black painted tin cans filled with water sitting in a panel acting as a shelf. Pivoted in the centre, the panel can be turned to face the room in the cool of the evening. The solar wall as illustrated at the Centre of Alternative Technology has a black painted surface on the outside of its mass, together with a glass sheet designed to encourage the input of short-wave solar radiation – the so called Trombe wall (see Diagram 3.18).

None of these rather extreme measures have yet gripped the imagination of the average British householder, but by the same token the whole technology of thermal stores in mass walling or floors seems to have equally passed unnoticed by the Building Regulations authorities. It is therefore not within the standard remit of conservatory companies, unless they are forcibly reminded of the need for an adequate concrete bed for the floor.

A development that could and should permeate British culture is the use of the hollow or perforated ceramic block, which is used all over the Continent to infill concrete frames. Their manufacture in Britain, where they were called 'poraton blocks', is limited by the fact that they

Photo 4.21 Ziegel blocks being used to line a stone wall in a new extension to a traditional building

were deemed to be of insufficient compressive strength, but as an internal lining to form a thermal mass they are absolutely ideal. Imported ziegel blocks are sometimes used in this way.

At this juncture it is pertinent to remind the reader of the three methods by which heat moves through building materials.

Conduction

Heat is passed from one molecule to the next because the latter is at a lower temperature, and energy which excites the individual atoms of the molecular structure is also transferred. In metals this heat transfer is very fast, but in dense materials such as stone or concretes it very slow, and this implies that the heat energy lost is also very slow. Materials such as urethane foam, based on cyanates, do not conduct heat and are therefore very popular as insulators. Conduction is one of the primary mechanisms of heat transfer in dense materials.

Convection

Heat is transferred through air. Warm air rises and warms a cooler surface including the internal surfaces of the pore structure of dense materials. Most convection in building structures occurs across cavities or across the two panes of glass in sealed units.

Radiation

Heat is radiated out from a hot surface to a cool surface, depending on the emissivity of the radiating surface and the absorptivity of the receiving surface. Low E glass is potentially one of the most advanced methodologies available to the sustainable builder that makes use of this scientific fact. It is designed with a metal film on the inside face of the innermost sheet of glass in the sealed unit. This inhibits the passage of some of the long-wave heat radiation out of the room, but allows into the room a goodly portion of the passage of short-wave solar radiation

193

via the outer pane and through the metal film. Most dense building materials are strong radiators of heat, especially brick and stone. Lean against an external wall built of either at the end of a hot summer's day and one can be literally scorched. Shelter nearby into the late evening and the heat can be felt radiating outwards. It is little wonder that stone and brick were always chosen for the central chimney breast and ingle in traditional yeoman farmhouses of the 16th and 17th centuries. Emissivity (E) can range from 0 to 1, the latter being perfect radiation, and unsurprisingly for most dense materials with lots of molecules packed together, E=0.9. Absorptivity equates to the E value for the majority of materials. Radiated heat can be lost across cavities as it moves from the inner surface to the outer surface of a wall, as well as across the cavity of a sealed unit. The use of foil-backed plasterboard and rigid foam panels has reduced this problem, as has the use of multiple layers of foil with a thin wadding of insulating material (trade name is Triso 10 multi-foil reflective insulation) used as roof insulation. This may be very effective in this respect, but completely ignores the need for the insulating material to breathe out warm moist air, particularly in a roof space. The result can be problems with condensation, the material acting as a vapour barrier.

U values and conduction, convection, and radiation

At this juncture it is necessary to get to grips with the concept of U value, which is the speed at which heat is lost across one square metre of the building element with a one Kelvin degree difference across the faces, measured in W/m^2K. Typical U values for well insulated houses verging on the passive house are:

- Roof 0.16
- Walls 0.35
- Windows 2.00
- Floor 0.25

Experiments to enhance the latent heat capacity of materials

Building on the knowledge that the amount of energy released or absorbed by a chemical substance during a change of state is called 'latent heat' (derived from the Latin *latere*, meaning to lie hidden), experiments have been conducted using both inorganic and organic phase change materials (PCMs). Simple examples of latent energy release are the melting of ice or the vaporisation of water by boiling. Energy flows when changing from one phase to another, that is, from a solid to a liquid and then to a gas. If water vapour condenses back to liquid on a surface, the latent energy absorbed during evaporation is released as 'sensible' energy onto the surface.

The aim of the experiments was to conserve energy and reduce dependency on fossil fuels, particularly at periods of peak demand. The basis of using materials such as butyl stearate $CH_3(CH_2)_{16}COO(CH_2)_3CH_3$, propyl palmitate $CH_3(CH_2)_{12}COOC_3H_7$ and paraffins is that as the temperature rises, the chemical bonds within the PCM break up, and the material changes from solid to liquid. The phase change is heat-seeking (endothermic) in the process and the PCM therefore absorbs heat. As the temperature cools the PCM returns to a solid phase and emits the heat it has absorbed. Salt hydrate type materials, such as Glauber's salt (sodium sulphate decahydrate), which also undergo a phase change have also been experimented with for incorporation into porous building materials, but rejected. Salts and dense materials are known to be an unhappy combination, as the salts mobilise in the pore structure on contact with water and have a tendency to expand in crystal form in the pore structure, causing its demise. In addition they are corrosive. Organic PCMs have been found to be the most effective within the human comfort temperature range of 16–25 °C, but are flammable and would generate very harmful fumes on combustion. They are also prone to oxidation and hence the break-up of their chemical bonds, increase the volume of the material they impregnate, and have an unpleasant odour. Primarily it would be difficult to make them fire retardant. Research has looked at the incorporation of organic PCMs into gypsum wall-board, concrete blocks, timber, wood particle board, and brick, as well as various aggregates. Gypsum board has been found to be the most effective as it can absorb up to 50% of its own weight, but the presence of calcium hydroxide $Ca(OH)_2$ in concrete being so strongly alkali can break up the bonds in organic materials. Containment is by means of surface tension. There have to be significant problems with the potential for major combustion, lack of breathability and volatile emissions. Set against this is the energy storing capacity of gypsum wallboard at 11 times that of normal wallboard through a 4 °C rise, and concrete which is in the order of 200–300% that of conventional blocks through a 6 °C change, the combination of pumice concrete block and paraffin being the most effective. The various concrete blocks tested include autoclave, regular, pumice, expanded shale, and OPC, with an energy storing capacity being 4 times greater than gypsum wallboard. This implies that impregnated wall board is more suitable for diurnal storage, whilst the concrete block gives a greater time load characteristic. It was concluded that PCM applied over a large surface area in a passive solar building is effective for the storage of solar gain, reducing room temperature by as much as 4 °C in the day and releasing this at night. Annual direct energy savings in the order of 15% are expected in a thermally moderate climate. At the time of the collation of the research on all of these aspects (1999) by Ruth Kelly at the Dublin Institute of Technology, the economic viability was low.

In conclusion – and in particular, as it is abundantly clear that such impregnated materials would have little capacity for dealing with warm moist air, a factor not even considered in the research – it is thought that they could only be useful in relation to the creation of a solar wall, dividing a sunspace from the main building (adapted from Kelly, 1999).

Ventilation and heat loss

From the discussion on hydro-carbon based materials it will be readily seen that it is extremely desirable to remove any possibility of emissions from these derivatives of oil-based products from the internal atmosphere of a building. In addition, it is necessary to disperse the carbon dioxide and vapour that humans breathe out, that great enemy of building materials, condensation, and excess heat when it builds up. This is the very crux of the breathing wall.

Unfortunately this is in sharp conflict with the equally desirable initiative to reduce air infiltration through cracks, voids, and porous building materials so that precious heat is not lost. Such infiltration is governed by pressure differentials inside and out. Ventilation is a more rigorous process whereby the amount of air allowed in is controlled manually, either by creating a purpose-made opening to take maximum advantage of pressure differentials or by mechanical fans which drive the air through ducts. Heat lost by both infiltration and ventilation is called 'ventilation heat loss'. The Victorians were firm believers in controlling ventilation manually, as can be seen at Tyntesfield House, near Bristol.

Traditional construction is often maligned with regard to ventilation rate (the speed at which air enters and leaves the building), which is regarded as too high for comfort. Ventilation rate is measured in air changes per hour (ac/h), and in a room with an open fire this can be as high

Photo 4.22 Manual ventilation controls at Victorian Tyntesfield House, near Bristol

as 4 ac/h, and in most traditional houses it is 1.5 ac/h. What is not generally realised is that in a traditional yeoman farmhouse the occupants never took off the majority of their clothing layers, apart from perhaps that which was designed to keep out rain. This is a long way removed from most young people today, who think that a comfortable temperature is that which allows them to walk around the house in a T-shirt. It should also be realised that this high ventilation rate enabled houses to breathe out moisture from rising damp at the base of walls and from floors, and any penetrating damp, by virtue of the chimney acting as passive stack ventilation (PSV). Nicholls and Hall (2006: 116) maintain that traditional PSV produces a heat loss rate of 2.9 kW, which is greater than fabric heat loss rate, and if the ventilation rate is reduced to 0.5 ac/h, this figure is reduced to just over one-third of the fabric heat loss rate.

There is no doubt that humans and animals, especially of the canine/feline variety, abhor draughts but the alternative, which is to cut down infiltration, can be detrimental to old or traditional building fabric (timber frames, stone, cob). What is required is a fine balance, which is not always best achieved by clever calculations but rather by usage and understanding of the fabric. It is therefore suggested that porous building materials should definitely be allowed to perform their maximum function, but that draughts via cracks in the join between window or door openings and the wall should be well sealed. The use of effective sealants around windows, such as brushes set into a rebate or latex strips, plus the use of heavy curtains is essential. Attention should be paid to removing altogether the tendency for a roaring gale underneath the floor joists of a suspended floor, by inserting insulation between the joists and underneath the floorboards. Loft hatches are a key area for attention, as are various ducts to the outside. In traditional buildings, however, there is a need to maintain a modicum of passive stack ventilation via chimneys and open flues, unless that latter are in use, when it is better to use a closed appliance rather than the open fire. Care must be taken when inserting closed appliances into buildings with thatched roofs. The concentration of hot condensing gases at a high level can create hotspots in the thatch surrounding the chimney. The result can be a slow smouldering charred mass of thatch that eventually explodes into a ball of flame. Many a thatched roof, and most of the building with it, has demised in this manner. Despite the move away from open fires to decrease ac/h, the presence of one open fire, with an appropriate flue, in tile or slate roofs is not an area of taboo whatever other proponents of sustainable buildings may advocate. A real fire has beneficial effects on mind and spirit, and that all-essential link with this most primitive force enjoyed by our forefathers may have an emotional benefit that far outweighs any saving in ventilation heat loss.

Passive stack ventilation can have very positive benefits to existing traditional construction, although it can make ground floors more draughty than upper floors, unless mitigated by a closed appliance. This is because warm air, heated by deliberation or by solar gain or heat from other domestic appliances etc., is less dense than cold air so it tends to rise upwards through the building. As it does so the internal pressure at the lower levels of the building starts to decrease, which results in a desire for equilibrium, hence the drawing in of fresh cold air through gaps in the structure at lower level. If these gaps, notably those around windows and doors, have been sealed and heavy curtains are in use across these openings, then this draught-inducing mechanism is cut down. In addition, internal draught lobbies can be very beneficial, and they are frequently found around entrance doors, which in existing traditional houses nearly always open from the main living room onto the street. That is where the majority of air infiltration occurs. Such devices are often found in buildings improved in the late 19th century. As the internal temperature of the lower level rises, due to say to an open fire giving radiant heat, the pressure difference increases, hence the propensity for open fires to create draught around the ankles. In essence this is because there is a greater pressure difference on the ground floor than on the upper floor. This is particularly noticeable on summer nights when the air movement on the upper floor is positively sluggish, even though windows on the ground floor may be open.

Deliberately contrived passive stack ventilation can be controlled by the use of dampers. The need for unhindered passive stack ventilation is also the reason why passive stack ventilators are above the height of any adjoining buildings.

Traditional buildings benefit greatly from porous building materials, such as stone and brick, removing pollutants and stale air as well as warm moist air that can cause moulds, and this should never be inhibited. This is in sharp contrast to sustainable new build, which is designed to have a completely sealed envelope or a very high insulation value, rendering them passive buildings. In these instances special care is taken to avoid the gaps that can occur naturally between differential materials such a concrete and timber. This necessitates the use of flexible fillers, of which silicon rubber is a typical example. In this instance porous building materials are taboo; the vapour barrier rules, even to the point of completely sealing the envelope, (a process best likened to putting the building into a large plastic bag), the use of large quantities of sealant together with designing out awkward joins between two differential materials. In addition, sealed combustion appliances are de-rigueur, and airtightness testing is the order of the day (see Chapter 5). Enveloping humans in this third plastic skin can surely not be healthy, and is the strongest argument for engaging in traditional building construction.

The problems of super insulation

The passive house, whereby the house is well insulated and free from air infiltration (and hence draughts), heated only by solar gain and heat from occupants and appliances for lighting and cooking, is everyone's dream of the sustainable home. Frequently called 'the passive house', it is the design standard to which building regulations are gradually progressing. If the electricity and any extra heating that may be required in a severe winter are produced from renewable sources, then it can be termed 'the carbon neutral house', and if the house can be entirely separated from any mains services then it can be termed 'the autonomous house'.

There are disadvantages to this enviable utopia. It is difficult to prevent thermal bridging of one form or another. Such problems can occur at the junction of the wall and the ground floor, which can only be prevented by ensuring that the floor insulation meets directly with the internal wall insulation, or if the wall insulation is within the cavity, the floor insulation is carried upwards and across the base of the wall so that the floor slab is independent of the wall. Thermal bridging can also occur where any internal element, such as a lintel, spans the full width of a cavity wall.

A further problem that can occur when a wall is heavily insulated on the inside, is that in freezing conditions the temperature near the outer surface of the wall will remain at a freezing temperature. It will not be able to benefit from heat gain from the internal environment as the insulation will block any such possibility. As a consequence water will freeze within the pore structure of the outer layers and will expand by one-tenth of its volume, thus bursting out the walls of the pores. This process can only result in damage to the external face of brickwork or stonework, as will be evidenced by the shearing off of whole angular portions of the material. The creation of vapour barriers can have a similar effect in that all externally derived moisture has to be held within the masonry, and cannot escape inwards. This encourages the mobilisation of salts within a more confined area of the masonry, and possibly more intense build-up of salt crystals within the pore structure, called 'crypto florescence'. This is particularly problematic for limestone, as the salt tends to be calcium sulphate caused by the chemical reaction between acid rain (sulphurous gases in rainfall) and the calcium carbonate of the limestone. Again the walls of the pores will be disrupted and the face of the masonry eroded, this time leaving a powdery residue. If by fortunate chance the majority of salts can migrate to the surface, the efflorescence of salts so formed will harmlessly be washed away by rainfall in exposed areas, although in limestone this can cause constant erosion of the surface. Sheltered zones on limestone will be more affected by acid-laced condensates to mobilise salts in a more limited fashion, although sadly this results in even greater catastrophe, a pattern of miniature volcanoes erupting on the surface. The effect can be seen in churchyards, where a many a treasured monument is no longer legible because of it.

Another form of super insulation is completely filling a cavity. Despite the many forms of clever polystyrene with vertical ridged faces to project water downwards, there is always a possibility of water tracking across. More cognisance should be taken of the exposure of elevations when making decisions about whether to include cavity wall insulation, south-west

Photo 4.23 Illegible inscription on a stone on a church wall due to calcium sulphate damage

elevations being most at risk from driving rain. Regrettably this is often not the case when renovating existing traditional buildings or when building new. It is the upper part of elevations, particularly gables, which are most at risk.

External wall insulation

One of the most effective forms of super insulation is that which is applied externally, particularly to elevations facing prevailing winds. On traditional stone or brick buildings this can very successfully done by weatherboarding over the insulation (itself a crucial matter – see discussion below), although the wood itself will require constant treatment. Here a wood oil/wax is much more successful than any applied paint layer.

Before getting too carried away with the concept of enveloping a whole building, there are other considerations, such as what this will mean for conserving the architectural quality of the building, be this vernacular or classically polite. A regional vernacular style of timber frame or stone will hardly be pre-eminent if all is enveloped in a rendered or boarded finish, and in polite buildings the gentle interplay between light and shade, not to mention the beauty of cornices, string courses, and raised window surrounds will not benefit architecturally by being masked with insulation. The positioning of rainwater gutters and down pipes, and window cills, are also adversely affected in a more practical but nonetheless equally problematic way.

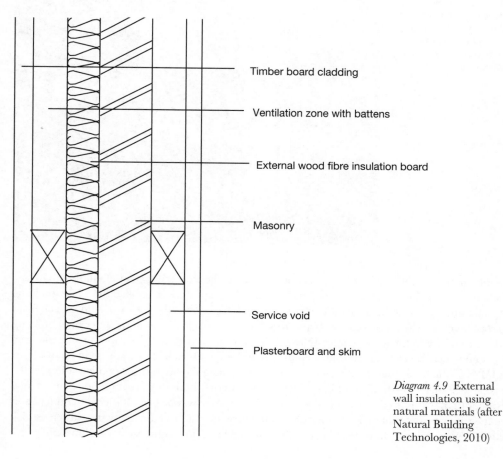

Timber board cladding

Ventilation zone with battens

External wood fibre insulation board

Masonry

Service void

Plasterboard and skim

Diagram 4.9 External wall insulation using natural materials (after Natural Building Technologies, 2010)

Photo 4.24 Boarded buildings facing down a valley in Guilsfield, near Welshpool – note the slate hanging used on the adjacent building in an earlier period to counteract the driving rain

Photo 4.25 North Wales vernacular – external insulation with a boarded finish would be detrimental to its vernacular appearance

The nature of the insulation used is critical. Just as the external face of a building material, be it masonry or timber, is highly dependent on being vapour permeable if the whole building is to breathe, so must the insulation material used. This inhibits the use of any closed-cell foam or polystyrene, which do not allow warm moist air (vapour) to do anything other than build up in the masonry in the form of interstitial condensation. Should such an impermeable material be used to clad an existing masonry building that does not have a damp course in the walls, or a DPM in the floor, then total havoc will result, including the build-up of condensation on the inside face, the growth of mould and increase in internal humidity creating poor air quality. Rather than the desired warm and cosy walls, the opposite will be true. It is vital to understand that existing traditional buildings perform in the way they do because firstly the walls are thick enough not to allow rainwater to get to the internal face, but also these walls allow any rising damp or warm moist air to pass to the outside. The only danger zone is if the site becomes saturated, which is where the French drain (see Chapter 7) comes into its own, wicking water away from the base of the walls, downhill to a soak-away, and thus reducing the degree of rising damp that the wall has to breathe out. The only successful materials that can used for insulation externally are those that are vapour permeable, such as wood or hemp fibre board, which can be lime rendered to further increase the capacity of the external face to transport moisture outwards.

There may be a natural desire to create a capillary free (totally waterproof) external cladding by tiling, slating, or even metal sheeting, or even inadvisably coating a render with a non-permeable paint (hydrophobic). Older handmade tiles have some capillarity and hygroscopicity, the firing of the clay into mullite (a glass-like substance filling the pores) being more limited due to the vagaries of the furnace. Later gas-fired modern tile has no such problem, so has very little capacity to take in moisture. Slate has traditional waterproofing qualities, and in Wales has been used throughout history for this purpose. The creation of such impermeable surfaces will inevitably not be conducive to dealing with the exit of internal moisture unless a ventilated gap is provided, and will result in more rainwater run-off. Directed at the base of the building this can permeate into foundations or linger on wooden window cills, causing decay. The more desirable scenario is a surface which can evaporate water, be it external rain or internal condensation, induced more evenly via a lime render or wooden boarding. The emphasis on lime render is deliberate. It contains a variable range of pore sizes that can help to buffer moisture content until external conditions are right for this to be wicked away by drying winds and sun.

Denser cement renders do not have this advantage, and in particular are more prone to cracking upon expansion or contraction of the backing material, creating crevices for rainwater to linger and spread into the substrate behind. This is bad enough where the substrate is solid masonry, but if it is an insulation material then this would cause it to lose its thermal resistance, due to filling the voids with water rather than air. It is for the same reason that there is a need for masonry surfaces to have lime mortar joints rather than cement mortar. Cement forces the masonry blocks to do all the work, resulting in mobilisation of salts or frost damage, while the

Photo 4.26 Lime rendering being applied (photo © Rob Buckley)

lime mortar holds the water until conditions are right, without any real threat to the masonry blocks.

Render on an insulation fibre board, or other less desirable material such as polystyrene, ideally needs to have sufficient depth to ensure that it can act as a buffer and deal with a large quantity of rainwater, as it cannot rely on the normal masonry backing for this role. In a dense substrate, particularly brick which has a high density of small pores and is very capillary open, any rainwater penetrating the depth of the render would normally be buffered until such time as the render could deal with its wicking outwards. This avoids any possibility of separation of the render and the brickwork, particularly that caused by mobilisation of salts on the surface of the brick by the water.

Weatherboarding has an ability to absorb internal moisture, although how much of it will escape to the outside, through the external grain of the surface if the latter is covered with a hydrophobic surface (oil or paint), is debatable. Moisture is more likely to escape via the end grain, where indeed it will be absorbed more rapidly (picture a bundle of drinking straws in a glass of water, each pulling up water rapidly via capillary action). This is why a cladding of hewn cedar or oak shingles is an attractive option. Although all shingles have a tendency to cup in extreme drying cycles (hot sun and wind), those that are hewn rather than sawn along the grain have a longer life expectancy.

Weatherboarding, like shingles, will always absorb some rainwater on its internal face, driven through the gaps between boards, hence the need for a vented space immediately behind.

The question will arise in most minds as to the effect of allowing an insulation like wood fibre board, hemp board, or straw board to act as a buffer for moist air travelling from inside to outside, in that it will surely reduce its thermal performance. It is thought that this is only a transient state in natural materials which will even itself out over time, whilst in mineral wool moisture will linger for long periods, causing a sustained loss of thermal performance. For this reason there is extreme scepticism about the performance of cavity wall insulation utilising closed-cell insulations with high vapour resistance. In effect, the insulation is acting as a vapour barrier on the inner leaf of the masonry, which if applied during the wet-trades process of construction will rarely give the inner leaf the chance to dry outwards. In addition, some gaps between panels will be inevitable, leading to condensation travelling into them, forming cold bridges. Of rather more significance is the fact that condensation generally cannot pass from inside to outside, and the penchant for moulds on internal walls is increased, with a reduced chance of balanced internal humidity and good internal air quality. The effect is similar to having closed-cell insulation on the internal surface of a wall, rendering the wall incapable of breathing out moist air (see below). Another similar problem is the use of closed-cell insulation between joists in a roof space, which is regrettably what many proponents of energy conservation put into operation in the relentless pursuit of ever more usable space. The result is that the timber rafters, being the only hygroscopic and vapour-permeable material in the insulation sandwich, will absorb vast quantities of warm moist air. The timber is only slender, 150 mm (6 in) at most, and the moisture content can reach 18% (some experts may put the figure higher), at which point there is a danger of dry rot. It is thus even more important when insulating roofs to consider the use of hygroscopic materials such as fibre boards or sheep's wool.

Internal wall insulation

This has been used for a considerable amount of time in older properties, generally to mask the problem of damp and uneven walls, so abhorrent to the pristine housewife of the 1960s

Photo 4.27 New matchboarding being re-applied to a traditional stone building

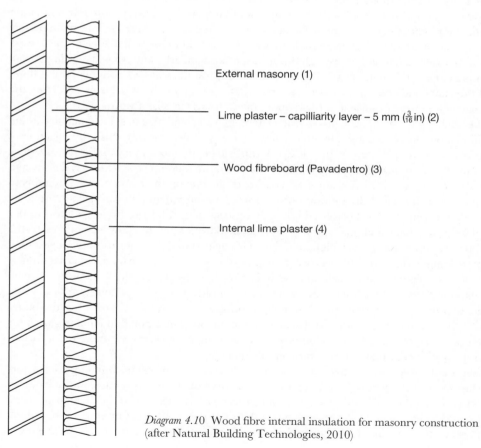

External masonry (1)

Lime plaster – capilliarity layer – 5 mm ($\frac{3}{16}$ in) (2)

Wood fibreboard (Pavadentro) (3)

Internal lime plaster (4)

Diagram 4.10 Wood fibre internal insulation for masonry construction (after Natural Building Technologies, 2010)

onwards. Even the Victorians were guilty of the same endeavour, battening out walls at dado level (below window cill to ground floor) and lining out with tongue and groove boarding. This was only ever designed as a short-term measure, the battens usually did rot, but over time it became an architectural entity in its own right and survives in many a cottage and farmhouse.

The Georgians were not averse to lining walls with hessian on battens in order to create a better surface for their expensive hand-blocked or imported Chinese wallpapers. Later versions involved battening out the wall, and using studs to support a polystyrene-backed plasterboard to create a warm dry surface. This created a form of cavity behind the plasterboard, which is all to the good as then some moisture escaping from the internal environment, via joints in plasterboard or around electrical sockets, could escape via the cavity to the outside. More pertinently in older buildings with no damp course, the rising damp from the ground floor or any rain penetration which reached the inside could also rise up through the vented zone, and back out through the wall.

Then came the tendency to fill in the voids between studs, using polystyrene or closed-cell insulation, with little thought as to how to deal with the build-up of moisture in the internal environment, or how to disperse any rising damp. Rain penetration on the external wall was quietly forgotten. The walls became even more saturated, with degradation of beam and joist ends as the result. This in turn gave old buildings a bad name. They were castigated and have often been demolished in great numbers as a result. This is a painful loss of a sustainable heritage in the name of an ill thought out repair regime.

By using natural fibre insulation the demands of both energy conservation and ventilation are met in a joint offensive, without the need for a vented cavity. Escaping internal moist air will be transferred towards the external environment or held in the buffer zone of the natural fibre until external conditions are right for its dispersal. Similarly, any rising damp in the wall and interstitial condensation will be able to utilise not just the wall but also the natural fibre, until external conditions allow for its dispersal via the oxygen-rich external air attracting the hydrogen of the water molecules. In addition, the wall surface is kept warm.

The reader may ponder why there is such attention being paid here to rising damp in existing traditional buildings, as surely the most efficacious way of dealing with this is by injection of silicone into the damp course. In brickwork this is true, due to a more uniform small-pore structure, but in stonework experience shows that the more random pore structure, density and pore size, do not readily get a universal coating of silicone on their inside face. Attempts to damp proof such stone walls and walls with a mix of masonry materials by injection are to be avoided, and where it has been tried is largely unsuccessful. A factor also forgotten with dry lining is the fact that traditionally stone walls benefited from a constant low heat offered by the open fire, summer and winter, night and day. This helped to drive outwards any penetrating or rising damp so that it became less of a problem to the decorative finish on the internal surface. In all events walls were historically coated in lime wash or distemper (chalk and water), which was vapour permeable, allowing moist air to travel outwards. Any penetrating damp of a virulent nature was evaporated off the internal surface of the wall, or held within the buffer zone of the internal lime render, and vented out of the building through the many gaps in its structure or via the large central chimney stack. To catapult such construction into a carbon neutral age using dry lining needs considerable thought.

Regrettably little constructive thought is ever given to this problem. Building contractors who favour an even more aggressive approach than dry lining, merrily remove the lime or clay plaster and coat the walls with two layers of neat cement as a form of vertical damp-proofing compound, thus obviating for ever the possibility of the walls functioning as a breathing membrane and reducing the building to road rubble in its next incarnation, as the stone will

Photo 4.28 Drylining being installed over a traditional stone wall – the timber should have been isolated to avoid the danger of dry rot

be irretrievable. Forcing so much externally derived water to sit in the wall, where it will only disperse slowly outwards and remain saturated for longer, reduces it thermal resistance, and reduces the effect of any insulation applied to the interior, not to mention greater mobilisation of salts.

Summary of options for traditional solid walls:

Traditional option – dry lining with a vented cavity: Vented cavity assists with the migration upwards and outwards of any escaping internal moist air and any rising damp. Use of vapour barrier to prevent internal moist air affecting the insulation resistance obviates any possibility of venting outwards internal humidity or any holding capacity for this (buffering).

Modern option – solid internal insulation with a vapour barrier: Escaping warm moist air around penetrations (e.g. sockets) infiltrates the insulation, reducing its thermal resistance. Rising damp and rain penetration linger, escaping only slowly via the masonry, mobilising salts and causing pore collapse, leading to stone decay. Joist and beam ends vulnerable to rot.

Preferred option – solid natural fibre – hygroscopic insulation: Escaping moisture penetrating the natural fibre insulation will be transferred outwards, as will any internal moist air, thus controlling the internal humidity, avoiding moulds and improving air quality. Any moisture in the walls will be buffered by the insulation until external conditions are conducive to its removal by increased oxygen activity (wind). Joist and beam ends have a better chance of survival as any build up of moisture in the vicinity can escape internally as well as externally. (The system does rely on good internal ventilation). Most pertinently there is no condensation on the interface between the insulation and the wall.

Other forms of hygroscopic buffering and insulation

At this juncture it is also pertinent to discuss the use of hemp plasters and clay plasters, which are not only beneficial in preventing condensation on previously cold surfaces but also boost the role of the wall as a thermal store. The insulating qualities of cob are legendary, keeping buildings warm in winter and cool in summer, and clay plasters will certainly emulate this effect. Clay plaster is similar to lime plaster as it is hygroscopic, and its use will avoid any possibility of condensation, the moisture being held within the plaster until such time as external condition or increased ventilation internally will result in the moisture being wicked away. Clay plasters are viewed with deep suspicion in the United Kingdom, being considered as potential sources of external damp ingress, but a thick clay plaster will be more successful at buffering rather than detrimental in respect of external damp penetration, and will also act as a thermal store.

Clay plasters have been used for centuries in various parts of Britain, and abound in Dorset, where sadly they have been the victim of over-zealous removal in recent decades. Buffering of moisture is particularly important in churches where the walls are not allowed to perform their natural function of acting as a thermal store, as churches are never given sufficient heat to balance out over longer periods. Instead they are subjected to intense periods of radiant heat, and short bursts of human occupation, both of which result in a sudden charge of internal moisture (warm air holds more moisture from human breath than cold air), which then proceeds to condense on the walls. Black mould can often be the result. Lime plasters to act as buffer zones by virtue of their more variable pore structure are essential in these buildings.

Conclusion

Traditional building construction must balance the need for walls to

- act as thermal stores
- have ventilation to ensure good air quality (including the use of wall structures to aid ventilation) and prevent the build-up of moisture-related moulds
- conserve heat and reduce CO_2 emissions.

This is a very tall order. It is little wonder that occupants of existing traditional buildings find themselves at a loss as to how to navigate their way through these demands, and that building control officers ignore the demands and needs of existing traditional buildings in favour of

Photo 4.29 Clay thermal store wall at the Centre for Alternative Technology, Machynlleth, Wales

imposing modern solutions not suited to their needs. It makes the construction of new traditional buildings using stone or traditional timber frame a difficult goal to achieve.

Of increasing concern to traditionalists is the demand to make buildings airtight in order to make them more energy efficient. It can readily be seen that airtightness should be more about controlling and eliminating unnecessary draught than about excluding ventilation altogether. In traditional building construction it is about allowing enough ventilation, not only by manual means of opening windows, but also by allowing walls to control vapour rather than trapping vapour within the structure. As any occupant of an existing traditional building will tell you, ventilation in the summer months has to be achieved in the most maximum way possible, with open doors and windows for much of the day. This again is how these buildings always performed. Hardy folk, in and out to their chickens or other tasks on the farm or cottage garden, wore the same clothes indoors as out, most of the time, and the old adage of 'put another jumper on' is very relevant to occupiers of such buildings. Even in the winter months most farmhouse and cottage doors remained open, although to suggest this to the average building control officer now would result in a very adverse reaction. Nonetheless, the whole issue of air permeability can be addressed by doing a fairly basic manual check on a windy day, as can the assessment of cold bridges. Hands are very useful temperature and draught gauges. 'Let the building do the talking, listen to what the building has to say' instead of making bald assumptions about where action is required, and never forget the role of shutters and heavy curtains. Our ancestors survived well enough with these devices, yet current generations turn their backs on this 'low-tech' approach. An altogether more interesting modern development is the invention of a solar driven venting system, which comes into its own for buildings and houses that stand unoccupied for long periods. This is the fate of many an attractive cottage in a remote location, not to mention whole villages consisting of holiday cottages or second homes occupied only at weekends.

There is a particular solar panel, roof or wall mounted, produced by a Danish company that heats cold air and feeds this pre-heated fresh air into a room or space of no greater than 70 m^2 (84 yd^2). More than one panel can be used for larger buildings. The literature (www.solarventi.co.uk) claims it is able to convert an outside temperature of 0 °C to 24 °C in good solar conditions. Thus not only is it providing a change of air, it is also providing heated air to give a modicum of winter heating. The ventilators can be controlled by thermostat, speed regulator, or manual switch (necessary in the summer). Why the idea has not been fully utilised is difficult to imagine. One of the main detractors will always be the aesthetic effect of large white boxes on traditional buildings with regional distinctiveness.

If insulation on the internal face of, say thick stone walls is deemed to be absolutely necessary, and there is a case to say it interferes with the role of the wall to act as a thermal store, then that insulation must surely be hygroscopic. The only way to achieve this is by using natural organic materials such as wood fibre, hemp, strawboard, hemp plaster, or by using mineral plasters such as lime and clay. All of these materials act as a buffer zone to absorb excess internal moisture, releasing it only when ventilation conditions are favourable. Such materials have certainly proved their worth, delivering buildings which have lasted hundreds of years, yet it is doubtful how long buildings using petroleum-derived closed-cell insulation materials will last, and even more doubtful how long their occupants will survive given an environment of trapped humid air, infested with toxins and carcinogens (volatile organic compounds, etc.). Of just as much concern is the practice of closed-cell insulation in lightweight buildings if it traps toxins in the heart of the building, although it has more of a role to play in large-scale metal framed buildings. In small-scale modern timber construction it will almost certainly lead to the rot in the timber, rendering the lifespan of such buildings very short indeed.

Research continues apace to try to prove the worth of traditional buildings, with the Building Research Establishment (BRE) being a leading player in the field. The Ministry of Justice (quoted in a joint seminar between BRE and English Heritage in 2007) had conducted research which showed that their pre-1900 building stock was more energy efficient per square metre – and thus deemed worthy of upgrading – than any of the rest of their buildings, with the worst performers in the period 1940s–1960s. This was deemed to be because the pre-1900 buildings were high mass construction whose walls acted as thermal stores, and which generally had better natural lighting and natural ventilation, whereas the later period buildings were of lightweight construction. Those of the period 1900s–1930s and 1970s–1980s were marginally less poor and could be improved with insulation. The latter period in particular had been constructed to rely on artificial lighting, oil heating, and air conditioning, as well as an open-plan layout that was more difficult to heat. These research results are thus a graphic example of how sustainable traditional buildings are.

As a result, the Ministry of Justice decided on a policy of retaining the older buildings but paying particular attention to increasing natural lighting via roof-lights, replacing outmoded heating systems with more efficient versions, and exploiting solar gain wherever possible, but using shutters and curtains for solar shading and night-time insulation (GHEU, 2007).

Sadly, such an enlightened approach is not universal and has not yet seemingly been grasped by developers and builders eager to meet the insulation requirements set down by government directives in general, and in particular by the Code for Sustainable Homes (www.communities. gov.uk/publications/planning).

There is a grave danger that the vast number of pre-1919 buildings will be damaged by the indiscriminate use of closed-cell insulation, on a scale that would far outweigh the quite dreadful damage done to buildings in the latter half of the 20th century by the use of cement mortar for pointing and rendering.

5

CLIMATE, SITE, AND
THERMAL PERFORMANCE

Siting and climate

The main problem with 'modern' development in the late 20th/early 21st century is that it has paid little or no heed to those elements of the earth that are essential for sustaining human life. Rather than buildings being regarded as a third skin, after natural skin and clothes, they have become regarded as distinctly unnatural elements to be traded for ever increasing sums. Their location and orientation in the United Kingdom is almost always determined by planning issues, many of them associated with amenity space, the provision of nearby facilities such as doctors, schools, and transport links, and employment possibilities. The availability of any empty plot of land in what is an extremely over-crowded island is the ultimate deciding factor, although development in main villages with existing facilities takes precedence. The inception and development of these villages in the period before or after the Norman Conquest has always been subject to an element of control, so in effect the modern planning officer replaces the lord of the manor. Their location within former open fields, and their subsequent development over many centuries, preserving their rural hinterland, has set the tone and constraints for requiring new development to respect this historical context, and rightly so. Brown-field development of existing redundant sites is another motivating factor.

This should not have precluded sensible orientation to sun, wind and rain, but in many instances it has, with an emphasis on frontages defining the street scene. On isolated farmsteads there is an historical reference to an orientation more suited to the elements, with few windows and a single door, often with these elements concentrated on elevations away from wind-driven rain.

Another problem that has beset development in the period after the Second World War is the desire for pastiche of earlier styles, using traditional elements in a non-traditional way. The use of mock softwood timber framing in the gables of post-war semi-detached houses, so-called Tudorbethan, was always going to pose a maintenance hazard. The gables invariably faced into the elements, were flimsy in their construction, and considerably out of access by normal ladders, thus being left to rot.

The desire for similar unfortunate constructions facing into south-westerly driven rain persist to this very day. Mock Georgian confections abound, with little thought to how their pseudo elements will perform, and rather than building for our time there is an unhealthy adherence to pastiche. The principle of life-cycle costing is well known to surveyors, whereby the whole lifespan of a building needs to be assessed, including its maintenance and eventual cost of demolition. Buildings need to return to being a barrier between humans and climate. Early man was in no doubt that his home was a shelter and a temple to the fire, that great moderator of cold air, and the source of all heat, light, and comfort. The current obsession with energy conservation is a factor of the deviation between external and internal environments. In the 1970s, cheap oil meant this deviation was of little consequence. Now it is totally fundamental,

Photo 5.1 Mock timber framing facing into the elements was always going to be problematic

both in terms of financial cost to householders and the cost to the environment. Heat loss through building fabric either by conduction through solid materials or by convection across air spaces, and by radiation from masonry surfaces, has become a major preoccupation. Even ventilation, which is actually just as fundamental to human and animal welfare as it is to the building, can result in heat loss by virtue of warm air being replaced by cold air. In this respect, possibly more is made of this than is actually good for human health. Modern buildings are required to be thoroughly sealed during the primary construction, and only after air testing are they then allocated a modicum of ventilation, via say trickle vents in the window frames. It is debatable whether this controlled ventilation is sufficient, however adeptly calculated, for real human health. 'Who can say what is sufficient? Have adequate studies been conducted of the health patterns of individuals in super-sealed buildings compared to those in traditional buildings whose very wall structure acts a medium for ventilation via its porosity? An anecdote springs readily to mind. When faced with an over-zealous environmental health officer in the 1970s who stated 'people are dying in these rural slums', one can recall replying 'yes, aged 97 usually'. Many rural cottages with somewhat over-abundant ventilation, especially in storm conditions, mitigated by only a single open fire, plus the most basic of sanitation and running water, housed occupants who were as sprightly in their nineties as they had been in their twenties. It would appear that ventilation is no bad thing, even though ventilation rates increase as wind increases. It is also abundantly clear that most of the post-war housing, particularly in the 1960s and 1970s, was poorly insulated, with massive heat loss (measured in joules per second, i.e. watts) and subject to little ventilation, resulting in cold corners infested with mould spores. These linked with the toxins from carpet glues and fabrics have possibly contributed to the current massive burden on the National Health Service.

Macro-climate versus micro-climate

The difference between these aspects was again well understood by earlier dwellers, as seen in the construction of courtyard houses in the 15th, 16th and early 17th centuries. These were an attempt to control the macro-climate by constructing walls to exclude wind and horizontal rain, thus creating a micro-climate. Why this appeared to cease in the 18th century is possibly a factor both of unreasonable classical influence, dictating the practice of bringing lawns up to the perimeter of the house, and the sheer maintenance workload dictated by so much built form in one place.

Macro-climate is that over which there is no physical control. Variations in temperature, humidity, wind speed, and direction, and most pertinently rainfall, occur without warning, and never more so than in the new 'global warming era'. Meteorological data is available for every month for the year, and has been for many years past, but is of little use in the face of so many changes and so little predictability.

Macro-climate is also modified by latitude and altitude. Latitude is the angular distance from the equator. Temperature decreases with increasing latitude, and even in the United Kingdom the temperature difference between north and south is acknowledged to be about 4 °C in the warmer parts of the year. Altitude is the height above sea level, and as altitude increases so the temperature decreases. Put more simply, it is colder on the top of a hill than it is in the valley bottom. Sadly, it can be equally true that cold air often sinks into valleys and condenses, causing fog. The presence of so much water in a gaseous state will lower temperatures in a valley. In addition, the west side of the United Kingdom is marginally warmer due to sea currents, but is also subject to moisture-laden westerly winds. In relation to hills, the orientation of the sun is a key factor. North-facing slopes provide inhospitable house locations, whilst the opposite is true of south-facing slopes.

Micro-climate is a man-made creation. Just as our canny forebears excluded wind by adopting a courtyard plan, so by the creation of walls, strong fences, hedges, and shelter belts of trees can we channel wind away from a dwelling, given enough peripheral space. Wind-chill factor even on a temperate spring day can considerably reduce comfortable temperatures, both inside and out. In addition, one of the key ingredients to creating a micro-climate is the creation of solar spaces achievable in conjunction with the creation of sheltered zones, or the deliberate creation of over-shadowing, usually by deciduous trees in the summer to create cool shade.

The sun in relation to creating a micro-climate site

The sun is pivotal to creating a comfortable life. The first thing one notices when visiting existing buildings of any antiquity is the extent to which the sun penetrates, and by doing so influences warmth and well-being. In traditional buildings, particularly in a farming context, the main rooms frequently face east, as morning activity such eating an early breakfast, saddling horses in stables, or commencing the threshing of corn, required maximum light levels. Farm labourers rarely welcomed the sun setting in the west as late as possible in the summer solstice and penetrating the house, as having returned from baking fields they welcomed a cool interior; but farmers certainly did in the farmyard, particularly regarding the orientation of over-wintering yards for cattle. Thus the orientation of the sun with regard to a new site for a domestic building, and its orientation with regard to the modification of an existing traditional building, is probably the most important factor to consider.

Sun in a kitchen on a cold winter morning can make an enormous difference to comfort levels inside a building. It is thus essential to find out where the sun is going to be placed in the sky at different times of the year. In existing buildings new owners are well advised to sit tight for a year, however uncomfortable their living conditions, conducting this careful monitoring exercise. (A further monitoring of enhanced areas of draught is further called for over a long period). Ingress of sunlight at various times of the year and day can be marked on ground-floor plans, and a careful correlation made in association with external land forms, trees and, other buildings. On a new site, intended site lines can be marked out using spray paint. This may entail visiting the site at regular intervals every day for a year, which may feel as if it is becoming a bore, but is well worth it in the end. The difference solar gain can make to an electricity bill is appreciable, particularly is attempts are made to store the heat in suitable fabric, i.e. dark ceramic tiles on the floor. Resorting to computer modelling of sun paths, so called solar-mapping to produce solar charts, is of course open to any individual with access to the Internet, but even if all the package requires are the co-ordinates for latitude and altitude, some understanding of the principles involved are essential. This revolves around an understanding of the real meaning of latitude and altitude, as well as the azimuth angle, in addition to the knowledge that the earth revolves around the sun once a year and around its own axis once a day. Latitude is the angular distance from the equator. Altitude is the vertical angle in the sky as defined by a triangle deduced from the position of the sun defined by the hypotenuse of the triangle and its position as defined by the height of the triangle. The azimuth angle is the position of the base of the triangle as an angular distance from the vertical. The angle so formed is the azimuth of the site. Numerous texts go into great detail about how this all works, but in reality there is no substitute for 'living the dream' and first-hand experience of where the sun is at any one time. The keeping of accurate logs on a daily or weekly basis is essential. Solar mapping requires a mathematical mind and the calculation of the azimuth angle as it moves across the site during the summer and winter solstices. It is especially important on the shortest and longest days, as are actual visits to the site. It is really only necessary for very complicated sites with many levels and potential obstacles to the sun.

Having done the solar mapping to get the best orientation, or having decided on the solar benefits or otherwise of the existing building, decisions about window sizes, overhangs to tame the angle of the sun, and ratio of glass to wall need to be made.

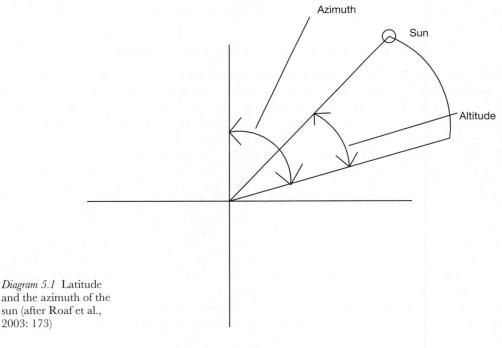

Diagram 5.1 Latitude and the azimuth of the sun (after Roaf et al., 2003: 173)

When planning for the sun, shadows created by the existing or intended building are crucial. The sun shining across the roof of a south-facing building will create a pool of shade on the north side. This may be welcome on a hot summer day, but in winter the low angle of the Sun will create an even longer shadow, putting much of a garden or even a neighbouring house in icy shade. This implies that the spacing between buildings is critical, as is the roof pitch. Keeping roof pitches shallow will mitigate the difference in the length of the shadow between summer and winter, but it should be noted that this conflicts with the need for a steep pitch on existing traditional building construction to remove rainwater. In addition, steep pitches are for this reason a traditional form suited to the English countryside. Such roofs may well have had their origin in the tent-like form of buildings that may have preceded the actual development of vertical walls. In addition, shallow pitch roofs of 10 degrees or less are often at the mercy of wind suction, across slopes or against wide gable ends. What can safely be done is to put other high buildings to the north of the intended domestic building, together with other non-domestic buildings such as stores and carports.

Closely spaced buildings can absorb considerable quantities of solar gain and feed this warmth into the air generally, thus pre-heating infiltrating air and causing uncomfortable conditions in high summer. This effect is noticeable in cities on a warm day.

Paving in the immediate perimeter of a house is capable of absorbing heat, which can then be absorbed by the adjoining walls. Dark surfaces absorb more heat than light surfaces, but by the same token paving in the middle distance needs to be light in colour to reflect light back into the interior of the building. More solar gain is possible from having dark floor tiles internally, illuminated by the sun shining through glass.

Solar radiation is one of nature's greatest free gifts. It results in:

- the gain of short-wave radiation and the limited loss of long-wave radiated heat through glass. This short-wave radiation heats internal building components. Even on a cool and breezy spring day the gradual heat gain in a fully glazed room can appear considerable.
- external air being warmed by the sun and deposited into the building via appropriately placed vents.
- increased light value that militates against the wasteful use of artificial electric light.

It should be noted that trees can filter solar radiation in both a helpful and an unhelpful manner. Trees generally come into leaf in mid May, and then allow only 18% of solar radiation through. Leaf loss occurs mid September to mid October, sometimes in a very wet summer as late as the first or second week in November. A single tree will then allow as much as 68% solar radiation through (Nicholls and Hall, 2006).

The remaining benefits of the sun revolve around the following:

- The need to prevent solar gain from escaping on the north side and through the ceiling of the area of maximum gain by using insulation, unless of course it is the intention to heat upper floors.
- The need to insulate ground floors against heat loss and provide a floor covering and base that retains solar gain. The ideal solution is a limecrete floor for new build and the repair of traditional buildings. A word of caution regarding floors in traditional buildings is made below.
- Careful planning is necessary on the location of rooms and their function (see section below).
- Paying particular attention to ceiling heights. The Victorians may have deliberately created rooms with tall ceilings secure in the knowledge that warm moist air rises and with it the

213

'ague'. This was a generalised concept of germ-infested air. They may well have been right but a high ceiled room is a cold room (hot air rises and is replaced by cold air), and correct ventilation can take care of warm moist air.

- Capturing solar heat via a roof slope, using the latter to reflect solar gain via a light coloured ceiling behind a clearstorey window into the room below (see Diagram 5.2).
- In existing buildings, constructing a solar collector room or sunspace. The south wall of an existing stone or brick building is absolutely ideal as a form of Trombe wall (see below and Chapter 3), a wall designed to hold solar gain which is locked into the molecular structure of the wall as well as warm air within the pore structure of the material. Brick has many small pores to facilitate convection of warm air from one pore to the next, whilst stone has a greater range of pore size for the same purpose, and both have a sufficiently dense molecular structure to act as thermal stores. Openings such as windows and doors can then release the heat through into main living rooms. In new build openings should be vents at high and low level, but in an existing building punching holes in the walls of historic fabric is not to be recommended. Existing doors and windows should be used instead.

Solar heat in floors

Some existing traditional buildings have brick paviors or massive stone flags on the floors. The latter are impossible to lift manually without contravening the CDM (Construction, Design and Management) Regulations 1994, and it is pointless to try and do so. Most rooms are too small for mechanical lifting devices.

If flags can be lifted without damage, and great care should be taken to establish this fact, by conducting a feasibility study and careful lifting of only one flag at a time, then they are ideal candidates for replacement on top of a limecrete floor in which underfloor heating has been incorporated. The molecular structure in the stone flags or the brick paviors hold the heat and releases it slowly.

Some Victorian buildings have floorboards on sleeper walls at ground level, and whilst wood cannot act as a solar floor it can be insulated at least on the ground floor. The cellular structure

Photo 5.2 The sunspace at the Centre for Alternative Technology, Machynlleth, Wales is designed to capture heat to filter into the rooms behind

of the wood can hold warm air by convection. Most of the time these boards can be lifted, the space between joists modified to take netting to hold insulation and the boards retained. If there is any danger of damage to old boards (and again a feasibility study should be conducted) it is better to provide an insulating floor covering instead. Floorboards of an earlier variety on upper floors should really never be disturbed.

Photo 5.3 Sleeper walls to hold joists for a boarded floor above at a Victorian school

Photo 5.4 Floorboards regrettably removed from an 18th-century house are a waste of a sustainable resource

All this aside, the principle of a floor absorbing solar heat during the day, and then starting to release it by late afternoon and evening, is one to be capitalised upon if at all possible. During the 1970s and 1980s there was an attempt to encourage sustainable builders to construct two particular sorts of features designed to capture solar gain. These were the rock store beneath the floor and the Trombe wall at the rear of a solar space or sunspace/modern conservatory. The rock store in the floor has never really captured the imagination, although a good many existing floors in sunspaces are solid concrete and this can have much the same effect. Limecrete also makes a good solar floor, but also inhibits rising damp by resisting capillary action and insulating against the cold. All sorts of experiments have also been tried with tanks of water beneath the floors. Water is a very efficient heat store, but storing large quantities of it in this way, in the somewhat bizarre water-filled wall or tank in the sunspace, is a recipe for problems. Water becomes very stagnant, attracts algae in direct sunlight (as any amateur keeper of goldfish will know to their cost), and starts to smell. The incorporation of a layer of small round stones within a floor immediately under the cement screed and on top of the damp proof course (in a standard floor) can act as a beneficial heat store. Many of these ideas must have been much sought after by the Romans, who doubtless hankered after their Mediterranean sun and substituted it instead with a fire that heated a stone or terracotta floor via a channel beneath to suck warm air underneath the floor and up through a vertical chimney.

Solar heat in walls

As can be seen, solar walls are a very efficient form of heat storage via solar gain. Again there have been experiments with the tin cans painted black holding water, installed against a solid board on shelves, the board being on a pivot and capable of being swung round to face an inner room during the evening. The idea is an attractive one in the sense that anything black holding water heats up very quickly; whole swimming pools in the Mediterranean can be heated via a single black hosepipe and pump with the maximum amount of hose pipe being exposed. The aesthetics of such a solar wall, however, are not very desirable.

A more usual form of Trombe wall is a solid masonry wall painted black and covered with a large glass panel. The heat stored in the wall during the day is then released into the interior room at night via vents, and cold air is sucked into the bottom, warmed as it rises, and then released into the room behind via a vent at the top of the wall. A variation on the solar wall is the blind solar window, again a masonry recess painted black and fronted by a glass sheet. Vents in the upper and lower portions of the masonry, controlled by miniature shutters, release stored heat into the room behind. The air between the wall and the glass panel heats up, rises, and is released through the upper vent during the whole of the day, to be replaced by cool air which then starts to be heated. A more novel idea is to create a long linear box of triangular form at plinth level, the box being filled with stones loosely packed, preferably black coated. A glass sloping cover then attracts solar gain, which heats cold air coming in at the base of the box, which is then warmed and fed into the house behind through a channel in the top of the box. A great deal of attention needs to be paid to meshing against insect and small animal ingress (adapted from Van Lengen, 2008).

Fireplaces

The art of the open fire goes hand in hand with the art of capturing solar gain. The former is a time-immemorial tradition and has as much to do with a sense of well-being, and being at

216

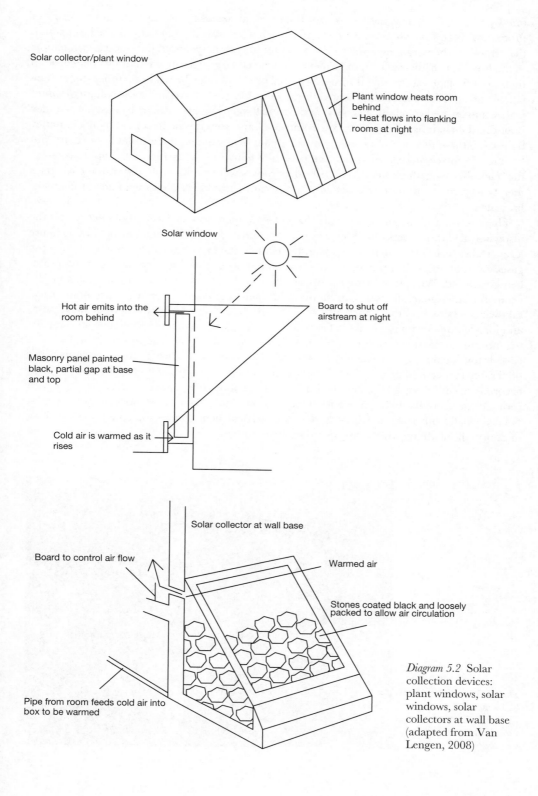

Solar collector/plant window

Plant window heats room behind
– Heat flows into flanking rooms at night

Solar window

Hot air emits into the room behind

Board to shut off airstream at night

Masonry panel painted black, partial gap at base and top

Cold air is warmed as it rises

Solar collector at wall base

Board to control air flow

Warmed air

Stones coated black and loosely packed to allow air circulation

Pipe from room feeds cold air into box to be warmed

Diagram 5.2 Solar collection devices: plant windows, solar windows, solar collectors at wall base (adapted from Van Lengen, 2008)

one with the world, as it has to do with the physical necessity for warmth and a means of processing food. It is discussed here as a macro-climate mitigation and a principal weapon in the armoury when creating a micro-climate. Our ancestors regarded their house as a temple to the fire. Post-medieval yeoman farmers constructed large open-fronted brick boxes around their fire (the ingle) surmounted by a substantial brick chimney breast (sometimes stone). The whole operated as a vast storage heater, the ingle a room within a room on the coldest of winter nights, and the bed with its blankets as curtains and tester (ceiling) drawn up against the chimney breast in the chamber above, to take advantage of and contain the heat held by the chimney breast. To these folk only these two rooms had any real validity, the ground floor room with the ingle fireplace always being referred to as 'the house' or 'houseplace', the upper room as the 'chamber over the houseplace'. All other rooms were peripheral, and in the early days merely workrooms until they eventually also became heated via their own ingle or chamber fireplaces.

There are many sustainability texts that refer to open fires as wasteful of energy and the instigator of carbon emissions. This is as may be, and of course it is infinitely preferable to use a renewable fuel source such as wood, and to contain the fire within a closed stove so that the gases given off in the burning of the wood are themselves re-incinerated, with only the water being expelled. What is of primary concern here is the position of the fireplace. Again, the yeoman farmer generally took care to place his ingle and chimney breast in the houseplace, sandwiched between other rooms in the centre of the building. Some earlier yeoman houses and fortified manor houses had the fire in a mural fireplace on the long wall, but this practice was not universal. It is therefore sensible to follow the lead of our wise forebears and site the closed stove against a wall which can store the heat, and upon a hearth that can do likewise, preferably in a central position in the core of the house. The dimensions of flues need to be in accordance with those advised by the Building Regulations (see Chapter 6). Tiling a flue with ceramic can considerably improve its ability to retain heat. Metal flue pipes should never be put externally but run internally to act as a vertical heat source, or contained within a masonry/tile cladding which can store their heat.

Photo 5.5 A smoke-blackened ingle – the room within the room in a traditional yeoman house

Photo 5.6 A newly inserted closed stove in a barn conversion – the wall behind is plastered with lime and leca (baked clay balls) and will act as a thermal store

Using a stove to heat water also is common sense, particularly if that water has been pre-heated by a solar panel, this being possible even in the depths of winter. Much nonsense is talked about the undesirability of mixing Rayburn and Aga water heating sources with solar heated water, but the main use of solar in the winter period is surely as a pre-heat treatment.

Sunshading

The converse of too much solar gain is discomfort. That which promotes heat gain can become intolerable in the heat of summer. There are a number of possible solutions to this problem:

- the physical creation of overhangs at eaves that work with the angle of the sun to prevent the unwelcome intrusion of the sun into interiors when it is high in the sky (see Diagram 5.3). If these can be manually manipulated so much the better, so as to avoid a possible unwelcome exclusion of the sun in spring when it has not reached its full height in the sky.
- the use of deciduous trees and shrubs strategically placed to provide shade in summer but not obstruct winter sunlight
- the use of physical and moveable shading devices
- the use of thermal store materials on walls and floors to act as buffers for stored heat.

Overhangs at eaves would ideally be moveable, but in practice this is rarely possible as a flat roof, for instance, would need to be thoroughly weathered at eaves. Complex mechanical systems are not advisable. First, they will be prone to mechanical breakdown in view of their vulnerable location in relation to the weather, and second, such systems demand energy usage.

External shutters such as the French variety, louvered in format, are another good option, as are deep reveals to windows, the latter being a useful by-product of an early Buildings Act (1774) to prevent the spread of fire across the elevation of buildings. External shutters appear not to have much precedent in English architecture, although internal shutters do, and the loss

219

Photo 5.7 Overhang at eaves and gables was traditional for weather protection but the idea can be modified for sunshading

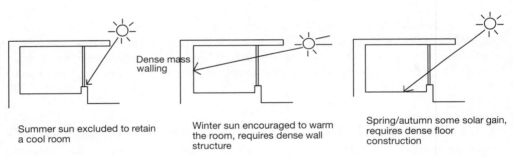

Dense mass walling

Summer sun excluded to retain a cool room

Winter sun encouraged to warm the room, requires dense wall structure

Spring/autumn some solar gain, requires dense floor construction

Diagram 5.3 Overhangs at eaves used for solar shading versus solar gain (adapted from Waterfield, 2006: 26)

of this device by the mid 19th century, or thereabouts, is something to be mourned. Quite how and why it occurred is unknown, as no better device is there to shut out the heat of the day, or to retain it in the evening.

A whole selection of modern blind systems is available. Louvres work well when the sun is low in the sky at the beginning and end of the day, and are thus good for east and west elevations. Horizontal shading work best on a directly due south elevation, where the sun is high in the sky.

Movable blinds are absolutely essential in a conservatory or sunspace. They not only shield the area from the harshest rays of the sun at its highest point in the sky, but also can encourage egress of warm air via the passive stack effect between the blind and glass, provided that a vent is provided at high level for the exhaust of this warm air. Most pertinently blinds retain heat on a winter's night, when such vents are firmly closed.

Various types of 'clever' blinds exist with their horizontal elements being formed from prisms or a section of a parabolic curve. These bend the light waves so that daylight is allowed through into the room but glare and the bulk of solar heat can be deflected outwards (adapted from

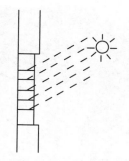
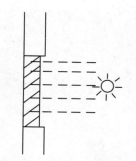

Diagram 5.4 Shading – moveable louvre style slats can be used to good advantage (adapted from Waterfield, 2006: 27)

Summer sun – south/south west facing elevation. Slats horizontal

Setting sun in the west. Slats nearly vertical

Waterfield, 2006) These are available mainly to the commercial market, whereas the average house owner really needs to concentrate upon the more readily available options. Once again, heavy curtains will provide much of what is required, as if the weather is that warm then occupants will surely be sitting outside.

Daylight – a factor of solar gain and orientation

Daylight is a function of solar gain. North-facing rooms may be conducive to the production of works of artistic genius, but as somewhere to engage in close work, be it reading, writing, or sewing etc., a sunny corner is admirable. The Victorians were conscious of this and introduced bay windows wherever they could onto Georgian buildings.

As human eyes age they demand increasing light levels to perform. Penetration of daylight into an interior is very dependent on planform (layout) and the depth of rooms. Whilst cubes may be very efficient in inhibiting heat loss, a long and narrow plan is more conducive to both maximum solar gain and daylight. Georgian rooms were frequently a cubic proportion, but benefited from tall sash windows. Intended to compensate for high ceilings, they also encouraged maximum depth of illumination. More traditional and early forms tend to have limited window area, due to the need to create minimum disruption in the forces within the thick walls, and enable maximum stability. This inevitably means dark corners, permanently lit in a modern age with 100 watts burning for most of the day and evening. Several dark corners can result at least 500 watts burning continuously. Lifestyle is very different to that of our forebears, who may have welcomed dark cool interiors on hot days and who in all events relied on the continuously burning open fire to provide much of their lighting requirements on winter nights. The burning of wood, in particular, necessitated the initial removal of moisture, which produces a bright flame for illumination. With the removal of these traditional fires, one of the best ways of counteracting the problem is cross lighting, if this can be achieved without loss of historic fabric. If this can only be achieved by cutting apertures into thick stone walls, then this is not recommended. It is possible to situate mirrors within the external and internal environs of a particularly dark room that will reflect a degree of natural light into the space.

A new building can take advantage of a design incorporating a clearstorey window, which immediately reflects light and solar gain into rear spaces. This is probably one of the most successful sustainable designs, together with the use of rooflights (of a design and size appropriate to the building) and sunpipes in existing buildings. These have liberated dark spaces in

Photo 5.8 Victorian
bay window designed
to create a sunny
corner in an otherwise
dark room

Diagram 5.5 Lighting
dark rooms with an
external mirror

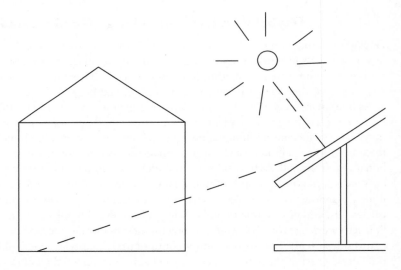

traditional buildings throughout the land, although protrusions on the roof take a deal of getting used to. Buildings of any antiquity traditionally have long uncluttered roof forms, and intrusions into such roofs needs to be carefully assessed.

South-facing glazing can help regulate solar gain in a somewhat converse fashion. The high angle of the sun in the summer helps the glazing to reflect away solar radiation, and this factor used in conjunction with tree induced shading can mitigate the most extreme effects of solar gain in a hot summer's day. In the winter the low angle of the sun allows more of the radiation to be transmitted through the glazing. In addition, the high angle of the sun results in a small area of floor being heated, whereas winter sun heats a larger area of floor. The importance of south-facing glazing in cutting down fuel bills cannot be over-stated.

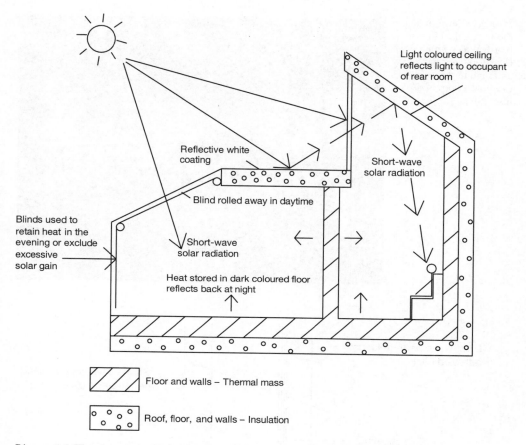

Diagram 5.6 The clearstorey format for maximum solar gain in new build (adapted from Waterfield, 2006: 29)

Daylight into a new build can and should be calculated with the utmost care. Natural light contains the full colour spectrum, although this is concentrated in the centre of the day, and the morning and evening sun is less intense and complete in this spectrum. A design for maximum daylight will do wonders to ease the burden of the portending electricity bill. In reality, of course, the increase in natural light via a large window has to be balanced against the consequent heat loss. Much can now be done, however, to retain heat by the use of low emissivity (low-E) glass. This incorporates a very thin layer of metallic metal coating on the outside face of the inner sheet of glass in a sealed unit. This traps in a percentage of the long-wave heat radiation, the basic heat within a room, but allows also a high percentage of the short-wave radiation from the sun to enter the room. The position of the low-E layer is deliberately sandwiched between the two glazed panels in order to protect it from wear and tear. This treatment can only be detected by a slight blue tinge to the glass unit.

Two types of low-E glass are available: soft- or hard-coated. The soft-coated glass cannot be toughened or laminated after application, so if this is required it needs to be done before the coating is applied. Hard low-E coatings can be applied after other treatments. There is a need for low-E glass to become more standard in its use for new sealed units. A word of caution is

Photo 5.9 New build at the Bala Lake Hotel, North Wales with a clearstory window, although in this example a floor is also inserted

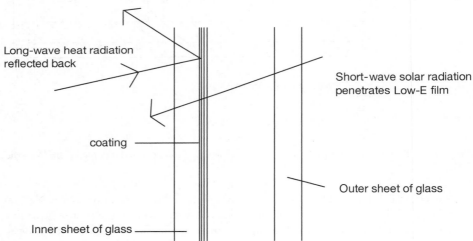

Long-wave heat radiation reflected back

Short-wave solar radiation penetrates Low-E film

coating

Outer sheet of glass

Inner sheet of glass

Diagram 5.7 Low-E glazing – section through a sealed unit (adapted from Waterfield, 2006)

needed instantly, however, for existing traditional windows. First, their automatic replacement with plastic windows, whatever the glazing used, needs to reconsidered if they have good solid traditional frames, be they sashes or casements, or even the more fragile and infinitely more precious metal casements on pintle hinges. This is an utter waste of a sustainable resource, and there is just as much benefit to be gained from secondary glazing, which can incorporate low-E glass, as there is from sealed units. Plastic windows are invariably ponderous attempts to emulate the architectural style of that which they seek to replace. Second, there is a myth about secondary glazing: it is universally described as ugly, yet there is no need for it to be so. There is also a possibility that it is frequently over-priced by the larger companies supplying sealed units in UPVC frames, a marketing ploy to stimulate demand for their products.

Photo 5.10 Historic metal casements with pintle hinges deserve to be retained and secondary glazed

The ratio of window to wall is closely allied to the development of an individual micro-climate. The typical ratio is 20%, with 60–70% of the windows being on the south side (Waterfield, 2006: 103) for solar gain. North-facing windows need to be kept to a minimum, sufficient for some natural lighting and associated room function of the most basic kind, i.e. corridors and bathrooms. Waterfield (2006) puts forward a very good rule of thumb of daylight being able to permeate into a room to a depth of around 2 or 2.5 times the head height of the window, with the greatest concentration opposite the mid point of the window. The reason for the popularity of Georgian and Victorian sash windows, long thought to be only an architectural desire, becomes immediately apparent.

More complex calculations can be achieved using the following equation, and in this instance it is highly recommended that this be done for new build. There is no greater joy than a well-lit room and no greater source of depression than a gloomy one, and whilst Horace Walpole at Strawberry Hill, Twickenham, in the mid 18th century may have thought his reborn Gothic 'gloomph' was admirable, suffers of seasonal affective disorder (SAD) do not.

Equation for daylight penetration in a single aspect room (Waterfield, 2006: 103):
L = depth of room
H = window height
W = room width
Rb = average surface reflectance at the rear of the room
$L/W + L/H$ = (or less than) $2/1-Rb$

Rb is normally about 0.5 for a light coloured room. Substituting figures in this equation, a room of about 4 m width (W) and with a window height of say 2 m (H) would need to be no more than 5.25 m in depth (L), otherwise the rear would be in a zone of poor lighting requiring artificial electric light. A recent building project with a full height window of around 3 m (H), a depth of around 5 m (L), and a width of around 3 m (W) yields a value of 3.2 for the lhs of the equation, which is less than 4 on the rhs of the equation (vis. 2 divided by 0.5), implying good daylight illumination. What has to be guarded against are high levels of light intensity in one zone of a room compared with a very low light level at, say the rear as this can produce a feeling of extreme discomfort.

Those existing buildings that are already gifted with windows on more than one elevation provide an element of cross-lighting. This will provide a more balanced intensity of lighting, but nothing can beat rooflights for providing a very even distribution of light within a room. This has to be balanced against somewhat higher levels of solar heat gain where it may not be required. Sun pipes or light pipes, a combination of a plastic dome on the roof leading to a highly reflective tube, can illuminate circulation spaces on upper floor landings and bathrooms. They are less successful for ground floor rooms, particularly as there is a need to pass through other spaces. Even standard bungalows have huge roof voids through which the pipes must pass before reaching the end diffuser.

A recent innovation to the market has been transparent insulation material (TIM), usually two sheets containing a gel or a honeycomb that allows diffused light through but resists heat loss. This can be a boon in a very contemporary design, as frequently large glazed areas in built-up neighbourhoods simply result in permanent curtaining by voile to preserve privacy, or even worse, plastic blinds. These have a habit of giving the dwelling an appearance more akin to a dentist's surgery or office. The role of curtains should not to be overlooked, however, in the fight to achieve maximum insulation and daylight. Whilst heavy curtains are of little use in the day, they can produce considerable benefit at night when the need arises to block off night sky radiation.

Reflective surfaces within rooms are also essential for the enjoyment of natural light. Light is reflected back into the room if the walls are white, and is systematically reduced with darker colours. If an existing room is so gloomy as to defy intervention, then it is a room best consigned for evening use, and dark bold colour will assist with the feeling of warmth. Again, surrounding the dwelling with light coloured surfaces, particularly walls immediately adjacent, upon which the sun may shine but not directly into the room, will assist and can be further enhanced by the use of mirrors. The proximity of adjacent buildings is often a problem which planners do their best to avoid in new build. It is thus a Town and Country Planning consideration. There is a recognised angle of daylight versus the distance from another building.

Existing traditional buildings will probably never provide enough natural daylight pene-tration into rooms, which is why the provision of a well designed sunspace, on the south

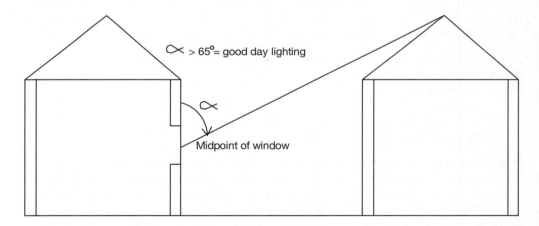

$\propto$ is measured from vertical, or if roof overhangs, from a line drawn between edge of overhang and midpoint of window

Diagram 5.8 Daylight availability in adjacent buildings (adapted from Waterfield, 2006: 106)

elevation, will provide not only adequate solar gain that can be fed into the main body of the ground floor, via the heavy stone or brick wall it abuts, but also an excellent well-lit space for close work. In addition, rather than pendent lighting which is so reminiscent of the solitary light bulb, task and mood lighting using low-energy lamps will inevitably be needed to supplement areas where daylight can never penetrate, and of course during the long dark winter nights.

Building shape and planform

The shape of a building is what will ultimately dictate the planform (room layout). Our ancestors were well versed in what will be shown to be the most efficient footprints, the circle and the rectangle. The former was associated with prehistoric man, and is to be found as a footprint in large numbers associated with the Iron Age. The latter has a too complex an origin to be discussed in a book on sustainability, but certainly appears with the post-Roman invasions and is commonly associated with the people then collectively known as 'the Saxons'. Footprints of large Saxon halls can be deduced from aerial photographs, and smaller farmstead clusters of buildings also show this rectangular format in a more modest form. The use of the rectangle dictated a linear arrangement of rooms, commonly referred to as 'cells' or 'units'. The central unit, the open hall, was given the only fire, and initially that fire was on the floor, with the location having some bias towards the 'high end'. Eventually the fire was contained in an ingle, which may block a former cross passage or back onto a former cross passage. The retention of the cross passage, which may have its origin in the use of one end of the hall as a cowhouse, and possibly also for threshing corn (utilising the through draught for winnowing), appears to be more symbolic than having an actual use in later forms. Certainly it was not an energy conservation measure, and its continued retention in the south-west of England in many of the ancient houses that remain is an energy conservation challenge. Never was there a draughtier feature in a house, but in its original form draught was part of the design. It was used to control the speed at which the fire burned. Room lengths varied from around 12 ft (3.6 m) for the very poorest yeoman farmer, husbandman, and cottager, to around 20 ft (6 m) for the very rich. Room depths were similar, but with a more modest maximum of around 16 ft (4.8 m).

Planning and internal arrangements

Different areas of a domestic building require a different approach to comfort levels. Our yeoman farmer or cottager ancestors frequently only heated the houseplace in their linear arranged house. This was the central room with an unheated parlour on one side and food store (buttery and pantry) on the other, the plan so created being three rectangular rooms in a line. The highest comfort zone was thus central and buffered on either side by unheated spaces. The parlour was a multi-functional room depending on social status, but was frequently a work room (spinning) or sleeping room, albeit unheated. High social status farmers installed an ingle if the room was to be used as a private retreat, away from the common folk in the houseplace. The buttery/pantry only gained heating when it became a kitchen, replacing the function of cooking in the houseplace. The central hearth remained a focal point, however, for much of life until the mid 20th century, and the chimney breast above the constantly burning fire acted as a vast storage radiator for the bed chamber above, as well as providing heating and lighting for this all-important central room. Servants and children, and even grandparents, must have dreaded the nightly exit from the warm hearth in the depths of winter, even though the windows were small to reduce heat loss. In 1779 the winter lasted from the beginning of November to the end of April, at temperatures well below freezing, so that the turnips froze in the fields and cattle were starving, as reported in *The Diary of a Country Parson* (Beresford, 1935). The diary records that the Parson did not go to bed, but sat propped up in chair by the constant fire. He makes no record of the fate of his servants or niece. By this period, in all events, rooms had taken on a more cubicular form, but the cellular format was obviously highly successful and is not to be spurned in modern-day layouts. Another factor to consider is that such traditional forms had main rooms facing east, away from the westerly winds but also because life in a farmhouse kitchen/houseplace started early so maximum light was needed. No longer tied to these requirements, unless by ownership of an ancient abode, in which case care should be taken not to foist unnecessary and damaging change upon such buildings, our modern living rooms can face due south for maximum solar gain.

The linear arrangement of living was extended to the development of the terraced house from the mid 19th century onwards, ranging from the most modest to grand Victorian villas, all cosily using the building on either side as a buffer zone against the cold. End terraces enjoyed an extra large plot of land on which to grow their vegetables, perhaps to offset the higher cost of coal, although this is not a recorded fact.

Even the ubiquitous semi-detached, a product of the 1920s/1930s, cut down the number of surfaces from which heat could be lost by dint of the party wall, although both forms of development were also a response to the need to cut down on building materials in order to produce cheaper homes. Terraces were built by cost-cutting mill owners and the like, and the suburbs for the new commuter (by train, rather than by car). Proud owners of such abodes still retained the practice of growing an abundance of vegetables in the long linear plot to the rear, a tradition established by their cottage ancestors. This would seem more like pleasure than toil after a hard day at the office.

Lessons to be learnt for new build planforms based on an understanding of ancient forms

Lessons need to be learnt from these previous plan types. So often in building construction designers are so anxious to break new ground, to make a mark on the landscape, that the

Photo 5.11 Rural terraced housing with large gardens, now ornamental, were designed to grow vegetables as a means of survival

previous lessons of history are totally ignored, to the detriment of sensible living regimes. By the same token it is important to create 'a building for our time' and not some pastiche or game of 'let us pretend it is the 18th century'. Some understanding of past domestic building form needs to be combined with 'life-cycle costing'. This implies a full understanding of the cost not only financially but also to society, in terms of the quarrying and production of the materials, the transport to site, the construction of the materials, the running costs, repair and maintenance costs over its projected lifetime, and finally the cost of demolition. In addition, there should be an assessment of what materials can be re-used and recycled. Only in this way can one be sure that what is being designed is truly long-term sustainable, as the yeoman houses so evidently were. Even Victorian houses with solid brick walls are eminently sustainable, despite being maligned on the grounds that the solid walls lack insulation. The fact that these walls act as breathing walls and thermal stores is often ignored.

For new build it is certainly sensible to realise that a streamlined format of building will reduce the impact of wind. Any sustainable new building will look to grouping the main daytime areas of occupation behind the south elevation. Ideally, the kitchen should face south-east to take full advantage of the morning sun, and to diminish chances of over-heating in the late afternoon/early evening when other heat sources being used for cooking could result in uncomfortably high temperatures.

Living rooms are best orientated towards the south-west in order to get maximum solar gain, and this might include the ubiquitous 'live-in kitchen'. This implies a room with both south-west and south-east orientation, really only possible in new build. Bedrooms above will benefit from solar gain to be enjoyed in the late evening if facing south-west. The problem arises from the modern need for a combined living and kitchen space. The former needs to be heated, the latter can get over-heated on a hot summer. What is not commonly realised, of course, is that the live-in kitchen is not modern at all, it is simply the houseplace reborn. It was the fundamental element of every medieval hall, post-medieval yeoman farmhouse, and later farmhouses, up until the middle of the 20th century. In many existing traditional houses the live-in kitchen has to be attached via a link.

Photo 5.12 A link detached extension is often the only means of obtaining the live-in kitchen currently so desirable

Northern orientations do not lend themselves to anything more than circulation areas, buffer zones, and storage areas, and some would say bathrooms, although a cold bathroom is somewhat reminiscent of life in the mid 20th century. One might also add garages to this list, although having such a unit as part of the main building always seems a waste of good floor space. Better that it should be a cartshed style of building in an appropriate location on the site. Bedrooms relegated to the north will not be able to enjoy the morning sunshine, so much a factor in the response of the body clock to the need to arise. If at all possible then, bathrooms and bedrooms need also to be on the south-east side. Gloomy northern corridors do not sound an entrancing prospect. They can benefit from borrowing light from the south, east, and west via glass bricks or be provided with a sunpipe, and a lower temperature of around 18 °C whereas living rooms need to be around 21 °C. In terms of lighting, circulation areas need only around 150 lux, whilst living areas need around 400 lux (Nicholls and Hall, 2006: 81).

Home-offices play a role in modern living. Those who are still forced to inhabit an office on a daily basis will have noticed how uncomfortable open plan offices are for a whole raft of reasons. Energy consumption is greater than for cellular-style smaller offices, but in all events such open plan offices are the hallmark of the late 20th century purpose-built 'filing cabinet for people'. These appear to become overheated at the slightest provocation, mostly due to the fact that there is one thermostat for the whole of the building, located in a cold zone. Such rooms are also notorious for their gloomy spots, demanding constant task lighting. Light is essential in an office environment as it produces a feeling of well-being and comfort. This of course conflicts with the demands of a computer screen, which requires a fairly low level of lux for effective reading. Home-offices are frequently consigned to gloomy rooms for this very reason, despite the fact that brain activity directly responds to serotonin stimulated by high light levels. The answer surely lies in a room which has a sunpipe to direct a high intensity of light onto a particular work surface, and sufficient low light levels in one area to render comfortable the use of a computer screen. None of this addresses thermal comfort, which may possibly be served by attracting heat into the room via a sunspace.

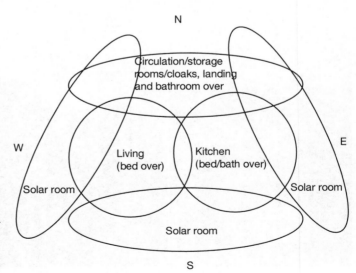

Diagram 5.9 Venn diagram of idealised new build layout with solar rooms on the south, west, and east sides

Volume and heat loss

Since the 1980s there has been a passion for building monolithic buildings in the countryside. Not only are they generally a blot on the landscape (despite best efforts, none of them has the grace and charm with which Robert Adam, in the mid to late 18th century, could imbue a large country house), they are also eminently non-sustainable. This is for the simple reason that the greater the volume the higher the 'ventilation heat loss rate', that is, the speed with which the warm air inside is replaced by the cold air from outside. The heat is lost in joules per second, measured as watts. There is a plus side which was well known to the Victorians, who knew that warm moist air could harbour germs, but because warm air rises it can be ejected from ventilation bricks in the upper parts of a room with a high ceiling. Thus larger volumes do increase the potential for ventilation, and conversely reduce the energy or cost of providing ventilation. In this country, however, the emphasis must be on reducing the volume in order to reduce heat loss.

Heat loss is in fact due a number of mechanical interactions:

- **Convection:** the upward movement of warm air currents
- **Conduction:** through the fabric (walls, roof, doors, windows)
- **Radiation:** heat lost via the outer surface of the building elements
- **Ventilation:** the replacement of warm air with colder air
- **Infiltration:** the penetration of cold air through cracks and small apertures in the building fabric.

Conduction has become an essential element of consideration as it is measured a by a quotient of 1 divided by the resistivity or total thermal resistance of all the elements of the fabric (roof, walls, windows, doors) to heat loss. This has become known as the U value and is calculated both for individual elements like windows and for the total thickness of the walls by adding together all those elements that make up the wall (see below on calculating U values). U values now dominate our lives as building conservators or building constructors. A low conductivity or U value for any element or all elements together means a high resistance to heat loss. U value must be considered in relation to surface area, which is in turn related to volume.

Photo 5.13 A large-volume new build in open countryside has a higher ventilation heat loss rate

As well as volume, and closely allied to it, heat loss is determined by air temperature outside and inside; so as to maintain a steady temperature within the house, heat loss must equal heat gain.

Volume is determined by the shape of the building, which in turn will dictate the ability of the building to channel away one its biggest enemies, wind. The wind-chill factor significantly lowers outside temperatures, and the greater the wind-chill factor the greater will be the difference between the inside and outside temperature. Infiltration on an extremely windy day with a cold temperature will produce the most frightening amount of infiltration, even around doors opening from the house into what is ostensibly a sealed sunspace. Volume will also determine the extent of overshadowing around the building, or the effect of this on other nearby buildings.

Volume also determines how large the layer of oxygen is, which is naturally adhered to the surface of all objects including the external face of buildings. The oxygen is part of the layer of air attached by friction to the surface, and in theory this air adds to the thermal resistance. Oxygen, however, is electro-negative and via the hydrogen bond attracts water. Water adsorption implies a rapid cooling of all external surfaces.

Volume is thus a key component of determining micro-climate, as the greater the volume the more surface area there is to lose heat from. The external skin of this volume must thus be regarded as a 'heat exchange unit'. The ideal is thus to keep the volume limited.

The most compact form that achieves this is the cube. The form was revered by our Georgian ancestors, in small country houses, rectories, and townhouse, for general room sizes and indeed for whole house plans, and when used in tandem with Count Rumford's fireplace improvements (Calloway, 1994) may have significantly improved comfort levels. The cube or square provides the smallest surface area for any given volume other than a sphere. Spherical shapes are not very practical for living unless it is in the form of a half-sphere or dome, although round houses were often used for country house lodges and roadside toll houses.

Earlier ancestors, from the post-Roman period into the late 17th century, preferred the rectangle. For a given volume this involves an elongation of the length and a reduction in the width. All of this involves an increase in the wall area. Nonetheless it was the form that dominated because it was easy to regard each rectangle as a single unit to which other similar units could be attached. Solid traditional yeoman houses of the 16th or 17th century often encompassed three such units, with only the central one being heated.

H-plan and L-plan houses are thus not sustainable forms unless the internal planning is so organised as to take account of the solar gain and mechanisms are put in place to shut off colder areas when the temperature drops. This is by far the best policy and too few houses are purpose designed with a 'snug'. Such planforms demand a complete understanding of those areas of the building best suited to retaining heat or capable of being cooled by ventilation as climate demands. Sometimes L-shaped or T-shaped plans are unavoidable because extending an existing rectangular traditional form cannot be done in any other way. A considerable number of traditional buildings can only be extended in a link-detached way, where the much longed for live-in kitchen/family room is linked by a glazed corridor to the existing traditional building. The aim here must always be to make the new build as thermally efficient/well insulated as possible and the old building function as a traditional thermal store. The error frequently lies in trying to apply the same rules to both types of construction. This is a recipe for problems.

The Elizabethan manor house (E-plan) was one of the most inefficient planforms ever to be devised, which although basically a thin rectangle, had high ceilings and narrow projections. There was considerable solar gain (often via more glass than wall), but also considerable heat loss through a large surface area, with on average a net loss of heat. Such houses rarely survive

Photo 5.14
A rectangular
three-unit yeoman
house, extended in a
later period and now
sadly lost due to the
'replacement dwelling'
policy, having been
devalued by plastic
windows

in anything like their original form, but where they do, the only sensible option appears to be to create rooms within rooms during the winter months. This can be achieved with heavy drapes on a frame, and one does wonder if this was not how the occupants coped with winter at the time of their inception, as well as clothing the walls in tapestries. It is analogous to the concept of the yurt. This structure, consisting of a felt cloth over a bentwood frame, was designed to house 10 or more people wearing their outdoor clothes permanently. Heat was provided by a central brazier as well as the body heat of the individuals. The structure resisted the infiltration of cold air from driving wind, and night sky radiation which reduced temperatures further was resisted with an 'early to bed under thick quilts' regime (Roaf et al., 2003: 25–26). The alternative for the modern dweller in an Elizabethan house is a monstrous energy bill. What has to be remembered, however, is that many of these houses started life as one side of a courtyard, their front elevations being sheltered on all sides by an enclosure that excluded wind and created a micro-climate for their main rooms behind the principal elevation. It is the absence of this courtyard, due to the heavy maintenance regime it generated plus the demands of fashion in the 18th century (lawns coming up to the perimeter of the building), which resulted in exposed elevations and eventual demise.

It is still possible to create such a courtyard format today when designing new build, but the concept of creating courtyard micro-climates appears difficult for architects to take on board. There is actually no difficulty in providing this, even with a simple bungalow format on three sides of a courtyard.

Bungalows are in fact problematic, as where in essence two rectangles of very large proportions are being joined together there may be some decrease in wall area, but the floor and roof area are doubled and considerable heat loss ensues. It is ironic that this is the building form favoured by so many elderly couples, who will find it difficult to pay energy bills on a pension.

A cruciform format was adopted for most historic churches. This evolved from liturgy and an amalgamation of the classical basilica and the Saxon form with side chapels, which became transepts. It is interesting to note that this involves a considerable increase in wall area, and is one of the reasons why churches are so difficult to heat. This is added to the fact that the walls are gigantic thermal stores, but in general there is limited possibility of them acting as such as heat is only provided for a short time, and it is easy to see why on balance the whole building acts as a thermal-loss mechanism. Either churches must be in constant use to warrant constant low-level heating, which is the only way the increasingly popular underfloor heating system can operate (one wonders if their proponents actually realise this), or the exigencies of their fabric must be accepted and with it the need to wear coats and operate a short burst of radiant heat, to heat the people rather than the fabric. The downside of the latter is rising damp and salt mobilisation that erodes the fabric. U value of walls must therefore be considered in relation surface area, which in turn is related to the volume.

Deep-plan buildings, typically shops in historic towns, formed from the development behind narrow trading frontage of former living spaces, are the most disadvantaged of all planforms. Even in their original usage, born from the high rentals of narrow frontages, living spaces must have been dark, cold, and gloomy. Retail outlets which have utilised the space by dint of demolishing all cross walls (a dramatic loss of historic fabric and thus in itself a non-sustainable activity) are disadvantaged as regards solar gain, natural daylight, and natural ventilation. Consequently huge sums are spent on energy consumption for heating and cooling, the most wasteful of which must surely be the warm air curtain as the door is opened, the heat energy wafting skywards.

There are several ways of reducing surface area heat loss:

- Insulating the ceilings of rooms rather than the under-side of roof slopes, so that the volume is more akin to that of a cube.
- Be constantly aware that any deviation from the cubicular form will result in heat loss. Any such extensions on houses should be capable of being independently heated or cooled, or kept deliberately free of artificially produced heat, because as soon as it is produced it may be lost without great attention to the U values of the areas concerned.
- When considering any other planform, utilise the rectangle with much of its long side facing south. The passive solar gain, which can amount to as much as 10% of space heating requirements, will mitigate any thermal loss by virtue of the increase of surface area.
- Use porches and conservatories as buffer zones against the cold rather than as heated rooms on their own account. Regard them as a means of transferring solar gain into a body of thermal mass behind, this body being part of the main build structure. Consider internal porches. They form a valuable buffer zone against the cold.

It is very difficult in practice to be disciplined about the use of conservatories. They are such attractive areas for working or playing that to abandon them when the sun is no longer providing free heat is almost impossible. All good books on sustainability will exhort that this must be so but have no guidelines on the alternatives. Closing the blinds in the day to conserve heat is not an option as this shuts out valuable daylight. Having a dark ceramic tile on the floor will certainly help, as will a ceramic tile on the wall surface forming the elevation of the house. All these surfaces will then absorb and radiate heat. Triple glazing with argon gas infill, and/or low-E glass, will considerably assist with heat retention.

Energy balance

The surface area to volume ratio needs to be particularly understood when dealing with small buildings, as the internal space will have a greater contact with the outside world, and will cool down rapidly when temperatures are at their lowest. This needs to be finely balanced by the ability to have the space in contact with more solar gain, and natural daylight, plus natural ventilation, via a south-facing glass wall. This is why the elongated format of building with much of its south side presented as a solar collector is so much more efficacious than a larger building with many of its rooms locked into the core. Whilst these rooms will lose less heat due to surrounding rooms acting as a buffer zone, they will not gain from solar energy, natural light, and ventilation. Unfortunately this balance between heat gain and natural daylight due to solar energy, and the design of the building to ensure minimum heat loss due to the surface area to volume ratio, is something that can really only be done by computer modelling. This will undoubtedly play a role in the future design of buildings to meet the Code for Sustainable Homes, but in the meantime old-fashioned commonsense must prevail. The Elizabethan manor houses discussed above were, in fact, often given mezzanine floors to divide up very tall spaces, thus rendering them more energy efficient. Using loft spaces for living in an existing traditional house or in a new build can mean that those on the floor below enjoy a more comfortable temperature, but it also means that the roof of such rooms needs insulation, which can never be as efficient as an insulated ceiling. Care also has to be taken in traditional buildings where ceiling insulation is desired. It should not be done to the detriment of a decorative plaster ceiling (ceiling rose and cornice) or the aesthetic quality of fine Georgian rooms.

If cooler rooms are required, and in the British climate this could only be on the south side of a building, then ceilings can be higher, allowing warm air to escape through high level windows provided for the purpose. The clearstorey window frequently associated with the need for solar gain to reach to the rear of a room, as already illustrated, can create high level ventilation that can encourage many air changes, which in a cold period would need to be treated with care as it can encourage down draughts. Draughts are the bane of many a traditional house dweller, and are more usually associated with increased air infiltration finding its way through the bottom of doors which have been modified many times in tandem with various floor coverings through the centuries (adapted from Nicholls and Hall, 2006).

The role of earth, wind, water, and fire

At its most primitive the elements of earth, wind, water, and fire, so beloved of astrologists, have dictated the layout and disposition of the domestic built form since time began. The ability of the earth to yield a building material, be it timber, stone, or the earth itself, or support the building, the basic human need for water, and the relationship of fire to draught, have always been the very basis of human dwellings.

Earth and topography

In an age where buildings made out of earth are few and cob (clay or chalk subsoil) is not so readily acceptable as a building material (see Chapter 6 for reasons why), it needs a brave soul to champion its cause. There is no greater a sustainable material, as having been removed from the earth and shaped into a wall, it can be returned to the earth. Sadly, so often it is treated as an impediment to an extension and simply removed. This is a shocking waste of a valuable material.

The most important factor for most new sustainable build, and extensions to existing traditional buildings, is the ability of the earth to support a new structure.

Dense clay is regarded as a good basis, depending on the plasticity index. All is well until localised deciduous tree cover, required for shading, proceeds to draw moisture out of the clay,

Photo 5.15 Wasteful chalk removal, made necessary to allow for an extension at Milton Abbas, Dorset – this is a modest example compared with most

causing the movement of nearby foundations. Very wet clay is certainly to be avoided as it is prone to shrinkage in very dry periods, causing problems arising from subsidence. Conversely it is prone to clay heave, which creates very grave problems for building foundations. Traditional timber framed buildings have been known to break their back in such a situation. An example springs to mind, that of a timber framed building located on a moated site. The Victorians in their infinite wisdom removed all but one side of the moat, and over time and with increasingly adverse wet weather conditions the clay expanded, and the centre of the building suffered clay heave.

Clay is, however, better than light crumbly soil. In very adverse climates, such as those found in North America, the frost line, the depth to which frost penetrates, is also a factor (Snell and Callahan, 2005: 109). It may start to be a factor in Britain if the present pattern of severe winters continues.

Although geological maps are useful, they so often do not reveal the potential for anomalies beneath the very plot required for the build. It is far safer to rely on the building inspector, who will have a working knowledge of the subsoil, and if necessary employ a soil engineer. Whilst this might seem an extravagance, it is far safer than making a very expensive mistake based upon the evidence of one trial hole. Some very basic data can be relied upon in order to make a judgement. Igneous or metamorphic rock in a sound condition is approximately 5 times more load bearing than sedimentary rock, and 20 times greater than dense sand. It is 50 times greater in its capacity for loading than stiff clay (Snell and Callahan, 2005: 110).

Earth will also have a bearing on the possibilities for access. It is pointless to have a grand design for sustainable living if the access to the site is dictated by a subsoil and topography that defies sensible access without the expenditure of non-sustainable materials on creating a track. In particular the initial construction will demand access for large transporters of building materials. Some existing traditional buildings are very remote, as in earlier periods farmers spent most of their time on the surrounding hinterland and the construction being vernacular did not involve transport of materials. Getting to market involved horses or donkeys with

236

panniers, if even a rudimentary track for a cart was not available, and in all events carts were more flexible than modern transport. If a site needs major excavation to make it usable, more thought is required. A recent scenario revolved around excavating a large tract of land around an existing cottage, in order to provide a double garage. Such an indiscriminate alteration of the contours can not only lead to problems with the water table and water flow, especially on chalk where the latter is acting as an aquifer, but also to a loss of integrity of the site. The cottage had indeed been created in the first instance by cutting into the medieval topography in the 18th century, and the house constructed of the material, but to have perpetuated this for what can only be termed a 'passing phase' would have been the negation of the very topography that gave the site its charm and regulated water flow.

For new sustainable construction the topography of the site will determine location of key buildings, including the house itself, and ancillary buildings. Topography will determine the unwelcome possibility of shading and loss of solar gain. The domestic building usually needs to be to the very north of the site, so that this building itself does not cause unwelcome shade. Additional outbuildings also will need to be to the north, so that they in turn do not shade the south elevation of the domestic outbuilding. Space for the delivery of building materials in the form of a turning circle and also for emergency vehicles needs to be anticipated. Turning circles for various vehicles, including emergency, can be found on the Internet. A key factor for a sustainable lifestyle will be the orientation of the garden, particularly the 'outside room' and with it the siting of concomitant outbuildings, particularly if buildings for small-scale stock rearing and vehicle shelters are envisaged. Note that garages are regarded as wasteful space – wet vehicles in all events need only sufficient cover to prevent rain ingress and maximum ability to benefit from drying winds, as did their predecessors, the farm cart. Areas for less than graceful structures such as the oil tank, which building regulations demand should be enclosed within a blockwork wall, need to very carefully thought about, and not left to chance. Many a traditional building for which this has by default been an afterthought have suffered visually from this unfortunate but sometimes necessary appurtenance.

All of this planning will probably need to be done in diagrammatic form for the planning authority, in all events, but physical setting out on the site with the aid of a compass, 30 m (100 ft) tape measure, hammer, stakes, and possibly a homemade sextant, will leave nothing to chance. Again a Venn diagram for less than precisely defined areas such as gardens will suffice.

Photo 5.16
An 18th-century cottage hewn out of chalk topography, now important for its setting

Water

Water is such a key element that it really does not demand extensive discussion, as all traditional building sites are really only ever located in relation to a water source. Water remains a fundamental component of decision making for the location of new build in rural areas. A turbulent source of running water could be utilised for turbines to generate a modest amount of electricity. Rainwater harvesting can be done on a grand scale, in which case it needs large tanks with possible attendant bacterial growth problems, but better on a small scale so that valuable water from roof-derived run-off is never wasted.

Buildings and water in close proximity are never a good mix, and whilst there is the odd existing traditional building that incorporates a stream flowing beneath, mills being the obvious example, such a combination is fraught with problems, not the least of which are vermin. Generally building materials and water are unhappy bedfellows, and a great deal of attention needs to be paid to creating a dry building platform. From the 12th century to the late Middle Ages, moated sites provided a good way of doing this. Then for some reason this technology fell out of favour, rather at much the same time as the ridge and furrow fell out of favour as a way of producing well-drained fields. Terracotta drainage pipes took the place of these more fundamental technologies, and in recent years the advocacy of the French drain to create a dry building platform has done wonders for existing traditional buildings. Its role in new build should not be ignored.

The interaction between building materials and water is the very essence of understanding built form. The primary role of materials is to resist the penetration of rainwater into the fabric and the interstices of the construction, usually by virtue of a satisfactory roof cladding. Dense materials such as slate, riven so that their face is delineated with numerous miniature gutters, are ideal at resisting rainwater and directing it into gutters where it can then fall to ground in designated areas to drain away into a soakaway or into mains drainage designed for this purpose.

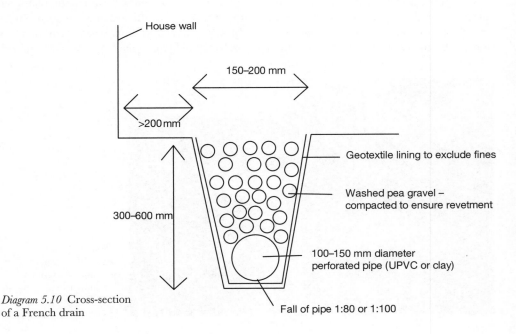

Diagram 5.10 Cross-section of a French drain

At this juncture it might be as well to indicate how damaging the introduction of internal plumbing paraphernalia can be to an existing traditional building. Underfloor heating systems have to some extent liberated new build, and some traditional build. Not only are radiators ugly, they are also inherently inefficient in their function, particularly when used in conjunction with a fuel-hungry boiler. This leads to them being used in an on–off fashion, resulting in the building always being heated up from cold, so that building materials that have the ability to act as a thermal store get no chance to do so, and in fact become woefully confused. No sooner have they heated up and moderately expanded than they are forced to cool down and contract.

Wind and fire

The dependency of fire upon the speed and direction of wind is an age-old relationship long ago forgotten by most building dwellers. Fires became very unpopular in the 1970s, leading to an unhealthy dependency upon cheap oil. What had previously been the very heart of the home, from early man onwards, and had been positively venerated by advocates of the Arts and Crafts movement, has become a distant memory.

> A massive stone chimney provides a welcoming hearth, the centre of the Arts and Crafts philosophy.
>
> (Massey and Maxwell, 1995: 88)

Most people only become reacquainted with the mystery and magic that is the 'fire' by visiting an open air building museum such as that at Singleton in West Sussex (Weald and Downland). The memory of opening shutters on the big dais windows, in order to better show students the dark and gloomy interior, and nearly causing a major conflagration and loss of the entire building, is one never to be forgotten. A more graphic representation of how the fire was controlled by draught would be difficult to find. This was also, of course, part of the role of the doors to the cross passage at the lower end. It is by examining this one aspect alone that one achieves a detailed understanding of the role of wind in controlling fire. Those more mundane

Photo 5.17 Massive chimneys on a building in Torquay, Devon, at the centre of the Arts and Crafts philosophy that the fire was the heart of the home

fireplaces that still remain in everyday use will have owners who are intimately acquainted with the need for up-draught in the chimney in particular to start the fire, and the everyday control of its burning rate so as to conserve fuel. Such skills, in particular those in demand for controlling a solid fuel Rayburn or Aga cooker, have all but been lost. Thirty years ago, every country farmhouse had one; it was the heart of the home and a very welcome focus on a bitterly cold winter's day spent in the yard. These machines are sometimes maligned by the sustainability movement and yet something which produces heat, cooking, and hot water in abundance is surely no bad thing.

Wind is sadly not predictable but the use of a simple wind sock for 12 months before commencing construction will give a reasonable indication of what is to be expected and when in terms of wind direction. That time in the caravan as the essential precursor to a new build or refurbishment of an existing building on the site is never wasted, however uncomfortable it is deemed to be. If a wind turbine is contemplated then something more technical will be required, such as an anemometer, a device for measuring wind speed. This is also applicable to an existing traditional house. One might not be able to influence its location or siting, but much can be gained from a complete understanding of wind direction and with it infiltration. Better to do this than launch into a grand masterplan that may involve removing large chunks of the building. The building should decide the scheme, not the scheme emasculate the building. The trick is to work out what the original logic was in the siting. Was the main room sited east not just because of the early morning sun, or a cool room to come home to in the evening after a day spent in hot sunshine, but because the site attracts south-westerly winds?

Pressure differentials are what determine the extent of air infiltration from wind. The pressure differential between the windward and leeward side forces air infiltration into the building from the windward side. Thus more heat energy is required to maintain the ambient temperature. As wind speed increases so the demand for heat energy is intensified. Cold winds normally come from the north and east, whilst warmer south-westerly winds are often saturated with rain, which in turn tends to cool down structures. It is often a difficult choice as to what elevation to insert windows as wind changes direction, but generally windows high up in a north or east gable will be very difficult to maintain. Wind-driven rain generally affects the upper portion of a building because wind speed increases with altitude. This is a lesson sadly not learnt by some traditional builders in the past, but then every village had a carpenter who welcomed making a replacement window as a relief from the tedium of making coffins. Windows are a serious problem on elevations facing into persistently stormy conditions. Heat loss through single glazing increases by 20% on exposed elevations (Nicholls and Hall, 2006: 63). Houses on the top of hills are particularly vulnerable to wind speed as wind is sucked up one side of the hill and forced over the top. Any building obstructing thus will be at the mercy of the full wind force. In addition, the area on the leeside may be subject to eddies, outside of a zone of still air, which can whip up debris such as leaves and small branches.

The corners of buildings are also very vulnerable. Corners act as natural ventilation zones for moisture drying out in walls so in this sense wind is very valuable, but it can equally well produce an excess of moisture concentration in gables facing into the wind. Photo 5.18 shows the application of a hemp-lime plaster to the interior of a gable facing into south-westerly rain soaked winds to try to counteract a constant degradation of a previous internal gypsum plaster. The evaporation of the moisture in the hemp-lime plaster is seen concentrated in the corner of the room, illustrating the capacity for corners to act as evaporation zones.

Having deduced where the major problem wind direction is located, the next issue is to slow it down. It is possible that nature in the form of hills or other rough terrain, or even other buildings, will do the job reasonably effectively. Inland sites do fare better than coastal sites in

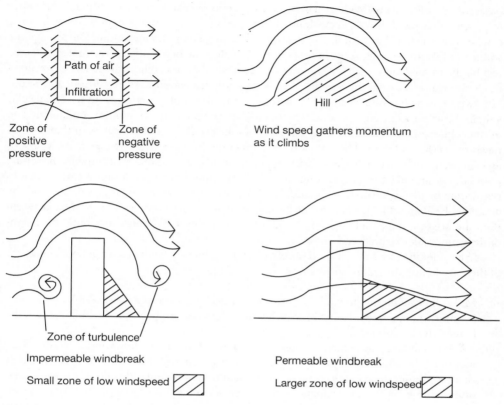

Diagram 5.11 Wind behaviour (adapted from Nicholls and Hall, 2006: 62–65)

Photo 5.18 Hemp-lime plaster drying out towards a corner evaporation zone due to wind action on the outside of a gable facing down a valley

this respect. Anything that produces friction will slow wind down, particularly trees with their bases infilled with shrubs. Walls can in most instances resist wind pressure, although it is better to provide them with some permeability, but fences generally cannot. Bunds or earth banks can be very effective. Permeability of the wind break is a factor in determining the force of the wind, so whilst some is desirable in solid objects, the less the permeability, the more protected will be the immediate lee side of the obstacle. Increased permeability will produce a zone of moderate wind speed on the leeside. Tree belts, acting as windbreaks, need to be deciduous otherwise they will produce excessive shade that may seriously interfere with potential solar gain in the winter. The gaps produced by leaf loss in winter can result in more wind infiltration in the colder zone of the year. This is typical of many building conundrums, in that there is never the perfect solution. There is a problem on virgin sites in that it will take many years for an effective wind break to develop. Trees are useful in all events as they can enhance the setting of buildings, and act also as a carbon sink, thus offsetting any expenditure of carbon due to renovation or new construction. In addition, birdsong is one of the great joys of life, and the creation of a feature that fosters wildlife habitats is of permanent benefit. The distance between the wind break and the building must be gauged, and needs to be around five times the height of the windbreak (Nicholls and Hall, 2006: 65).

Further modifications to the wind problem can be achieved via the construction, and many of these tricks are utilised in existing traditional buildings:

- Make the wall on the prevailing windward side as heavy and thick as possible.
- Site the building behind a bank, ensuring that there is a gap between the building and the base of the bank for drainage.
- Site the building with a long sloping roof facing into the prevailing wind direction. This was most definitely the most favoured option traditionally, and whilst the outshots that sheltered under these long sloping roofs were almost always utilised as dairies, brewhouses, or other messy and cold occupations, this usage being the primary factor, the combination made eminent common sense. These workrooms acted as buffer zones to the warmer, wind-sheltered living rooms.
- Avoid the use of any windows on an elevation facing into the wind, or use only the smallest windows. Dairies traditionally faced north and were normally only given a slit window with a central bar and perforated fly mesh covering.
- Site entrance doors on an elevation away from the prevailing wind, although it has to be said that this did not occur in ancient hall houses, where it must have been intended, at least initially, that the wind should be pulled through by pressure differentials from one side to the other, if the cross passage was to be used for farm purposes or control of the fire.
- In new build emulate the practice of using subsidiary buildings to shelter the domestic building. The workaday nature of most traditional houses, with numerous outbuildings ranging from the smallest coal shed up to the largest threshing barn, meant that the domestic building could be shielded, thus preventing it facing into a tunnel of wind. These buildings were inevitably staggered. The threshing barn invariably stood to the north of the yard to shield animals from cold north winds and encourage maximum south-facing exposure. The farmhouse often faced the barn, with its front elevation facing south also, at least by the 18th century, although earlier formats have the front elevation facing onto the yard.
- Create sheltered courtyards to form wind-free micro-climates.
- Modern methodologies revolve around the almost impossible task of totally sealing the perimeter of all of the building. The concept of air pressure testing (see below) revolves

around this principle. The fact that it is virtually impossible to secure complete enclosure in a building envelope, or the fact that this could actually be undesirable in terms of human health, is a matter of debate.

Other principles involve ensuring that the surface presented to the prevailing wind is streamlined in order to reduce the pressure differentials that result from presenting a large surface area to the wind, creating more potential for warm air to be sucked out of the building. This principle may well be utilised for cars in an effective manner, but it is more difficult for buildings. The long narrow building with its main elevation facing due south will create a naturally narrow west gable that will assist in this streamlining process.

Other good modern principles are:

- Avoidance of flat roofs, or if absolutely necessary, the considerable insulation of such roofs, to counteract the upward suction of wind which can draw warm air vertically out of the building.
- Avoidance of high rise buildings altogether, which can present a vast surface area to prevailing winds.
- The use of hipped roofs with pitches of 30 to 45 degrees encourages smooth air flow. This produces yet another building conundrum as steeply pitched roof are more efficient at removing rainwater, and indeed are the most traditional form of British roof.
- Shorter walls facing the prevailing wind to encourage a slowing down of the wind. Wind whipping against a long wall will gather speed.

(Adapted from Nicholls and Hall, 2006)

Airtightness testing

The modern phenomenon of airtightness demanded of new buildings is closely allied to the whole concept of minimising air infiltration, the chief cause of which will be wind. It is something to be approached cautiously with existing traditional buildings, be they vernacular or polite, as they will have been constructed in a totally different way with an expectation of draughts and to some extent of rising damp. No technology was available to counteract these problems at the time of construction, so ways of living with them were contrived and need to be continuously developed without detriment to the building (a typical example being the Victorian match-boarded dado to cover the signs of rising damp). Consequently, these old buildings are complete living and breathing mechanisms (see discussion of the breathing wall). Raging draughts, such as those that develop during extreme wind and storm conditions and gale-force winds, should not have to be tolerated. The most common-sense action is to choose a windy day or night and using the hand test, a hand across any gap around doors and windows, above skirting, below doors etc., it can be readily discerned where a draught-proofing device needs to be installed. Early forms of the latter were often strips of leather or fabric tacked to the inside of the door frame. Modern equivalents of closed-cell polystyrene will do the job but rapidly becomes dirty and compressed, thus ceasing to function. It is not aesthetically pleasing and the old ways seem to work best. A more solid latex product made in Germany is also satisfactory.

If, however, a new sustainable construction is deemed to have good air-tightness performance, then it is deemed suitable to undergo an air pressure test. This is usually carried out following initial construction and prior to installing any controlled ventilation, such as trickle vents. In domestic buildings the front door is replaced by a panel containing a fan. The fan is operated at a set speed until a constant internal pressure is achieved inside the building, usually 50 Pascals (Pa) above atmospheric pressure, which is normally 101 Pa. At this constant pressure it is deemed that the volume of air escaping from the building via leakage is equal to the volume of air being drawing into the building by the fan; 50 Pa is deliberately chosen as being deemed to be close to normal pressure on a non-windy day. The air permeability of the building is then measured in cubic metres of air per hour, per square metre of the envelope (m3/h/m²). The division by the area of the envelope is necessary to achieve this calculation. The result is compared against a known table of values for best practice and is useful for comparison between different buildings.

The ventilation heat loss rate, measured in air changes per hour, at standard room temperatures and pressures is a more fundamental requirement. A known volume of tracer gas is injected into each room so that it can then mix with the air in the room, and its concentration is then measured using a gas analyser. This is compared with its concentration after a set time, when some of the gas has left the room and been replaced by even more fresh air. The rate of fall after a period of time is the ventilation rate. This result is set within the context of internal and external temperature differences and external wind speeds. Actual zones of the room where the gas may be finding an escape route is best deduced by a smoke test, so that the smoke can be seen to be escaping through gaps in the external envelope. A rather better technique is thermal imaging, which will be able to deduce escaping warm air from a building on a thermographic image.

The conclusion of these tests, which one would like to believe are super efficient but which seem to be somewhat over the top as regards basic common sense, is that three strategies must

Photo 5.19 Air-tightness equipment in position

then be adopted in accordance with Part L of the Building Regulations, to provide controlled ventilation, as discussed in Chapter 4:

- Trickle ventilators over the top of new windows, to provide a controlled path for ventilation of the building.
- Purge ventilation, usually in the form of windows that can be opened.
- Extract fans or passive stack ventilation (infinitely preferable) to deal with moisture to remove potential for mould spores and remove odours.

From the previous discussions and this discussion of air testing it can be seen that the method of ventilating a building is the key to its survival. In addition, the amount of ventilation is going to be determined by ventilation heat loss rates, which are in turn determined by wind, a totally unpredictable force of nature.

Understanding U values

This is another phenomenon of modern life and modern building construction. Like ventilation, it is an attempt to measure the possible heat loss from a building, but is marginally more successful because it is less dependent on the vagaries of nature, although the values ascribed to various materials appear to pay little heed to what actually happens to external materials when they are cooled down by rainwater. Driving rain fills the pore structure (miniscule voids) of masonry materials with water, thus decreasing the number of pores holding air. This air helps to resist heat loss and is one of the keys to understanding the role of masonry in contributing to the overall thermal capacity. As well as resisting heat loss, the air in the pore structure can become warmed by heat within the building, this heat being passed into the molecular structure, and thus contributes to the heat store, holding this heat until it can then be released back into the building. Driving rain, normally the type that forms an angle with the vertical, can thus increase the thermal conductivity. A building material becomes more dense as the pores are filled with water, and this gives a greater path for heat to be conducted across the depth of the material. This thermal conductivity is measured as k, where k is the measure of the rate of heat conduction through one cubic metre of a material with 1 K temperature difference across two opposite faces, thus measured in Watts per metre Kelvin (W/mK). Very dense materials, such as slate, a compressed mudstone, and sandstone, have little porosity and have a naturally high density (measured in kg/m^3). This high density is also due to the fact that the molecules are packed tightly together. Thus they conduct very fast and have a high value of k. Wood, which has a linear cellular structure, with the only density lying in the walls of the linear cells, has a low density. Once dried, the linear cells hold a lot of air, and thus have a much lower value of k.

It becomes readily apparent also that when water filled the pore structure of masonry it makes the masonry wall very cold. If this moisture travels towards the interior face (in fact an unusual occurrence), cooling it down, then this will cause condensation and black mould. Every device should be utilised in building design to avoid masonry walls from having to suffer the full blast of driving rain. Overhangs are particularly useful in this respect, and this device was well known in the early 19th century, with even modest town houses being given extended cornices supported on modillion eaves. More modest vernacular houses were given outshots for workaday activities, creating a long sloping roof directed into the full force of the driving rain, and clad with slate to direct the rain away from the walls beneath. The living areas were sited on the other side of the building and thus protected from this extensively cooling rain. A belt

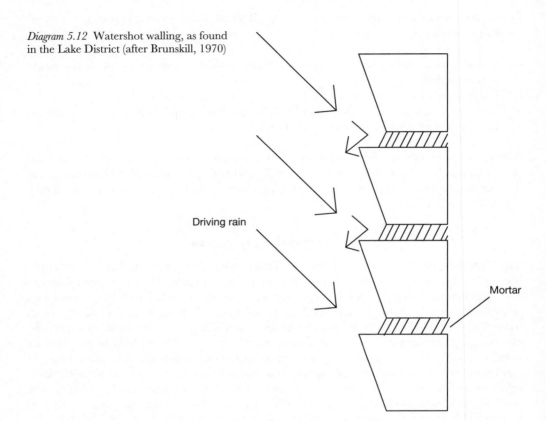

Diagram 5.12 Watershot walling, as found in the Lake District (after Brunskill, 1970)

Driving rain

Mortar

of trees in the path of driving rain will also mitigate its force. In the Lake District watershot walls were devised for the very purpose of deflecting rain downwards.

When all else fails, cladding the exterior with a thick lime 'overcoat' is a failsafe methodology. It is essential to apply it at least three layers in thickness (see section on lime renders for mixes). Lime has a graded pore structure if it has been prepared using well-graded sand, and the outer zone copes with the saturation until the pore structure can breathe it out again.

Summary

Thermal conductivity (k) increases with density. The larger the value of thermal conductivity, the faster heat is transferred through a material. It is measured in W/mK.

Actual U values are measured in terms of thermal resistance (R):

R= 1/k (m²k/W) where 1 is the thickness of a 1 m cube of material.

R is in fact the ability of an actual thickness of material to resist the passage of heat. It follows that increasing the thickness of a material will normally resist the transfer of heat, but this can also be done by decreasing the value of the conductivity k. The lower the value of k, the higher will be the figure for 1/k. The figure for k is lower if a material is used which has less density and more porosity.

In practice all the materials across a wall have their own value for R. Thus a typical wall with

- an internal plaster finish is given a value for R_p
- a blockwork outer skin is given a value for R_{bl}
- a layer of insulation attached to the outside face of the blockwork is given a value of R_i
- the cavity has a value of R_{cav}
- the outer skin of brickwork is given a value R_b
- the still air attached by friction on the outer face is given a value of R_{so}
- the still air attached to the inside face is given a value of R_{si}

$$R_T = R_{si} + R_p + R_{bl} + R_i + R_{cav} + R_b + R_{so}$$
Where R_T is the total thermal resistance
The U value is calculated as $1/R_T$
The U value is thus lower the higher the total thermal resistance

In reality the issue is somewhat more complicated as the total R has to take into account thermal bridges, a pathway of high conductivity in an otherwise low thermal conductivity layer. Timber studs used to mount plasterboard are regarded as a thermal bridge, although the logic is somewhat confusing as timber is an insulator in its own right if the linear cellular structure is air filled.

This means identifying a maximum heat loss and minimum heat loss flow through the structure, adding the thermal resistance for each together and dividing by 2 to get the average thermal resistance (for more details see Nicholls and Hall, 2006: 102–106). Fortunately the thermal resistance R is known for most standard building materials.

Typical thermal resistance for the above example might be:

External resistance = 0.04
100 mm brickwork = 0.14
Cavity of 50 mm = 0.18
Blockwork = 0.16
Plaster (gypsum) = 0.03
Internal resistance = 0.13
Total R = 0.68
U = 1/0.68
U = 1.47 W/m$_2$k

that is, heat is conducted in this standard form of construction at the rate of 1.47 watts per square metre for every degree of temperature difference measured in Kelvin, between the inner and outer surface of the walls. This is a very low U value, and is below the maximum value of 2 W/m^{2}K demanded by most building regulation requirements.

Typical requirements now are:

Floors – 0.25
Solid 9 in external brick walls – 0.35
Roof with loft – 0.16
Roof with insulation at rafters – 0.20
Flat roof – 0.25
Windows – 2

(Griffiths, 2007: 45)

All thermal elements such as these must meet the standards set out in Building Regulations, Approved Document L1B. In addition, in an existing traditional or polite building the requirements need to be met whenever a new thermal element is altered. The only exceptions are where the payback is deemed to be less than 15 years, or the floor area would be decreased by more than 5%. There is also the possibility of trade-off, which is something that needs full exploitation with existing traditional buildings be they listed as being of special architectural or historic interest or not. It is thus possible to achieve a higher insulation value on one element where another, such as an exposed stone wall, is required to be retained (see also discussion on breathing walls). This frequently occurs with farm building conversions, and a typical untouched example where this trade-off needed to be achieved is shown in Photo 5.20. Here the main insulation is proposed for the roof, which in all events needs to be repaired extensively. The best approach is always to upgrade those building elements that are going to be disturbed in all events.

There is a far better option for this traditional stone building as it will allow the stone walls to act as a thermal store, which would not occur if masked by insulation to achieve the desired U value. In addition, the material used may have been one that had a dubious health record (see discussion in Chapter 4 on this aspect). This more than anything else must be of the gravest concern, particularly to those engaged on the repair of an existing traditional building whose walls are not only capable of acting as a thermal store but also of breathing out noxious gases by acting as a form of ventilation. Why can this aspect not be taken more seriously by building regulation authorities?

U values for the passive house

It is possible to correlate the total U value for all the elements of a building so that the heat loss (1/total thermal resistance) is balanced out by heat gain from nothing more than 'heat gained from heat energy need for cooking and lighting, from the heat of the bodies of the occupants, and from solar gain'. Such buildings are normally given heat ventilation recovery systems (HVRS) so than any heat from exhaust air is recovered.

Photo 5.20 Proposed barn conversion – untouched roof ideal for incorporating high levels of insulation above the rafters without disfiguring the roof construction

Photo 5.21 Almshouses in Milton Abbas, Dorset, now have a heat ventilation recovery system to remove condensation and improve air quality

Such an aspiration is difficult to apply to an existing traditional building, although this does not preclude HVRS being installed to remove stale air created by previous unsatisfactory cement coatings on internal walls. In the almshouses of Milton Abbas, Dorset, the architects have installed such a system which has massively improved the internal environment and air quality. Previously the stale air could almost be cut with a knife; such is the impact upon traditional building construction of rendering walls incapable of breathing.

Insulation introduced into these buildings is not only going to mask their abilities to breathe out exhaust gases via the structure, it is also going to compromise the floor area with its build-up against the existing wall. It may also compromise ceiling heights, or the character of existing doors and windows, upon which the house depends for its integrity. This is not to say that such insulation is never possible, but it needs very careful consideration. Wood fibre boards will best allow a masonry wall to act as a breathing wall and a thermal store. In all events, calculating the U values of existing masonry walls is virtually impossible, due to the impossibility of catering for voids, so designers over compensate by calculating the thermal resistance on the basis of the insulation only.

Flooding

Flooding is an issue which is frequently becoming a problem for both old and new build. It is difficult to see how new build, especially if constructed on the timber frame principle, can survive such an onslaught, ditto traditional timber framed buildings. Traditional buildings with thick stone walls clearly have a better chance, but even they will be subject to salt mobilisation so that crypto florescence will occur at lower levels, and with it exfoliation due to damaged pore structure. This salt, once mobilised, will always remain a problem unless attempts are made to poultice it out. Such attempts are rarely one hundred per cent successful. Salt mobilisation is a factor never mentioned in the press or by insurance assessors, but is nonetheless very real.

It is rare for traditional buildings to be sited near to rivers that burst their banks regularly unless the water management regime in the vicinity has altered over time and stimulated a propensity for flooding. Nonetheless such events do occur and care has to be taken when

Photo 5.22 Flooding at Marnhull, Dorset, in November 2009 – this now appears to be an annual event

specifying insulation, with a particular mind to the nature of the existing site. Insulation that is cellulose molecule based, for example wood fibre boards, will deteriorate beyond recall, and it would be even more dangerous to use a closed-cell material which would not allow the base of the wall to dry out either internally or externally. Where severe flooding is a constant threat, then traditional buildings need all the porosity they can gain, both in walls and floors adjacent to walls, inside and out, for example the French drain, to encourage egress of moisture. This must take precedence over the calculation of U values.

6

THE ROLE OF THE BUILDING REGULATIONS IN THE SUSTAINABILITY OF TRADITIONAL BUILDING CONSTRUCTION

The origin and need for building regulations

There has always been something of a love–hate relationship between building regulations and existing traditional buildings, especially when the latter are being modified in any way. Much depends on how any individual building inspector views traditional building construction and how much discretion can be allowed (with the exception of fire safety). That there should be regulation of new build is without question, as a recent appraisal of a Georgian house attests. A perfectly solid early 18th-century house was seemingly updated with larger early 19th-century windows by removing much of the front wall and replacing it with brick. Unfortunately this new wall was not tied into the stone end and cross walls, so was virtually wafting in the breeze. The resultant cracks that had persisted for much of the period since c.1800 were discovered only when every corner of every room was opened up. A more telling need for a building inspector at the time would be difficult to find. Sadly this story is not unusual in the frantic building activity that characterises classical revival in the 18th and early 19th centuries.

Building regulations have their origin in the Great Fire of London in 1666, when it was realised that some regulation was needed to counter the rapid spread of fire across elevations. Building Acts, mainly to curtain the spread of fire, were passed in 1709, and again in 1774 as the first had little effect in the rural areas where sash windows flush with the external wall persisted until much later after their London counterparts. It was not until after the Industrial Revolution, which led to appalling living conditions, including whole families living in cellars and the spread of cholera, that very serious consideration was given to a government-led reform. The result was the Public Health Act of 1875. This was revised in 1936 and 1961, culminating in the Building Regulations 1965. This document has had numerous revisions, but the Building Regulations 2000 were particularly designed to meet the requirements of the Building Act 1984. This was designed to ensure that the health and welfare of persons living in or working in or nearby buildings was secured. The primary purpose of the Building Act 1984 was, however, to boost the potential for the conservation of fuel and power, and it is this requirement that provides the greatest challenge for existing traditional building construction.

The Building Regulations are supported by the Approved Documents (sometimes referred to as 'Parts' of the Building Regulations) which correspond to the areas covered by the Regulations. The purpose of the Approved Documents is to provide technical and practical guidance on ways in which Schedule 1 of the 1984 Act (the actual Building Regulations) and Regulation 7 of the same Act, can be put into operation, as well as reproducing the actual requirements contained in the Building Regulations relevant to the subjects of the individual documents.

An important point, which must form the basis of new sustainable construction or modification of existing traditional construction, and one which is very often not mentioned, is that

Photo 6.1 This Georgian house under repair illustrates the need for building regulations: removal of the render showed that the front wall was not tied to the gable walls

there are alternative ways of complying with requirements. The key is persuading the building control officer that there is an alternative way of doing so.

The current Approved Documents are:

- Part A – Structural
- Part B – Fire Safety
- Part C – Site Preparation and Resistance to Moisture
- Part D – Toxic Substances
- Part E – Resistance to the Passage of Sound
- Part F – Ventilation
- Part G – Hygiene
- Part H – Drainage and Waste Disposal
- Part J – Combustion Appliances and Fuel Storage Systems
- Part K – Protection from Falling, Collision, and Impact
- Part L1A/L2A – Conservation of Fuel and Power in New Dwellings
- Part L1B/L2B – Conservation of Fuel and Power in Existing Buildings
- Part M – Access and Facilities for Disabled People
- Part N – Glazing: Safety in Relation to Impact, Opening, and Cleaning
- Part P – Electrical Safety

Of all of these, Parts C, E, F, and L1B often provide the greatest challenges to maintaining existing traditional building construction for domestic buildings. With regard to new traditional building construction, Parts A, C, E, F, J, and part L1A provide considerable challenges.

The aim of the following discussion is show where the pitfalls are for traditional construction of existing domestic buildings and to indicate any possible pitfalls for new sustainable traditional domestic construction. *It is not intended to be anything other than an indication of the scope of the Regulations and their impact on traditional building construction and cannot be used as a substitute for consulting the actual regulations as set out above.* Other building types will be touched upon, but in the main it is domestic buildings that provide the greatest challenge, or buildings such as those designed for use by the general public.

Materials and workmanship – traditional building construction

Perhaps the most technically challenging for traditional building construction is Regulation 7 – Materials and Workmanship. It requires that work shall be carried out with adequate and proper materials which are appropriate for the circumstances in which they are used, are adequately mixed and prepared, and are applied, used, or fixed so as to adequately perform the function for which they were designed. Finally, the building work so resulting should be carried out in a workmanlike manner. That these materials can include stone, timber, and thatch is not in dispute, but the regulation is largely geared to materials that can be assessed for fitness against the list of standards given below. The standards are very much geared to modern and largely unsustainable materials, such as cement, for which calculations are relatively easy. One must not forget also that asbestos, in its day, was a material that at the time of its use provided for ease of calculation, but has subsequently proved untenable in relation to human health. Unfortunately it is all too easy to forget the heinous mistakes of the past in the face of mass enthusiasm for a modern material that appears to offer many advantages. The standards are:

* British Standards (BS)
* European Standards (EN)
* National and international technical specifications from European Community member states
* CE marking (Construction Products Directive 1994)
* Tests and calculations (for example, those that conform with UKAS Accreditation Scheme for Testing Laboratories)
* Independent UK product certification schemes

All of these can be very rigorous and pose problems when new build in, say straw bales is attempted. Lime putty, the basic ingredient of all traditional mortars, has suffered dreadfully in the past. Only with the innovation of hydraulic limes, some of which have obtained the necessary European certification, has the situation eased. Lime putty remains to many 'a black

Photo 6.2 Lime putty is not a black art but a fundamental material for repair

art', particularly those in professions such as structural engineers whose measure of 'fitness for purpose' largely revolves around materials for which calculations can easily be made, such as steel and concrete.

Another test for fitness can, however, include 'past experience of the material as an indication that it is capable of performing the function for that for which it is intended', and it is this that most advocates of traditional materials rely. It is also possible to ask the Local Authority to sample test the material, but chronic under-funding of Local Authorities makes this a less satisfactory option. Tests are normally only conducted on sewers and drains.

The question of recycling materials is also a vexed one. Whilst Regulation 7(0.2) says that environmental impact can be minimised by careful choice of materials and use of recycled materials not having any adverse implications for the health and safety standards of building work, this is not always recognised. A classic example of where this concept has not been embraced is a recent example of the re-siting of 17th-century windows in a new blockwork wall, the latter being the result of the need to widen the building. The building inspector involved was unhappy with the use of old windows in new walls, as it was felt that they could not be adequately draught-sealed around their periphery, and this despite the liberal use on most building sites of urethane foam sealants.

Another criticism frequently levelled against traditional materials is that they are short-lived or subject to change, again despite the fact that many petroleum distillate products are even more grievously subject to change by UV radiation. The regulation states that such materials should only be used in works where these changes do not adversely affect their performance. As will be seen from the following, the very basis of most traditional materials is that they are subject to change, but that the change is eminently more reversible, the material normally reverting to its original format: clay swells and shrinks; if lime cracks it self-heals, filling the crack again; and cob can be re-constituted using the very same material taken from the building. All these materials are considered to be short-lived, yet their very advantage lies in the fact that they are recyclable and renewable with no extra environmental cost. The immense prejudice against the use of materials that are eminently renewable or recyclable but have what is deemed

Photo 6.3 Window removed from the old crosswing to house in 6.1

Photo 6.4 Window re-sited in the new crosswing to house in 6.1

to be a shorter lifespan is a hurdle still to be conquered in the minds of the regulators. The fact that some lime renders date from the 17th century, and cob buildings, clay mortared, and plastered walls frequently from the same date, is never regarded as being part of the equation. Why the lessons of history can be easily disregarded is one of the mysteries of mankind.

Clay and the Building Regulations

Materials for which reliance has to be placed on proving previous capability of performing the function for their original intention include clay for use in mortars and plasters.

A high proportion of buildings in the United Kingdom, located in areas where clay is the predominant subsoil, have utilised the resource instead of lime, for both bedding stones and for producing a haired clay plaster, sometimes with an element of lime included. Lime mortar in these areas was restricted to the pointing of the building externally and for external renders. The reasoning is obvious: clay requires no processing or modification, but limestone requires burning at a high temperature, around 800–900°C, before it will dissolve into water to form a 'putty' and then have an aggregate added to form a mortar/render. All of this is hard work and requires capital input, such as kilns and a workforce, whilst clay needs only to be dug from the ground and put through a pug mill to remove stones. In modern parlance, clay is an eminently sustainable material, having no carbon emissions. The battle to try to use it as a bedding material for stone, in the face of the Building Regulations, would only incite numerous examples being given by the building inspector concerned of existing traditional buildings which move with the seasons, and even with the variation of climatic conditions in any one day. Such buildings abound in North Dorset, but their owners have accepted the odd crack as part of the daily and seasonal norm. Clay plasters have a better chance of being more universally accepted in new build and as replacement plasters in existing traditional buildings, but only if they are used on internal walls. Clay is an eminently hydrophilic (water-loving) material, and if used as a plaster will absorb condensation caused by modern living, and will thus curtail interstitial condensation, but its use

Photo 6.5 Clay
mortared joints

on the inside of external walls will depend on much more acceptance of the concept of the breathing wall. This is a concept very analogous to that of the wall acting as a thermal store. Neither of these concepts have, as yet, any credence in the Building Regulations, an omission which is the crux of the problems encountered in traditional construction.

The use of clay traditionally revolved around the use of very thick walls. Precipitation was largely absorbed by the outer portion of the wall, and by the lime mortar joints specifically designed to be more resistant to rain penetration, or by a lime-based render, acting as a generous overcoat. In respect of the latter two methodologies for wall protection, it was the porous nature of the lime (discussed in detail elsewhere) that catered for a certain level of absorption in a wetting cycle, then a quick release of this moisture in a drying cycle. Without this, clay mortars and plasters are not tenable. Proving this is, of course, a difficult matter. In the majority of the areas where clay abounds as a binder and internal plaster there is actually little cognisance of its existence by building contractors or building professionals generally. Contractors frequently mistake the plaster for an old and crumbly lime plaster, remove it, and rake out the joints to gain a key for the dreaded application of two coats of neat cement 'to keep out the damp'! This is followed by an application of a coat of gypsum plaster to create a good finish for decoration. If the traditionalist should be fortunate enough to intervene, it is nearly always too late to save the clay plaster, its malleability making it a ready victim for the wheelbarrow and the skip. To persuade the contractor that it should be replaced at the very least by a lime plaster takes a great deal of personal charm! To attempt to persuade its replacement like-for-like will probably meet with derision, but this is not to say that one should not persevere.

Cob and the Building Regulations

Cob, which also abounded in areas of clay and chalk, is a similarly maligned material, not so much regarding its use in existing traditional buildings but rather for new build. Set against this is the knowledge that it has been used for thousands of years in many cultures all over the world, as a material which as the natural subsoil is readily available, and has the unique benefit of acting as a thermal store. Its whole basis for construction is the thermal flywheel effect, whereby temperature fluctuations are flattened out and buildings are kept cool in summer, but in winter absorb the heat of the fire so that the building acts as a vast storage radiator. Its sole problem is that it requires a certain level of moisture to remain intact, the water molecules acting as a form of binder between each particles of clay or chalk. Too much and the wall from which it is constructed turns to porridge, too little and the wall turns to powder. Frequent is the story told throughout villages where cob is the basis of the local vernacular, of the occupants of a bed finding themselves in the street overnight. Such occurrences although rare do still happen, prompted by the misguided use in the past of cement renders and plasters (see Photo 6.6 – fortunately the house was empty).

Photo 6.6 Partial collapse of a building at Milton Abbas, Dorset

Photo 6.7 Material from the collapse was later broken up and reconstituted with water to rebuild the collapsed wall – notice how it resembles coarse hempcrete

The use of cob for new build could be regarded as contrary to Approved Document C (site preparation and resistance to moisture), in particular the provision of adequate subsoil drainage to avoid the passage of ground moisture to the interior of a building and water-borne contaminants to the foundations of the building (Part C1). Also in Part C2 there is a very specific requirement for the floors and walls to adequately protect people using the building from the harmful effects caused by ground water. All of this creates a requirement for a solid or suspended floor to be constructed to prevent undue moisture from reaching the upper surface of the floor and a wall which prevents undue moisture from reaching the inside of the building. The solid floor would instantly drive moisture up the cob wall, and the latter is simply not possible with a cob wall for the reasons stated above. The advantage of the cob wall acting as a wick for the vast amount of warm moist air which modern life generates is rarely regarded as part of the equation. Further problems with cob walls will be discussed below in the section on walls. Major problems revolve around the difficulties in calculating the thermal performance of cob walls, and it is this inability to provide exact calculations that further inhibits the use of this most sustainable of building materials. Some research has been conducted by the University of Plymouth, long associated with pioneering the re-use of cob. The research was conducted on existing cob buildings, using thermal probes to assess the impact of the moisture in cob on thermal properties, in order to address the lack of data for the material. The research concluded that in-situ testing varied from laboratory values currently used in thermal engineering calculations (Pilkington et al., 2008).

Gernot Minke has written the pioneering *Earth Construction Handbook: The Building Material Earth in Modern Architecture* in 2000, which supplies valuable data.

Hempcrete and straw bales and the Building Regulations

A material very similar in its nature to cob is coming very much to the fore. It is a mixture of lime and hemp stalks (hurds) to form 'hempcrete'. This possibly has a comparative in the old tradition of mixing straw with clay or chalk subsoil, and indeed they are very similar in appearance. This was done to provide a form of internal corset and to prevent massive shrinkage, but it also helped to boost the thermal performance. Hempcrete forms a solid, non-load bearing insulated mass wall, but the difficulties lie in providing thermal calculations. Much is dependent on the ability of the individual traditionalist to convince the building inspector that the ability of such walls to act as a thermal store far outweigh any disadvantages of thermal loss. Woolley and Bevan (2009) have experimented with building a house with hempcrete walls in County Donegal, Ireland and are testing U values. Cited in their article are tests on a house in Suffolk by Plymouth University giving a U value of 0.36–0.37 W/m^2K for the walls. In addition, hempcrete blocks do not rely on the need for moisture to remain intact, so they can cheerfully be given a damp-proof membrane below walls and floor, as can straw bales. The bales, used as load-bearing walls or as infill to a timber frame, are becoming increasingly popular. This form of construction appears to be able to surmount the legislative hurdles posed by the Building Regulations, and in fact a text on the subject (Jones, 2002) states that a typical bale of straw has a U value of 0.13. Straw timber panels are now manufactured, complete with a sheath of OSB (orientated strand board) and a wooden frame, which can be lifted into position in the same way as other less eco-friendly engineered panels. They are deemed to exceed the current Building Regulation standards for insulation for heat and sound. In fact a whole variety of cellulose materials can be used to mix with lime or even cement to improve insulation, to include flax, coir, jute, and woodchip. The technology is not particularly new as the wood-wool slab, so beloved of timber frame repairers in the 1970s, can be regarded as a precursor.

An additional hurdle for straw bales is compression testing when they are used in the raw for mass walling construction. The requirement of Part A (A1 – loading), that the construction should transmit to the ground safely and without causing deflection, combined dead, imposed, and wind loads, comes into play. Coupled with the need to satisfy Part B on fire safety, notably that the external walls of the building shall adequately resist the spread of fire over walls, and from one building to another (having regard to the use and position of the building), plus the need for internal walls to maintain stability for a reasonable period, the challenges for straw-bale construction are many and varied. These aspects of compression and U value can, however, be tested on site with a sacrificial prototype building without huge detriment to the overall cost of the project. The material itself is low cost, and the need to expend some costs to allay a building control officer's concerns is a relatively simple exercise. Peter Walker, of the BRE Centre for Innnovative Construction Materials at the University of Bath, has conducted considerable research into straw-bale construction (Lawrence et al., 2009).

Something for which the Building Regulations do not legislate also needs consideration. It is a problem with many existing traditional buildings and could be a problem with hempcrete and straw bales. It is the autumn invasion of mice and even rats. Many existing traditional buildings have a cement-render skirt to provide a barrier to this yearly problem, and straw bales or hempcrete blocks similarly need a good lime render, the bottom metre of which is a stainless steel mesh (personal communication, Rob Buckley DCRS).

Photo 6.8 Straw bale in construction (photo © Rob Buckley)

Photo 6.9 Straw bale construction nearing completion (photo © Rob Buckley)

The one disadvantage which cannot be overlooked is that, with the exception of straw, the majority of these materials are imported and thus can incur sea miles. In fact there is no reason why hemp cannot be grown in this country, simply a morbid fear of growing the wrong form of hemp and ending up with an unwanted supply of cannabis. This appears to be the principal inhibitor.

With all the materials such as cob, hempcrete, and straw bales, the issue of foundations that will satisfy Part A (A2 – ground movement), is an important issue to contemplate. There is a need to show that the building so formed can be constructed so that ground movement caused by swelling, shrinkage, or freezing of the subsoil, or landslip and other forms of subsidence, will not impair stability. Cob is traditionally a material founded on little more than a gravel-filled trench and a plinth constructed of stone, flint, or brick. Rebuilding cob structures because the walls have developed cracks is a frequent scenario. This is possible because cob enables the existing material to be simply be chopped up and mixed once again with chopped straw and water to reform itself, all without incurring any issues concerning CO_2 emissions either from transport or manufacture. Illustrated in Chapter 2 is a hapless example where the owner was not even prepared to contemplate it, so a typical c.1850 cottage was erased from the landscape to be replaced with a concrete block look-alike (Photos 2.6 and 2.7). The question will always arise as to whether reconstituted cob is new build or a repair. There will be a tendency for building control officers to regard it as the former, so that the whole weight of Part A will be brought to bear. There are isolated examples in Dorset where an enlightened building control officer has allowed the original gravel bed to be utilised (personal communication, Rob Buckley), showing an understanding of the need for cob to retain an element of moisture, albeit that the exact amount cannot be proved. More unorthodox methods, such as tyres filled with gravel or concrete acting as a foundation for straw-bale buildings, will probably require even an even more enlightened approach by the regulators concerned.

Stone and timber versus the Building Regulations

Stone and timber are much more acceptable materials with regard to satisfying Part A (structure), Part B (fire safety), and Part C (resistance to contaminants and moisture). Stone is as old as mankind, has a pore structure that copes with precipitation, and often with the resultant salt mobilisation, but is more valued by traditional builders for its strength in compression.

Photo 6.10 Cob barn prior to reconstruction using the existing material (photo © Rob Buckley)

Photo 6.11 A more substantial section of barn complete with joinery of its period, where only the wall tops need reconstruction (photo © Rob Buckley)

Photo 6.12 Church Cottage, Droop, Dorset, rebuilt in 1939 in an age before building regulations

Unfortunately the whole ethos of using every scrap of the quarry stone for building a wall in random or regular uncoursed construction has long since vanished, replaced with the less satisfactory and eminently more wasteful practice of using stone as a rainscreen cladding to an inner blockwork wall. The latter appears to demand a slavish adherence to using ashlar stone in a regular formation, incurring considerable wastage at the quarry. The residue is frequently consigned to being crushed for roadstone. Stone has further suffered from the practice of small quarries being bought up and deliberately being shut down by large conglomerates, a practice which had also plagued small brickmakers and lime burners (at the hands of the cement industry), in the decades following the Second World War. It has taken the building conservation movement some time to pioneer the re-opening of small stone quarries in order to meet a shortfall in the material required for repair of important listed buildings. Even so there is rarely enough for new build, and thus much stone used in new build is imported from China and other Far Eastern countries. Stone that has been transported half way round the world can never be regarded as a sustainable building material, not only in view of the carbon emissions accrued but also because of the social inequalities which afflict its production.

Timber has even greater connotations in this respect, even if it is Forest Stewardship Council (FSC) accredited, the hallmark of a well-managed sustainable source. This cannot obviate the emissions expended in transport. Nonetheless it is a renewable resource, and was a key component of those buildings which now characterise the formerly wooded areas of Britain, usually on clay lands, the clay also being a key ingredient of the daub for the infill panels. Existing timber framed buildings inevitably have no membrane beneath the sill beam, which sits on a masonry plinth, and nearly all timber framed buildings need the sill beam and lower portions of the vertical studs replaced. The practice of putting a lead tray beneath the sill beam to prevent rising damp has been tried and deemed to be a failure, as the lead is prone to attack from the mobilised tannic acid in oak. The result is a steady attrition of the lead into ever-deepening ruts following the run of corrosion, culminating in total loss of the lead beneath. The approach may well satisfy Part C initially, but the long-term prognosis is poor. Better to use Hyload DPC or similar beneath the timber wall plate, although by doing so the ability of the timber to lose moisture into the masonry is inhibited, and this can lead to enhanced decay of the timber itself. Such a situation is characteristic of many building materials. Whilst solving one problem, the solution is creating another.

Another area of weakness in timber framed structures is the face plate of joints so formed by the creation of the mortise behind, but fortunately the frames are still regarded as having longevity. This avoids the requirement under Sections 19 and 20 of the Building Act 1984 not to use short-lived materials (materials that are, in the absence of special care, liable to rapid deterioration). Doing so can lead to the rejection of the intended building form, or the imposition of a clause for limited use, (after which time the building will have to be removed), or the imposition of a clause specifying restricted use.

Timber frames used in new build rarely utilise this ancient technology of mortise and tenon, although a recent building in Herefordshire (see photos 6.13 and 6.14) shows that it can and is still done. New build timber frames tends to use sheathing to the frame, providing a diaphragm to avoid any potential problems with joints between components. The principal problem with traditional frames is, in fact, more associated with Part L (conservation of fuel

Photo 6.13 New timber framed building in Herefordshire during construction

Photo 6.14 Detail of treatment of panels in the new traditional timber framed house, showing sheathing to take render, also lead as a damp course which may well be leached due to tannic acid from the oak

and power). The builder of the Herefordshire building illustrated pronounced that this was the last timber building he would be able to construct where the frame remained visible internally and externally, because of ever-increasing restriction on the U values allowed. The increasing demands of energy regulations stipulating a U value of 0.20 W/m²K will mean that internally the walls will require considerably more insulation. The example shown of this traditional form of construction utilises Styrofoam panels set in mastic, fronted by a mesh to hold a lime render, and making full use of mastics in an attempt to seal in the panel.

In concluding this section on materials and the Building Regulations, there is underlying all of the above the need to establish adequacy of workmanship. This is normally established by using the British Standard Code of Practice, as detailed in BS 8000 (workmanship on building sites). The aspects most relevant to sustainable building materials are:

- Part 3: 1989 Code of Practice for Masonry
- Part 5: 1999 Code of Practice for Carpentry and Joinery
- Part 6: 1990 Code of Practice for Slating and Tiling of Roofs and claddings
- Part 10: 1999 Code of Practice for Plastering and Rendering
- Part 11: 1995 Code of Practice for Wall and Floor Tiling (Section 11.2 1990 stone tiles and Section 11.1 1989 for ceramic tiles)
- Part 12: 1989 Code of Practice Decorative Wall-coverings and Painting

Parts 3 and 6 are undoubtedly the most useful to the sustainable builder, as the Codes of Practice are largely geared to modern building construction.

The Building Regulations and existing traditional building construction – material change of use

Conversion of non-domestic buildings to habitable use involving a material change of use, including the conversion of former outbuildings into domestic use adjacent to an existing

Photo 6.15 A Dorset house and barn about to be joined together to form a major complex, an exercise which will require stringent satisfaction of the Building Regulations

domestic building, invoke regulations which are seemingly at odds with sustainable concepts. Even a traditional cottage or farmhouse that has been out of use for some time or which requires a major overhaul may well fall victim to an over-zealous approach.

It is thus important to discuss the proposed scheme with the building control officer well in advance, putting forward ideas such as retaining thermal mass, using traditional haired or hemp-lime plaster for insulation, and retaining existing joinery and glazing. The retention of joinery is as much from a non-wastage point of view as any desire to retain historic artefacts, although clearly previous generations were equally motivated by the desire not to be wasteful, hence the survival of joinery from the 17th century. Retained thermal elements have to be upgraded to meet higher values. Again this can mean the loss of existing traditional windows and doors in timber, and this can be counter-productive to sustainable building, particularly if there is any degree of craftsmanship involved. Older windows tend to have intricate mouldings and be in wood that is slow grown with a closely set pattern of growth rings. This dense concentration of winter wood, dark in colour, is a store of resins or tannins that helps to preserve the wood, and in addition the close-ringed format enables the cutters to produce a crisp clean moulding. This is rarely possible with fast-grown softwoods, with a preponderance of summer wood filled with sap and the reason why so many modern window mouldings fail to please. In all material change of use an existing rooflight or roof window which has a U value worse than $3.3 \text{ W/m}^2\text{K}$ will require replacement. This does affect survival of Victorian cast iron rooflights, whose design has inspired the much vaunted conservation rooflight. It is regrettable to lose such features, but it has to be said that they perform badly in dwellings where so much warm moist air is now generated, and attract so much condensation that it drips onto the floor. The degradation of the ironwork is also more rapid, all in sharp contrast to an era when bedrooms were rarely heated, thus not holding moisture within warm air, and when bathrooms were an unknown quantity in cottages, the steam from the tin bath in front of the fire wending its way up the chimney. This change in lifestyle and its effect on the fabric of traditional stone and brick buildings is not fully appreciated by those who aspire to be in charge of their future, be they owners, architects, or surveyors.

Photo 6.16 A former 18th-century panelled bedroom in use as a bathroom with all the attendant problems that will be caused to the fabric by condensation

The building components largely affected are set out below. Generally there is a requirement to upgrade walls, floors, and roofs to meet current standards, with particular reference to insulation, air-tightness and control of airborne sound. These certainly pose problems for existing traditional construction, particularly thick stone walls designed to provide thermal mass and breathing walls, as the intention is now to provide them with insulation and render them airtight. Internal insulation invariably creates problems with any features projecting from the walls, such as the dados, dado panelling, and cornices associated with polite architecture.

Then there is the matter of consequential improvement. If a building has a floor area of greater than 1,000 m² (1,196 yd²) and the proposed building work includes an extension, or initial provision of any fixed building services, or an increase to the installed capacity of such, then the energy efficiency of the whole building consequentially must be improved. This can include replacing all existing windows or doors which have a U value worse than 3.3 W/m²K.

Photo 6.17 Modern extension following the form of the existing farm building which has been converted to domestic use – this change of use has to satisfy strict energy conservation requirements

Walls

All traditional buildings tend to have solid walls, either 225 mm (9 in) or 325 mm (13 in) if in brick, or considerably thicker, sometimes up to 600 mm (24 in) if in stone, often built without a damp proof course. This is something of a conundrum, as clearly rain is not a new phenomenon. In a period, however, when many roofs were thatched or even stone tiled, a generous overhang beyond the plain of the wall surface was the norm. This catapulted water into a drainage channel dug around the building, which if gravel-filled coped with the drainage of this water. Even in towns before the modern age it was common to have a leet down the centre of the roadway into which water from roofs was directed down a camber, helping to wash away the detritus of occupation. Unfortunately all of these concepts seem to have been lost in the mists of time, so that there is a somewhat unrealistic reliance on gutters and down pipes. These are often no longer robust cast iron, but are constructed of plastic that expands and contracts on the joints with changes in temperature, goes brittle and breaks with UV radiation, and is also often of an inadequate profile to cope with current rainfall related to climate change. Add to this the preponderance of many layers of tarmac on pavements in towns, steadily creeping their way up the plinth of buildings, completely disallowing any possibility of drainage away from lower courses of masonry. All is a recipe for disaster, and whilst one would not want to advocate the rebirth of the leet, there is an urgent need for porous paving in towns and for the construction of French drains around rural detached buildings. This is an eminently sensible device with its perforated pipe set in pea gravel (essential to create multiple mini-waterfalls) and leading down from one corner to a soakaway, and is by far the most effective means of creating a dry building platform for existing buildings.

Some traditional buildings are timber framed in construction. Found in areas where the clay subsoil created ideal conditions for oak forests, which were in turn cleared to produce agricultural land, they can date from the medieval period to as late as the early 1800s. The scantling of the uprights and cross rails steadily diminished in size as the timber became less abundant throughout this period.

The main requirements of the Building Regulations are that such buildings should now meet standards for:

- basic requirements for stability
- conservation of fuel and power
- airborne impact and sound.

There is usually little difficulty with stability, especially as regards masonry structures. Their external and internal walls readily form the required 'robust 3-D box structure in plan, with walls adequately connected by masonry bonding', but a great deal of misconception still reigns regarding the binder between the building blocks and its ability to contribute to stability. There are still factions that condemn clay binders and even lime mortar, both of which are eminently sustainable in that they have not required carbon-inducing processes to any great extent.

Conservation of fuel and power, closely allied to thermal performance, is a vexed issue with traditional buildings, which is why a whole chapter has been devoted to it. Traditional masonry buildings perform in a wholly sustainable way, in that their walls are constructed of a porous material. In the case of brick there are many small pores, and with stone there is usually a variety of pore structure. This means that such walls act as a thermal store, each pore holding heat in the inner part of a wall, to release it at the point when rooms are cooling down. Of even greater importance is the ability of such walls to absorb warm moist air and wick it away via a zone of interstitial condensation that exists deep within the core of thick masonry walls.

In contrast, the Building Regulations require that reasonable provision be made for limiting heat gains and losses through thermal elements or other parts of the building fabric by providing insulation, usually on the inside face of the wall. This normally consists of a cavity to give a break in the path for moisture, a breather membrane, followed by a sheathing board and framework to hold insulation, completed by a vapour control layer. The latter is deemed necessary as soon as insulation is applied in order that the latter does not have its pore structure disrupted by moisture ingress. All this is very laudable but it immediately negates the function of the actual masonry wall as a heat store or as a means of wicking away internal warm moist air. External insulation is of course possible, but this often means that the character of the building is obscured completely or certain of its elements, such as window and door openings, are set within deep recesses, creating a very different ambience of light and shade. It is particularly problematic with Georgian buildings in this respect, creating all manner of hiatus in an otherwise unblemished elevation.

Other aspects of fuel conservation are easier to comply with, such as limiting heat gains and losses from pipes, ducts, and vessels for space heating/cooling, and hot water services. Providing and commissioning energy efficient building services with effective controls, and providing the owner with sufficient information to enable the building to be operated efficiently, is eminently sensible. It is these measures, as well as providing enhanced insulation in areas where breathing walls will not be affected, that should take precedence.

Airborne impact and sound seems at first to be an almost negligible issue, particularly as the regulations state that the amount of sound resistance provided by a wall depends on its construction. Traditional buildings almost always have thick walls, even dividing the internal layout, which should readily absorb sound yet the regulations state the following. If the existing wall is masonry and has a thickness of at least 100 mm (4 in) and is plastered on both faces, then it requires an independent frame (gap of 35 mm, 1½ in) clad with mineral wool as an absorbent material (tightly compressed), with two layers of plasterboard. The independent panel should not be in contact with the existing wall. Again, such walls treated this way are unble to absorb warm moist air and act as a thermal store, thus negating their traditional role as breathing walls, and much negotiation with the building control officer will be necessary. There are instances, of course, when there is a need for such a treatment, for example when fire risk is paramount. If a wall is common to two or more buildings, then it must adequately resist the spread of fire between the buildings and the spread of fire over these surfaces.

Photo 6.18 This rare early 18th-century house, with a façade designed to create light and shade, would be impossible to insulate externally, and in all events the thick stone walls act as a thermal store

Ceilings and upper floors

A first floor constructed of wooden joists and wooden floorboards, supported on solid main beams, is both a joy to behold and eminently sustainable. In the medieval period to around 1700 such timber would have been home grown and travelled very little distance. After that time it would have been imported from the Baltic States, so having travelled perhaps thousands of miles it is even more incumbent upon all later generations to ensure that the wood remains.

Traditional buildings relied upon natural processes such as hot air rising to heat upper floor, bedroom fireplaces being utilised only for the sick and infirm. Gaps between boards may have allowed this to happen, but generally wood is an excellent insulator and helps to keep warm air trapped in the ground floor rooms. Of greater significance in heating upper floors was the massive brick chimney breast, which is to be found in most traditional master bedrooms; heated constantly by a fire which never went out, summer or winter, night or day, it acted as a vast storage radiator. So significant was this that many such bedrooms had the chimney breast as the headboard, and encapsulated this warmth in a bed hung with curtains and a fabric ceiling (tester). Servants and children were considered to be hardier souls, requiring very little heating in sleeping chambers.

Two aspects cloud this romantic picture, that of resistance to airborne and impact sound, and the need for compartmentation in respect of fire spread when change of use is contemplated.

Again the regulations state that the amount of sound resistance will depend on the combined mass of the existing floor and any independent ceiling, plus the amount of absorbent material. This also depends on the isolation of the independent ceiling and the airtightness of the whole construction. Thus traditional buildings can be doomed to failure, as their very qualities of breathability are what ensure that they survive. It is a basic requirement of all traditional building construction that it is able to wick away warm moist air, and replace it with less contaminated air, this latter action normally being achieved through the operation of the traditional fireplace. This achieved something in the region of four air changes an hour. Fortunately where historic buildings are undergoing a material change of use it is realised that it is not possible to improve sound insulation to the standards set out in Approved Document E, particularly if the building has very special characteristics, such as decorative wall features. In these circumstances it is understood that the aim is to improve sound insulation to the 'extent that it is practically possible'.

Two types of floor treatment are recommended by the regulations. If the existing floor is timber, the gaps between the boards should be sealed by overlaying with hardboard or filling the gap with sealant. Unfortunately most sealants are by their very nature non-reversible and can only result in long-term damage to boards. There is even a suggestion that the floorboards can be replaced with boarding of minimum thickness of 12 mm ($\frac{1}{2}$ in) with mineral wool laid between the joists in the floor cavity, resulting in the loss of floorboards, a sustainable resource.

Some traditional buildings of the 18th and 19th centuries have independent ceilings, which allows mineral wool (why not sheep's wool?) to be laid in between the joists (minimum thickness 100 mm/$\frac{1}{3}$ in), minimum density 10 kg³/22 lb³), but this has to be accompanied with two layers of plasterboard with staggered joints. Should this be insisted upon in relation to a ceiling with decorative ceiling rose and cornice, then the resultant architectural havoc is not difficult to imagine. That there have been such instances in the name of fire protection in change of use of country houses to elderly persons' accommodation is well known. The practice has been standard, regrettably, since the 1970s. Independent ceilings can of course be created below an existing ceiling, leaving the latter archaeologically encapsulated, in which case they can be fixed by independent joists attached only to the surrounding walls. Alternatively, the regulations state that additional support should be provided by resilient hangers attached directly to the existing floor base and the perimeter of the independent ceiling should be sealed with tape or sealant. Neither of these solutions is in any way sustainable, and the latter can be problematic in historic buildings if it adheres so tightly that removing it at a later date will cause damage to the plaster. A problem arises where a window head is near to the existing ceiling, as is often the case in grand Georgian rooms. The regulations state that the independent ceiling can be raised to form a pelmet-recess, but this is not always possible as the regulations also state that there should be a 25 mm (1 in) gap between the top of the independent ceiling and the floor base above.

Another way of dealing with the problem is to create a platform floor above the decorative ceiling with absorbent material. Again mineral wool is recommended between the joists with a floating layer over the boards of minimum thickness of 8 mm ($\frac{1}{3}$ in) fixed together (spot bonded or glued/screwed) and laid on a resilient layer of mineral wool (minimum thickness 25 mm/1 in), paper faced on the underside and with joints tightly butted and taped. Gaps between floating layer and skirting are again recommended for flexible sealant (see comment above). There are in fact a number of patented systems on the market for the resistance of airborne sound in floors. Most use manmade materials such as polystyrene, and few have green credentials.

The conclusion can only be that if change of use is going to lead to measures that will result in the loss of intrinsic merit, and potentially the long-term demise of the building because it cannot deal with warm moist air and the resulting sustainability being heavily compromised, then a rethink of the whole scheme is urgently required.

Fire resistance, compartmentation, and sound resistance

The upgrading of buildings, either for conversion of non-domestic to residential use, or even buildings which are very run-down and require extensive rehabilitation, may well fall victim to the arduous requirements associated with fire compartmentation. The aim of these regulations is laudable – the saving of life over buildings – but the effect on the building's ability to function as a traditional building as described above can be severely compromised. The requirements revolve around the need to sub-divide the building into compartments separated from one another by walls and floors of fire-resisting construction, whenever possible. If parts of the building are occupied for different reasons, this is essential. In particular the wall and floor between the garage and the house must have a 30-minute fire resistance, with any opening having a 30-minute fire door. Flats and maisonettes have even more rigorous requirements for compartmentation between each unit and refuse storage chambers. The tables in Approved Document B need to be consulted for appropriate fire resistance. This is very problematic for existing country houses being converted into flats. Either the floorboards need to be sealed beneath with mineral wool and fireline boarding, or the gaps between joists packed with fire-resistant material and the ceiling (which may have decorative plaster ceiling roses and cornices) undercloaked with two sheets of plasterboard or fireline board. The resulting appearance is a grave disappointment in a once fine house Georgian house displaying elegant ceiling/cornice decoration. In the case of a substantial yeoman farmhouse that has a spacious roof space previously used only for storage which needs to be brought into domestic use, then the requirements become very rigorous. The floors need to have a 30-minute fire resistance, and all load-bearing elements, to include structural frames, beams, and floor structures, need to have a minimum standard of fire resistance as set out in Approved Document B. Junctions between a compartment wall and an external wall need to be restrained to reduce the movement of the wall away from the floor when exposed to fire. This will almost certainly involve lifting floorboards, an exercise which frequently results in their damage. It has been known for boards to be totally scrapped in order to make way for chipboard in such situations! In addition, compartment walls should be able to accommodate the predicted deflection of the floor above, by either having a suitable head detail between the wall and the floor that can deform but maintain integrity when exposed to fire, or the wall redesigned to resist the additional vertical load. There is also a need for the creation of a protected shaft from the attic floor down to the ground floor exit (see below for fire doors and corridors). None of these is a good option when dealing with historic fabric of great integrity and eminent sustainability. Also the use of materials such as fireline board and mineral wool, or even plasterboard, can hardly be described as 'green'.

There is a good case for examining carefully the possible harm of conversion of a building to flats, especially the loss of decorative features or sustainable building materials such as lime or wood which will never be created again (close-ringed softwoods are very difficult to source, and existing lime plaster walls are a sustainable resource), against the sustainability argument for making better use of space and satisfying a need for housing demand.

Ceilings generally in conversions can retain their lath and plaster cladding providing the building control officer deems it satisfactory to satisfy Part B (fire safety), but again it is normally a requirement that it should be upgraded with two layers of plasterboard with staggered joints, total mass per unit 20 kg/m^2.

One of the greatest losses in existing buildings being converted to flats or elderly persons' accommodation is internal doors. For generations of conversions their loss has been total, having been replaced with the ubiquitous fire door, involving not only the loss of their historic integrity and craftsmanship but also the green material of wood. Even door frames are not spared, as in all such situations they need a good perimeter sealing, although intumescent strips can be inserted around door frames and doors. In addition, doors can now be encapsulated in a form of intumescent wallpaper, which bubbles up to form a seal around and across the whole door surface when heated in a fire situation. The problem is compounded by the need for doors in a flats-conversion situation to meet sound insulation standards, inevitably meaning even more intervention in the case of thin panelled doors. The loss of such doors is non-sustainable.

Historic staircases in existing buildings can range from the simple newel stair to the elaborate dog-leg complete with decorative balusters and handrails. Sound insulation on doors to cupboards under is also a requirement for staircases, particularly if they are ascending from one functional area to another. The resistance to airborne sound depends on the mass of the stair, the mass and isolation of any independent ceiling, the airtightness of any cupboard under the stair, and finally the stair covering. Here at least is something which is non-invasive and can be done using a thoroughly green material such as jute or sea-grass. It must, however, be a minimum of 6 mm (⅜ in) thickness, be laid over the stair treads completely, and be securely fixed so that it does not become a hazard. Gluing is the example given, but glues are inevitably engineered from strong hydrogen-carbon compounds, e.g. formaldehyde, and often laced with nitrogen. They are disruptive to human tissue, particularly if they off-gas. In all events no one should seriously countenance gluing a stair carpet to an historic staircase as it is non-reversible. Far better to resort to the old established method of carpet tacks.

Cupboards under stairs are a traditional form of storage in what is otherwise a dead area. Regulation demands that they be lined with plasterboard together with an absorbent layer of mineral wool, immediately below the underside of the stair, with the wall similarly treated with plasterboard. If there is no cupboard, then an independent ceiling needs to be constructed under the stair, almost impossible if the existing stair has a curvilinear format already lath and

Photo 6.19 Wreathed handrail

Photo 6.20 17th-century staircase, a problem regarding insulation and airborne sound due to the very nature of its form and construction

271

plastered. A strong case needs to be made for retaining the existing traditional lime-plastered underside.

Staircases are seen to be joining one separately functioning area from another within a fire-protected shaft so that all doors opening onto the staircase need to be fire doors. Again, intumescent coatings on existing doors are eminently more sustainable than replacing doors with purpose-made fire doors.

Junctions also have a considerable role to play in containing airborne sound and fire. Junctions with abutting construction, i.e. the perimeter of a new ceiling, junctions with external or load-bearing walls, and junctions with floor penetrations such as piped services (excluding gas pipes which can be contained in a separate ventilated shaft) and ducts all need to be surrounded by sound-absorbent and fireproof material. The use of sealants and mineral wool abounds in the advice. The impact of these materials on the building and the project needs to be anticipated before making decisions on the wisdom of conversion.

In conclusion, a material change of use of a building is a multi-layered beast. It can cover many scenarios, including change of use from non-domestic to dwelling, conversion to flats, boarding house, hotel, institution, public building, and conversion to more than one dwelling, conversion to shop, bar, restaurant, or any other form of public building. In addition to the approved documents referred to above and those listed below, any change to public use will incur a requirement to satisfy Approved Document M (access and use). This will mean attention to how access is gained from the site boundary and any on-site car parking space, plus sanitary conveniences suitable for the disabled. Of greater import to existing buildings is the requirement that change of use brings for thermal elements, in particular joinery. Any thermal element needs to be upgraded, and this includes existing windows (and rooflights) or doors separating the inside from external space. If these elements have a U-value less than 3.3 W/m^{2}K, then they will have to be replaced. Even more careful thought needs to be given to weighing the advantages of a new use against the potential loss of sustainable building construction, and in particular wooden windows.

Generally, material alteration to domestic buildings such as substantially replacing a thermal element (especially windows), renovating a thermal element, or making an element part of the thermal envelope (where previously it was not), now need to be addressed. Also included is providing a controlled fitting or service, for example central heating. This is all under the aegis of Part L. Immediately existing windows, however historic or interesting, are affected and will end up being scrapped, and existing breathing walls, very much part of the thermal envelope, will require insulation, thus removing their ability to act as breathing walls.

In addition, all the requirements of the following Approved Documents must now be taken into account (only those having an impact on building construction of domestic buildings are listed here):

- Part A (structure)
- Part C1(2) (resistance to moisture)
- Paragraph B1 (means of warning and escape)
- Paragraph B2 (internal fire spread – structure)
- Paragraph B3 (external fire spread)
- Paragraph B4(2) (external fire spread – roofs)
- Paragraph B5 (access and facilities for the fire service)
- Part E (resistance to the passage of sound)
- Part F1 (ventilation)
- Part L1 (conservation of fuel and power: now divided into L1A – new dwellings, and L1B – existing dwellings)
- Part M (access to and use of buildings)

In practice much of the above means that existing buildings, the retention of which is eminently sustainable, as are many of their components, such as windows, will not be able to be retained. The windows will need to be sealed units and existing single-glazed windows cannot be adapted to hold them. The glazing bars of traditional windows are very thin. This became a fashion in the 18th and 19th centuries and was also related to the thin muff (cylinder) and crown glass that was available at the time. This hand-blown glass also becomes the victim of removal, to be replaced by sealed units that have a minimum U value 3.3 W/m²K.

Many of the above requirements now apply to all existing buildings where a modification is being made to one part of the building in order to meet the requirement to generally improve thermal performance.

The Building Regulations and new construction

Foundations and floors (Approved Document A)

All new buildings must meet the following construction requirements:

- The combined dead, imposed, and wind loads are sustained and transmitted to the ground safely without compromising stability.
- Ground moisture caused by swelling, shrinkage, or freezing of subsoil, land slip, or subsidence will not affect stability of any part of the building.
- Precautions are taken against contaminants on or in the ground covered by the building.
- Adequate subsoil drainage to avoid the passage of moisture to the interior and damage to foundations via waterborne contaminants of a corrosive, flammable, or radio-active/toxic nature.

The normal requirement is for a damp proof course (DPC) of bituminous material, polyethylene, engineering brick, or slates in cement mortar, continuous with a damp proof membrane (DPM). The former needs to sit at least 150 mm (6 ft) above ground floor level. For a cavity wall, the cavity should be taken down at least 225 mm (¾ in) below the level of the lowest damp proof course, with a damp proof tray to prevent water getting to the inner leaf.

This raises a whole host of issues for the sustainable building construction.

One of the most popular forms of flooring in this realm is the limecrete floor. Whilst it is frequently accepted in the replacement of an old floor in an existing building of historical significance, it is often taboo for new construction. It is simply thought too insubstantial or even 'hocus-pocus' to meet the stringent requirements of Part A for foundations set out above, although it now has Local Authority Building Control Approval. What is frequently forgotten is that lime is essentially the same type of material as cement, except that it does not contain clay and that it is burnt at a lower temperature in order to release carbon dioxide. This carbon dioxide is then re-absorbed on setting, so reducing the carbon footprint. Cement, in contrast, requires intense burning so that it becomes highly reactive. Instantly one can see why lime is so much more beneficial to the environment. There is also mystification as to why the under-layer consisting of hard clay balls or foamed glass is damp proof. The answer is simple: there are no fines to encourage extensive wicking upwards of moisture, and too many voids for water molecules to be attracted upwards, but this relies on a good understanding of the nature of the water molecule and in particular the strength of the hydrogen bond.

The analogy is as previously discussed in Chapter 4 – that of placing test tubes of various diameters in a bowl of water. A fat test tube will not result in as high a water level within as a thin test tube. This is because the insides of the tubes are oxygen-rich. The pull of the oxygen molecules, lining the internal surface of the thin tube, upon the water molecules, via the hydrogen bond is much greater than the pull of those lining the fatter test tube. In the latter there is much more free oxygen above the level of the water, which means that there is less pull from the walls of the tube.

In addition, most limecrete floors utilise hydraulic lime, formed by burning a limestone rich in silicates and aluminates, the properties usually of a clay content. The burning process renders the latter more reactive and a close bond is formed between the calcium content and these constituents (calcium silicate and calcium carbonate). This means that they in effect become reactive with water when taking on a cure, giving some of the properties of Portland cement (OPC), including a faster set. This is always useful when laying a floor, and has an analogy with the ancient practice of adding sour milk to a traditional lime and aggregate floor to make a lime-based concrete. The milk contained casein which induced a bond between this and the calcium (calcium caseinate), which hastened the setting process.

The limecrete floor is the ideal choice for sustainable new build utilising materials such as straw, wood, or cob, materials for which vapour permeability is essential as it creates a breathing floor capable of flexing with the building (see Chapter 7 for a detailed description).

Another area of potential conflict with the requirements of Part A for foundations and floors is the whole issue of using unbaked earth. Cob, adobe (unbaked earth blocks), and rammed earth all avoid the whole issue of cementitious products that require burning. Unfortunately, the very basis of such walls requires a level of moisture content so that the power of the hydrogen bond can actually exist to hold together the particles of earth in order to avoid the walls turning to powder. In addition, a certain level of natural drainage is required in the foundations of such walls so that the walls are not completely saturated and turn to 'porridge'. The normal cement strip foundation would thus be taboo, as would any adherence to the principles set out in Part A. It is thus little wonder that there are relatively few unbaked earth buildings, apart from the odd children's play shelter or bus shelter. All are deemed buildings for infrequent human use. Traditional cob buildings in Britain were given a foundation of stone or flint or brick or some mix of these, which continued upwards into a plinth and provided just the right qualities of ebb and flow of water to keep the wall at the right moisture content.

Straw bale buildings offer somewhat greater flexibility in the choice of foundation, including local stone, concrete blocks, bricks, or even tyres filled with crushed stone aggregate. The possibilities are endless, and the whole situation less fraught than with cob. With such a highly insulated wall there is further encouragement to make sure that the plinth walls and the floor are also well insulated.

Timber framed buildings are as traditional as any known to mankind, and in this respect are on a par with unbaked earth. Early buildings (post-Roman) utilised misshapen boughs tied together with vines, but the technology became rapidly sophisticated and by the Middle Ages the mortise and tenon was the basis of joining together timbers on the ground in the form of A-frames etc., to be then raised from the ground. Other cultures specialised in round pole construction to make log cabins, which depended on notched joints for their stability. In all these situations it is fatal to induce too much rigidity into foundations. Traditional rubble foundations using stone and lime, as opposed to a concrete strip foundation, will actually perform better, if the building control officer can be persuaded.

All of this contrasts sharply with the normal requirements of the regulations for floors set out in Approved Document A. The normal expectation is that of a ground floor of concrete or a suspended timber floor built on sleeper walls. The former has a damp proof membrane, normally a polythene sheet, the latter has the sleeper walls built on over-site concrete between the brickwork foundations and plinth. Timber sleeper plates rest on each wall, supporting the joists for the floor. The ends are commonly suspended on metal hangers in a modern build, but in the 19th century were let into the masonry, a fatal practice which can lead to dry rot. The practice of using beam and blockwork floors has largely superseded this. Floorboards are now rarely used, having given way to chipboard, a material which is saturated with formaldehyde. Materials such as cement, burnt at over 1,000 °C, create CO_2 emissions, and polythene is derived from petroleum distillates. All contrast unfavourably with natural materials.

Size of residential buildings (Approved Document A)

There is a requirement that the maximum height of a building for residential purposes, measured from the lowest finished ground level to the highest point of any wall or roof, should be less than 15 m (2 yd). In addition, the height of the building should not exceed twice the width of the building, and there are also rules governing projections, which in effect also cover extensions. The import of these regulations on constructions made of cob or straw bale is of little relevance, as they tend to be one- or two-storey at the most. Timber framed buildings are also traditionally similar in their scale and proportion, although so-called American balloon framing, employing an element of pre-stressing in every floor, can be three storeys.

Floors (Approved Documents A, B, and E)

Suspended timber floors on sleeper walls in theory require to be sealed, leaving their original ventilation of sub-floor function in jeopardy. In reality it means that these floors now need to have insulation between the joists held in place with plywood sheets, or plastic mesh, vented on the underside to prevent moisture inducing decay. Suspended floors also need to satisfy Part B for fire safety and have cavity barriers for the same reason.

Floor detailing can be self-certificated by consulting the website for Robust Details Ltd (www.robustdetails.com), developed in 2004 as an alternative to the need for compulsory sound testing to manage, monitor, and promote robust detailing to satisfy sound testing requirements (pre-completion sound testing) and ensure compliance with the Building Regulations. Intermediate floors need to resist airborne sound, such as speech, musical instruments, loudspeakers, footsteps, and furniture being moved about.

Many ground floors will be intended as solid floors, and thus must be constructed so that the thermal transmittance U value does not exceed 0.7W/m2K at any point. In fact, ceramic tiles absorb solar heat during the day, hold it like a storage radiator, and release it at night. Solid floors lose heat more rapidly from the perimeter than they do from the central core; this is because the ground in the centre of the floor heats up over time. It thus follows that perimeter insulation is advisable. It is also pertinent that a large ground floor area with an open-plan layout is a better than several small rooms cold-bridging with the ground.

Intermediate floors have a requirement to be attached to the external walls by tension straps, so that they provide horizontal diaphragms. All such floors must meet the minimum levels of fire resistance, as per Approved Document B, usually 30 minutes. Joists are hung from the inner leaf of masonry to avoid penetration of the inner leaf that would create an air path for infiltration of cold air.

Walls (Approved Documents A, B, and C)

The Building Regulations allow for a range of interpretations of possible structure to include the older style solid walls of 225 mm (9 in), cavity walls with two leaves braced with metal ties, or modern timber frame. Nowhere is straw bale or earth mentioned. The internal wall of cavities can be load-bearing brick or blockwork, or non load-bearing lightweight blocks, manufactured boards, or timber studding.

Materials used for internal linings must restrict the spread of flame and the amount of heat released. In truth a lime plaster does perform this function well, but the standard advice is to use gypsum. A wall common to two or more buildings has to be designed so that it adequately resists the spread of fire between the buildings. This is only common sense. External walls must be so constructed that they have a low rate of heat release, and thus be capable of reducing risk of ignition from an external source and spread of fire over the surface. Some building control officers may argue that a coating of lime render on straw bales may not adequately resist the spread of flame. Weatherboarding may also be suspect here unless it is backed by a rather less green material such as fireline board that can resist the spread of flame.

The main need for walls is to protect the people who use the building from the harmful effects of ground moisture, precipitation, and wind-driven spray, as well as interstitial and surface condensation. It is the need to satisfy all of these requirements that renders green materials such as stone and cob at odds with the requirements. Stone does a good job generally but cob needs not to resist ground moisture, the lack of which can threaten its very existence.

The Building Regulations fortunately realise that the use of materials such as urea formaldehyde foam, used for cavity wall insulation, should not be allowed to penetrate occupied parts of a building where it can become a health risk to persons in the building, by virtue of its concentration. What is not revealed is how this can be gauged. Does one have to wait until someone is ill? Urea formaldehyde is in theory rigidly controlled. It should be manufactured in accordance with BS 5617:1985 and installed in accordance BS 5618:1985 and the Approved Installer Scheme.

Walls also have to be designed so that the noise from domestic activity and the transmission of echoes from an adjoining dwelling or other parts of the building is kept to a level that does not affect sleep, rest, and engagement in normal activities (Parts A and E). This requirement can surely be reasonably met in sustainable construction, such as straw bales and the use of lime, hemp, or clay plasters. Existing stone or brick buildings with lime plasters should also be regarded as superior in this respect, but much depends on the view of the building control officer. New build is almost entirely geared to the requirements for the conservation of fuel and power (Part L), involving lining out walls which limit the ability of lime plaster and stone walls to act as a breathing walls. Masonry provides thermal mass so that brick, stone, and earth provide passive cooling systems in the summer and thermal stores in the winter. Foil-backed plasterboard actually stops masonry from performing these vital functions. A standard requirement appears to be to introduce yet another internal layer of insulation, complete with a vapour barrier, thus negating the ability of the walls to take in moisture and send it outwards. In addition, the ability of the walls to act as a heat store, by virtue of storing heat in the molecular structure of the internal surface, is also severely compromised. They are regarded as transmitting all the internally gained heat to the outside, which is simply not the case. In this respect the requirement that reasonable provision be made for limiting heat gains and losses through thermal elements can be a problem for sustainable building materials. Other requirements include the more reasonable guarding against heat losses through pipes, ducts, and vessels used for space heating and hot water services, and providing and commissioning energy efficient building services with effective controls (although where this leaves the humble solid fuel Rayburn, Aga or wood-burning stove, whose control is invariably the human hand, it is difficult to say). Of key interest is the requirement to provide the owner with sufficient information regarding the building services and maintenance requirements, so that the building can be operated in such a manner as to use no more fuel and power than is reasonable (Part L). The latter is surely the aim of most sustainable building constructions, the behaviour of the materials being the key factor in achieving this aim.

There is a general requirement set out in Approved Document A (A1/2 Section 2C3c) that walls should satisfy BS 5628:Part 3:2001. This refers to the Code of Practice for masonry materials and components, design and workmanship and, interestingly, thermal properties.

Approved Document A includes basic requirements for stability. These include the need to form a robust 3-D box structure in plan, with internal and external walls adequately connected by masonry bonding or mechanical fixing. Residential building height should be no greater than 15 m (16 yd) and in residential buildings up to three storeys, and all walls be they external, internal load-bearing, compartment and separating walls, must extend up to full height. Thickness of wall is more flexibly couched in terms of roof loading, wind speed, and floor area, type of materials, loading, opening and overhangs, etc. This could thus be applied to a wider range of sustainable building materials. There is a requirement that solid external walls and compartment walls in coursed brickwork or blockwork should be one-sixteenth of the storey height in thickness. Some concession to ecological build is evident as there is a mention of construction in uncoursed stone, flints, and clunch (a type of soft limestone) needing to be not less than 1.33 times the thickness required for brick or block walls.

Solid external walls can still be constructed, this being the normal form of traditional sustainable building construction. There is a requirement for them to hold moisture arising from rain and snow until it can be released in a dry period without penetrating inside of the building or causing damage (Part C). Solid external walls exposed to very severe conditions should be protected by an external impervious cladding, or built with stonework/brickwork at least 328 mm (13 in) thick, or dense aggregate concrete blockwork at least 250 mm (10 in) in thickness, or lightweight aggregate blockwork at least 215 mm (8½ in). Blockwork requires a two-coat render with a total thickness of 20 mm (¾ in), with a mix of 1:1:6 (lime: cement: well-graded sharp sand) unless the blocks are dense, in which case the mix is 1 cement:½ sand. It is noted that there is no mention of traditional 1:3 (lime:well-graded sand mix), or of natural hydraulic lime (NHL), a regrettable omission. Solid external walls can be insulated on the outside or on the inside. Where the insulation is on the inside, a cavity should be provided to give a break in the path for moisture. Where the insulation is on the outside, it should be well shielded from any moisture ingress by a rain screen cladding to ensure that there is no degradation in the properties of the insulation as a result of water ingress.

Cavity walls are required to be at least 90 mm (3½ in) for each leaf thickness, with a cavity of 50 mm (2 in) wide, with wall ties having a horizontal spacing of 900 mm (35 in) and a vertical spacing of 450 mm (18 in). Parapet walls have equations which need to be satisfied to calculate the wall thickness. Vertical restraint can be provided by intermediate buttressing walls, piers, or chimneys dividing the wall into distinct lengths within each storey.

Chimneys, a seemingly long-forgotten feature of all domestic buildings, are fortunately still catered for and should measure at least twice the thickness of the wall, measured at right angles to the wall. They can act as passive stack ventilation.

Openings and recesses, determined in terms of their number, size, and position, should be adequately supported. Most building construction utilises concrete lintels, which are manufactured to an exact specification. Timber appears to pose particular problems with calculations, so unless this sustainable material is actually part of a timber framed construction it is usually eschewed. Interestingly, there is a requirement for no openings in walls below ground floor, except for small holes for services and ventilation. This appears to preclude the construction of the traditional cellar entered externally, a key feature of sustainable building construction reliant on the thermal flywheel effect for cold storage and for below-ground heat stores. A key factor is that joints between walls and door frames or window frames must resist rain penetration in order to prevent damage to any part of the building. This is only right and sensible, but the materials invariably used are not in terms of sustainable building. They are invariably mastics of various petroleum-based types, or urethane foam. The latter appears to

Photo 6.22 Sustainable masonry building but with glue joints, which therefore cannot be regarded as totally sustainable

Photo 6.23 New brick townhouses in Oswestry, Shropshire, showing the form of sustainable building which will ultimately permeate English townscapes

Photo 6.24 Another part of the Oswestry complex, fronting onto and fitting into the historic street scene

have taken the building construction world by storm, but once present has no qualities of reversibility and adheres to building components for dear life, rendering them totally useless for recycling. The role of traditional lime putty mortar or linseed oil-based putty acting as a mastic needs to be investigated as a good substitute.

DPCs and cavity trays must be provided to ensure that water around lintels, under cills, and at abutments between wall and roof drains outwards.

There is definitely a problem with mortars, however. Many sustainable materials rely heavily on the use of lime mortar as an equally sustainable material, but there is a directive to use 1:1:5 or 6 (cement:lime:fine aggregate). Not only is cement not a sustainable material due to the intense burning it requires, but mixing cement and lime together is introducing both a chemical set with water and a chemical set with carbon dioxide into a mix where these different chemical reactions may well interfere with each other. The use of a fine aggregate is also questionable.

Photo 6.25 Purpose-built passive stack ventilation on a sustainable construction

Lime mortars require grades of aggregate which tessellate together, each coated with lime to the thickness of a coat of paint. Not only does this make for a mortar capable of taking a loading, but the presence of the larger particles of aggregate in the surface of the joint provides a robust weathering surface. The remainder of advice about mortars is the standards associated with cement, such as BS 5628 and BS EN 998–2. Stone itself is specified as squared and dressed, which may help to explain the vast acreage of quarry waste sent to be crushed for road stone, when in reality this stone could be used for random uncoursed construction in time-honoured tradition. Various British Standards are quoted in order to ensure minimum compressive strength.

In all new construction, gable ends are required to be strapped to roofs by tension straps of galvanised steel, also to strap walls to floors above ground level at intervals not exceeding 2 m (6½ ft). This heavy reliance on galvanising is questionable as in time it will be corroded away, being a sacrificial element, as will the steel, which must also have a limited lifespan once the galvanising has failed and the steel meets oxygen and water together. Such metals set up an electrolytic cell that will in time ravage the straps on which the construction depends.

Framed external walls are increasingly becoming the way forward for sustainable building construction. They are heavily dependent on insulation to minimise heat loss, and can be given a cladding such as weatherboarding which is eminently sustainable. The cladding needs to be separated from the insulation or sheathing by a vented and drained cavity with a membrane that is vapour-open but resists the passage of water on the inside of the cavity. The normal modern construction consists of a brick outer skin, then a cavity, then the frame itself, clad with a sheathing board coated in a breather membrane, with insulation between the frame members, then a vapour barrier. Sustainable construction frequently replaces the masonry skin with renewable wooden boarding and the insulation with sheep's wool or a similar naturally formed product capable of breathing out the internal warm moist air, thus dispensing with the need for a vapour barrier. The main disadvantage of the latter is that it traps warm moist air within the structure and does not allow the building to breathe, plus the fact that it is frequently

Photo 6.26 Steelwork inserted into a traditional building possibly has only a limited lifespan compared to the already great longevity of the existing building.

punctured during the construction process, thus creating paths for moisture to infest the insulation and render it useless. Sustainable construction utilises a stud framework sheathed externally with a wood fibre sheathing board, and internally with similar, with the core being Warmcell (cellulose) or sheep's wool. The whole is constructed to ensure exchange vapour transfer. The same is true of loft spaces. The use of sheep's wool or Warmcell removes the fear that warm moist air will seep into more conventional insulating materials, such as rock wool, via such apertures as loft hatches, and degrade the insulation. With natural materials the warm moist air will simply continue along its path out through the eaves' ventilation.

Cladding systems for timber framed structures are very much to the fore. The regulations state that cladding systems should resist rain, be joint-less or have sealed joints, or if designed to have overlapping dry joints should be impervious or weather-resisting. In addition, there is a requirement for a backing material which will direct any precipitation towards the outer face. These forms of cladding suggested, to include metal, plastic, glass, and bituminous products, do not pay much heed to sustainability, unlike the timber boarding.

There is a ready market for plastic look-alike boarded cladding, and realistically it is difficult to resist as it has greater longevity than wood. Alas, it suffers many of the same problems, becoming coated with green algae, which is difficult to remove, and will eventually pit and degrade with UV light, as well being lacking in breathability. It fares better than wood on the overlapped joints, but at what cost the environment, particularly when disposal has to be contemplated? Metal cladding is used for all portal framed constructions on industrial estates. Such buildings frequently do not need heating, but metal is hardly suitable for a domestic building that needs to retain heat. Weatherboarding is eminently more sustainable as it is renewable and allows buildings to blend in with the landscape. Of greater benefit are weather-resisting claddings, such as natural stone (although the wastage discussed above is relevant) and slate hanging, the latter being one of the most attractive and serviceable claddings known to man, with a time-honoured pedigree, albeit that it eventually suffers from nail sickness. Tile cladding is in the same category as wood, and recyclable. Very few buildings can accommodate

Photo 6.27
Modern
timber frames
in the form of
sheathing
cladding a
lightweight
frame are
often given
brick cladding
to resemble
masonry
buildings

Photo 6.28 A new timber framed building near Weymouth, Dorset with strong eco credentials complete with sunspace, all in renewable timber

Photo 6.29 Another view of the boarded timber frame building near Weymouth during construction

Photo 6.30 Traditional timber framed and boarded building at an educational recycling centre in Somerset

Photo 6.31 A barn-like timber clad sustainable build on the Dorset/Hampshire border

Photo 6.32 New sustainable building at Guilsfield, Welshpool also has barn-like connotations

jointless or sealed claddings, another category favoured by the regulations, as they do not accommodate structural movement due to expansion and contraction of the frame, and in all events such materials are frequently petroleum-distillate derived.

Of crucial importance is the form of the dry joint between the cladding units, which has to be designed so that precipitation will not pass through or will be directed towards the exposed face without penetrating beyond the back of the cladding. All materials, such as slate, tile, and wood, have what is called the 'angle of creep of rainwater'. Water has a unique ability to 'creep' due to the supreme attraction between the hydrogen of one molecule and the oxygen of the next molecule (the hydrogen bond). The amount of overlap is thus important in order to accommodate the angle of creep. In slate and tile, the water can mobilise carbonates and iron within the surface structure of the underside of these elements and cause degradation. In wood the water, if not given sufficient ventilation within the body of the overlap, will set up wet rot. Guidance on claddings and the fixings thereof is to be found in BS 8000–6:1990.

A problem does exist with windows and openings in external cladding of frames, as there is a need to ensure resistance to rain penetration and the rampant damage which may happen to a frame should such rain find a route of ingress. It is very important to provide a damp proof course underneath cills, above lintels, and on reveals of windows and doors. Techniques employed include: the setting of frames behind masonry claddings; the use of checking rebates to stop the water molecule in its tracks, this being a time-honoured tradition particularly used around sash windows in Georgian buildings; and the use of insulated cavity closers. Also of importance is the construction of such walls with interstitial condensation in mind. This is a zone of moisture which forms when the warm moist air escaping from the interior interacts with the descending wall temperature as the moisture travels nearer the exterior of the wall, to the point where the moisture rains out. It can of course be prevented altogether by the use of a vapour barrier, but trapping warm moist air in rooms inhabited by humans or animals is a recipe for rampant ill health. Undoubtedly in a frame construction the best materials for coping with the total removal of this zone of condensation are insulating materials that are naturally based, such as wool and cellulose-based products, and claddings which have a good pore structure, such as tile and wood. Thus not only is there a meeting of the requirement for sustainable materials but for healthy human beings as well. In addition, surface condensation on cold walls is avoided and with it the potential for mould growth. A requirement on this aspect is that the U value of any external wall of window/door reveal must not exceed $0.7w/m^2K$ at any point. There are some sensible regulations, such as that for fixings being corrosion resistant, ensuring both vertical and horizontal restraint when fixing the cladding, and the cladding capable of taking the fixings of any other external loading such, as handrails, that is an imposed load.

Internal linings for framed buildings must be capable of resisting the spread of fire. Such linings are classified on tests as set out in BS 476 and are set out in Appendix A of Approved Document B. Fortunately wood fibre boards are not seen to adding significantly to any fire hazard. Small rooms with a an area of not more than $4 m^2$ ($4 \frac{3}{4} yd^2$) in domestic accommodation are Class 3; other domestic rooms are Class 1, as are circulation spaces within buildings. Other circulation spaces, particularly the common areas of flats and maisonettes, are Class 0. It is particularly important to ensure that the wall and any floor between the garage and the house shall have a 30-minute standard of fire resistance, and all load-bearing elements have a fire resistance in line with the above Appendix A.

Compartmentation is an integral part of fire protection in all new build, sustainable or otherwise. The aim is to prevent the spread of fire by creating fire-resistant subdivisions. The degree of subdivision depends upon the use of the building and its fire load, the height of the floor of the top storey in the building, and the availability of a sprinkler system. Walls in semi-detached houses or houses in terraces should be constructed as a compartment wall and be the full height of the buildings to the apex and underside of the roof in a continuous vertical plane. The junction so formed should be covered with fire-stopping materials. The compartment wall should then extend up through the roof for a height of 375 mm (15 in), or a 1,500 mm (59 in) zone on either side of the wall should have a fireproof covering. This contrasts severely with practices in the past, as late as the 1950s, whereby such dwellings frequently shared a common loft space, so that a fire in one building could rapidly spread via this area to adjoining dwellings. It is particularly important to ensure that the junction of compartment floors with external walls is restrained at floor level to reduce the movement of the wall away from the floor in the event of a fire. Walls should be continuous at eaves. The key issue for sustainable building is that doors in such walls have to meet the same fire resistance as the wall, and sadly fire doors generally available can in no way be considered sustainable, being composed of distinctly non-green materials. The construction of compartment walls should be able to accommodate the predicted deflection of the floor above by having a suitable head detail between the wall and the floor that can deform but still maintain integrity, or can resist the additional vertical load from the floor above as it sags under fire conditions. Trusses need also to be designed so that failure due to fire in one compartment wall will not affect the stability of a truss in another compartment.

The question arises as to what would be the effect on, say straw bale construction. The answer is surely that such construction has to be confined to low single-unit dwellings. Still relevant will be the fact that the external wall of a building should not provide a medium for fire spread if it is a risk to health and safety, although it is still permissible to use timber lintel, windows and door frames, and timber staircases. This is good news for sustainable construction.

Fire resistant materials which comply with the above requirements are also short on green credentials, and there is no mention of lime. Cement, gypsum, mixes of these with perlite or vermiculite, glass fibre, blast furnace slag, and ceramic products, and intumescent mastics are the order of the day.

Airborne sound is now a major issue in new wall construction, particularly walls that separate one dwelling from another, but also walls between habitable rooms. This has resulted in something of an obsession with potential air paths, even those due to shrinkage. This is mirrored by an equal obsession with airtightness to ensure maximum heat retention. Both require careful design, avoiding obvious transgressions such as ill-fitting doors and windows, but still ensuring an amount of controlled ventilation for healthy living. Trombe walls, already discussed in Chapters 3 and 5, provide such a source of controlled ventilation and can provide alternatives to fan or heat exchange units. Nonetheless the incessant testing of new dwellings for airtightness can be a real downside for the creator of the sustainable building construction. A particular regulation, which is that of requiring porous materials to be sealed, strikes real fear in the heart of those on the quest for sustainability. Yes, it is important to seal gaps in the joints of a structure, and to ensure that sound is not spread through roof voids, but porous materials should surely be left to perform their invaluable functions as discussed. Such materials are the very heart and soul of the sustainable building. There are onerous tables set out in the building regulations with which all new walls with a separating function have to comply, particularly those adjoining a communal space, and rigorous sound insulation tests performed. Such tests have to be carried out without a soft covering on the floor. Floor joists have to be supported on hangers and not traverse underneath adjoining walls, with joists separated by timber blocking if they are at right angles to the walls. The principle aim is to block the transmission of impact sound such as speech, musical instruments, footsteps, and furniture moving.

Plasterboard is much beloved by conventional builders, or rather a dry laminate of plasterboard and mineral wool for enhancing the ability of walls to resist sound transmission. Green builders would favour using sheep's wool behind lime plastered onto a traditional lathe and stud subframe, or one of the panelvent or Diffutherm boarding systems. As these are effective for insulating against heat loss, they must surely be as equally effective in resisting sound transmission.

Wall construction has a number of general requirements in addition to the arduous requirements for fire and sound resistance set out above. Mortar is discussed but is generally assumed to be cement, when in fact lime mortar will do many walls more good in the long term. Of particular interest to sustainable builders is the need for a separating wall to be bonded to an external wall so that it contributes at least 50% of the bond at the junction. Again there are exhortations to use mineral wool, for example at the closure of a cavity at eaves level. Cavities are always assumed to be of recent construction, for example dense aggregate concrete block with a 50 mm (2 in) cavity, when in fact late 19th-century traditional brick buildings may have a cavity. At junctions between an external masonry wall and a timber frame inner leaf there is a requirement for a cavity closer, but more pertinently the requirement to use two layers of plasterboard where there is a separating floor, and the sealing of joints with tape, all unsuitable materials for the green builder.

Lime plaster on expanded metal lathing is surely an alternative, plus the fact that the ability of lime to seal cracks and self-seal any cracks which open up, is an added advantage.

Framed walls with absorbent material between the frames are frequently desired by the sustainable builder. There is a suggestion again that layers of plasterboard, independently fixed, be used for the internal wall finish, and a need to be able to calculate the mass per unit area of the leaves, the distance between the frames, and the absorption of the cavity, for airborne sound. There is also a strong steer to mineral wool batts. Again sheep's wool will surely perform just as well. A need to ensure that walls adjacent to fireplaces are of solid masonry construction cannot be disagreed with, and measurements are given.

Extensions

Extensions to existing buildings attract a number of requirements, and new build must meet current standards especially regarding insulation to walls, floor, and roof. The build must also have efficient heating, hot water systems, ventilation, and lighting. The wall, doors, and windows between the existing building and the extension should be insulated and draught-proofed to at least the same extent as the existing building. In addition, the U value for each element in the new build should meet the current standards of having U values of 0.35 for walls, 0.25 for floors, and 2.20 for windows, rooflights, and doors (all in W/m²K). Exemptions for listed buildings and buildings in Conservation Areas still apply, but this may not be for much longer.

As well as that for a low rate of surface flame spread (Class 3 for small rooms, Class 1 for others including garages), airborne sound is also an issue, with independent ceilings with absorbent material favoured over plasterboard laid on top of sound absorbing bars plus absorbent material, and with plasterboard on timber battens being the least favoured. Again

Photo 6.33 Extensions to traditional buildings need to meet all new Building Regulations

Photo 6.34 This extension is aimed at maximum solar gain

there is no mention of wooden lath and lime plaster, although if it already exists in the main building it can be retained as long as it is deemed to satisfy Part B on fire safety. If not lathe and plaster existing ceilings need to be upgraded with at least two layers of plasterboard with staggered joints giving a minimum total mass per unit area of 20 kg/m2 with an absorbent layer of mineral wool.

All new construction needs to be independently supported on independent joists fixed to the surrounding walls, or on hangers.

Sustainable building makes frequent use of rooflights, with light wells finished in a thermoplastic beneath, and sunpipes. In sustainable construction, wired glass would be a better option, in all events. The use of thermoplastic for any glazed element would be anathema to any sustainable builder, although it is so much the norm in modern building construction that the Regulations assume its use. Referred to as 'lighting diffusers', these cannot be used in fire protecting or fire resisting ceilings unless they meet the same fire resisting requirements as the ceiling, normally Class 1.

Roofs

Traditional roof construction consists of rafters fixed to a ridge board and supported by purlins, struts, and ties fixed to wall plates bedded on the top of walls. In ancient buildings the truss, consisting of heavy principal rafters and a hefty tie-beam, will do the main supporting work. Such roofs are clad with thatch, stone tiles, slates, or clay tiles.

Needless to say, the function of a roof is to resist penetration of rain and snow. Rainwater is automatically assumed to be carried away by a drainage system but in many existing buildings and in new build in the open countryside this may not be the case, and in all events in sustainable buildings there is a need to ensure that rainwater is put to good use, such as flushing toilets and being stored to use on gardens. The Building Regulations still do not deal with this.

In addition, roofs need to resist fire from an external source, external fire spread of flame, and the risk of fire spreading to another building beyond the boundary. Internal fire spread depends on fire-separating compartments, already dealt with.

Photo 6.35
A traditional roof has
lost its meaning in a
welter of new support
timbers and is in
danger of needing to
satisfy Building
Regulations for a new
roof

An instruction to ensure that all softwood timbers in roof construction are treated to prevent infestation by the house longhorn beetle, especially in the home counties, will give concern to sustainable builders. Such treatment is invariably solvent based and can damage human tissue. One of the issues that the regulations do not seem to address is the angle of creep of precipitation underneath slates in particular, which mobilises minerals within the body of the slate, notably calcium carbonate and pyrites, and results in degradation of the underside, and eventual demise. It is not countered by the requirement to use roofing felt, now universally superseded in all events by breathable membranes. Ventilation is a key issue in roof voids, and is allied with the need to limit heat gains and losses via the roof void to the rest of the building. This is where sheep's wool or hemp laid in batts between ceiling joists really comes into its own, despite what

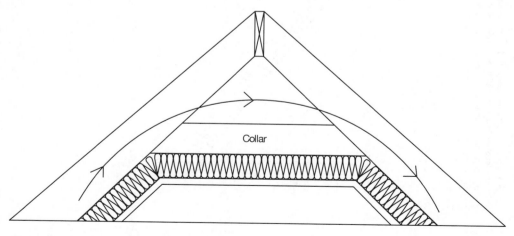

Collar

Diagram 6.1 Ventilation at apex is recommended for warm roofs (adapted from Griffiths, 2007: 50)

In a 'warm roof', the insulation can be laid above, between, or below rafter level, or in a combination of all these positions. This form of construction is generally chosen when the roof space is to be used for habitation

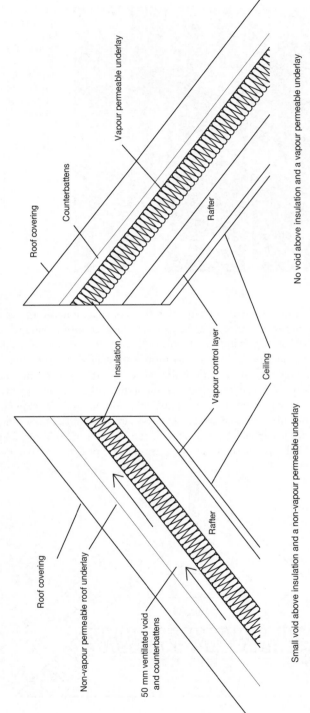

Roof covering

Non-vapour permeable roof underlay

50 mm ventilated void and counterbattens

Rafter

Insulation

Vapour control layer

Ceiling

Rafter

Roof covering

Counterbattens

Vapour permeable underlay

Small void above insulation and a non-vapour permeable underlay

No void above insulation and a vapour permeable underlay

Diagram 6.2 'Warm roof' options for ventilation (adapted from Marley Eternit Ltd, 2008)

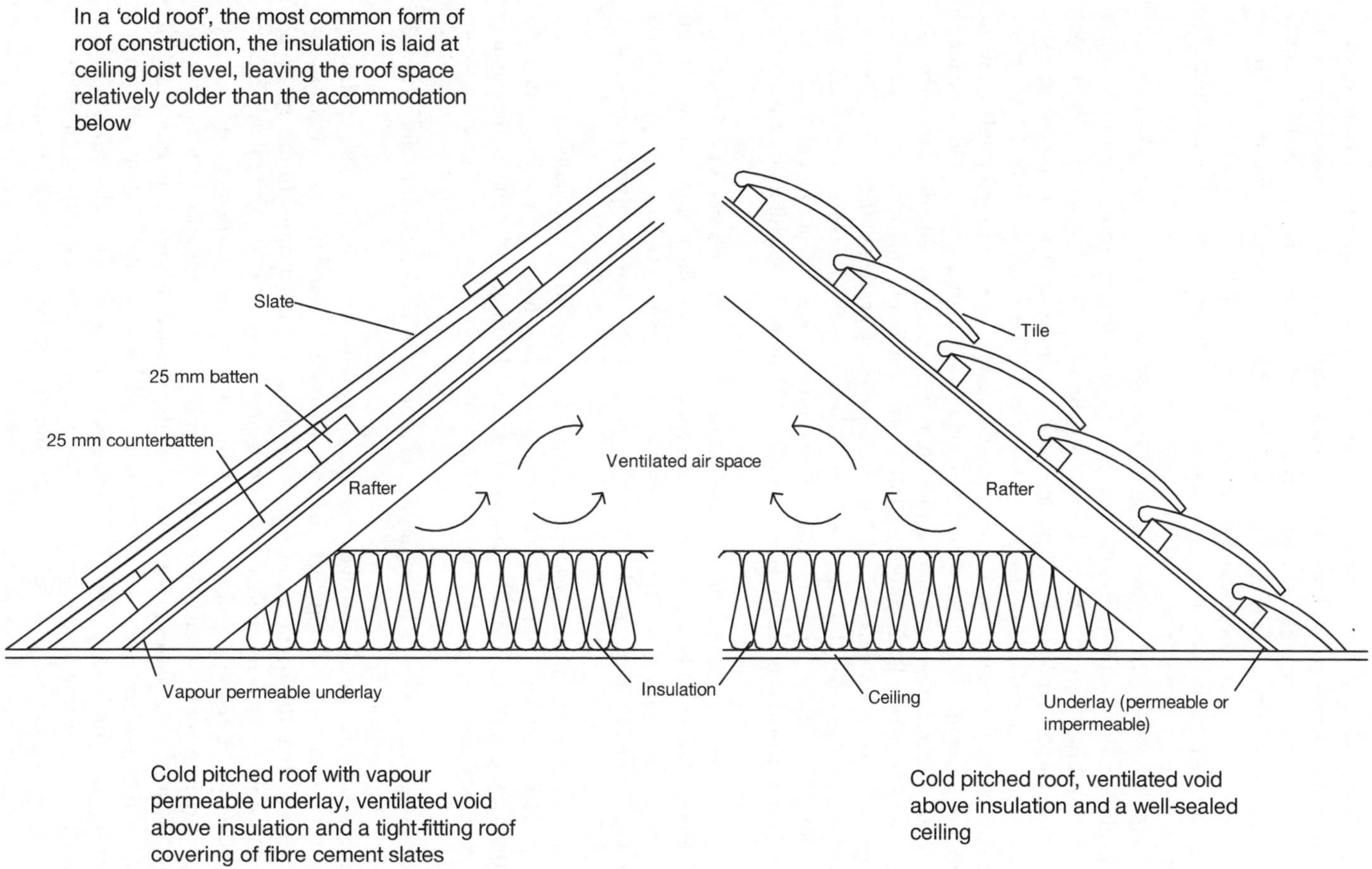

In a 'cold roof', the most common form of roof construction, the insulation is laid at ceiling joist level, leaving the roof space relatively colder than the accommodation below

Slate

25 mm batten

25 mm counterbatten

Rafter

Vapour permeable underlay

Ventilated air space

Cold pitched roof with vapour permeable underlay, ventilated void above insulation and a tight-fitting roof covering of fibre cement slates

Tile

Rafter

Ceiling

Insulation

Underlay (permeable or impermeable)

Cold pitched roof, ventilated void above insulation and a well-sealed ceiling

Diagram 6.3 'Cold roof' options for ventilation (adapted from Marley Eternit Ltd, 2008)

is always shown in the standard text books. The diagrams shown here reflect the latter, but it is a simple matter to substitute natural materials.

Ventilation is also assisted by the use of clay tiles with their minute but prolific pore structure and stone tiles, the latter being sadly prohibitive in cost unless a local quarry has been re-opened for the purpose of their production. This has been promoted in rural areas in recent years in order to meet the demand for building materials to repair historic buildings. Natural stone, slate, clay tiles, and wood shingles are classed as weather-resistant materials in the regulations and contrast sharply in their performance for the holistic benefit of the building with what are described as 'moisture resistant materials', such as metal, plastic, and bituminous products. BS 8000–6:1990 is the guidance for fixing roof materials.

Flat roofs suffer from interstitial condensation if they are of the cold deck variety, that is, where the insulation is beneath the deck, and can be affected by warm moist air rising up through the building. It is thus essential to ventilate above the insulation. Warm roof construction where the insulation is above a waterproof membrane, and in turn protected again by another waterproof membrane under the immediate decking material, is by far the most popular form of flat roof construction, but invariably involves the use of rigid urethane foam panels capable of bearing a degree of loading. A prominent manufacturer of this material claims, on their website (www.celotex.co.uk), that they have eco-credentials, with an independent assessment by BRE covering full lifetime impact, including raw material extraction, manufacture, distribution, and usage, to include an A rating in the Green Guide to Specification, and credits earned under the materials section of The Code for Sustainable Homes and BREEAM. Much of this thinking is based on the longevity of the material, but in all events it will not provide the vapour permeability that much of sustainable construction requires.

In apex roofs the issue of whether to have a warm or cold roof is an issue. A cold roof with both the insulation above and between the ceiling joists, all above the ceiling below, is often better for the roof rather than a warm roof with the insulation below between the sloping rafters of the roof itself. There are fewer problems with a cold roof in terms of vapour permeability affecting the rafters.

Flat roofs are frequently used as a *means of escape* in a fire situation and should lead to an external escape route and have 30 minutes fire resistance. Guarding may also be needed at a height of 800 mm (31 in). Where escape is from a window in the roof, the bottom of the opening may be 600 mm (2 ft) above the floor. Normally flat roofs used as escapes connect to another means of escape.

At the other extreme, roofs which have a pitch of more than 70 degrees have to be treated as walls in terms of insulation.

One of the major problems with roof voids is the degree of warm moist air rising up through the building and causing the potential for mould growth on rafters and trusses, as well as more serious problems such as dry rot and underside lead corrosion in flat roofs. In theory, simply laying mineral wool in between ceiling joists is not going to allay this problem; if anything, it will have its efficacy reduced by the absorption of this water, unless enclosed in a plastic quilt. Using sheep's wool, in contrast, allows this moisture to pass through and if the void is well ventilated the moisture then passes to the outside, via this ventilation and via porous materials such as stone and clay tiles. This may be in conflict with the regulation that the thermal transmittance (U value) does not exceed $0.35 \text{ W/m}^2 \text{K}$ at any point, but makes for a healthier roof void capable of resisting mould growth. This is a prime example where the needs of traditional construction can be jeopardised by the Building Regulations. By the same token, there is a clear regulation that roofs with a pitch of more than 15 degrees need ventilation at eaves equivalent to a continual strip of 10 mm ($\frac{1}{2}$ in) wide. A mono-pitch roof should have both the eaves ventilation of 10 mm ($\frac{1}{2}$ in) and ventilation at the junction of the roof and the wall abutment of at least 5 mm ($\frac{1}{4}$ in) wide. Roofs with a pitch of less than 15 degrees, and the flat roof (cold roof) discussed above, need a continuous strip 25 mm (1 in) wide in two opposite sides to promote cross-ventilation, with a void of 50 mm (2in) between the roof deck and the insulation. In many instances of sustainable build the insulation follows the pitch of the roof and is between and under the rafters. This requires ventilation at the ridge at least 5 mm ($\frac{1}{4}$ in) wide. In many traditional existing buildings the bedrooms are in the roof space, with the ceiling following the pitch of the roof. In these instances it is best to treat the roof as a warm roof, removing the external cladding to fit this insulation. If insulation is inserted under the rafters, then the whole needs to be treated as a flat roof, with the required ventilation and spacing. Roofs with a span exceeding 10 m (11 yd) may require more, totalling 0.6% of the roof area. Such ventilation openings need to be fitted with fly-proof screens.

Passive stack ventilation is increasingly favoured by sustainable builders. Such ducts need to securely fixed to a wooden strut that is itself securely fixed at both ends, with the duct being insulated with at least 25 mm (1 in) of material having a thermal conductivity of 0.04 W/m2K. Rigid insulated ducts need to be used for any external projection (to protect against condensation), and if projecting from the roof more than 0.5m from the roof ridge, needs to extend to at least the height of the ridge. Typically used in wet rooms, these ducts need to be separate from any other passive ventilation introduced into the building. The crucial point is that terminals should feed condensation down onto the roof slope, and the best way of doing this is to have the terminal emitting from the roof ridge, which also protects against buffeting by wind.

The structural capability of roofs is also considered. It is thought that a traditional cut timber roof consisting of rafters, purlins, and ceiling joists generally has sufficient resistance to wind forces. Diagonal bracing is needed for trussed rafter roofs (traditionally called 'coupled rafter roofs'), for which BS 5268:Part 3:1998 should be consulted for guidance. Horizontal and lateral restraint to walls at roof level is an important issue.

Compartmentation to guard against fire spread is also an important issue with new build and has been discussed earlier in this chapter. Concealed spaces through which smoke and flame can spread are an issue, and existing traditional buildings have suffered considerably from total destruction due to this problem. Roof claddings can change, particularly where it is essential to extend the useful life of the building. This is a situation frequently encountered with thatch, where it is now common to use water reed rather than wheat straw. This in itself does not often lead to extra weighting, but certainly over time the weight on a thatch roof does tend to increase, and can result in deflection of roof members. Something that is not commonly realised is that introducing strengthening work or replacement roof members into a roof in a situation where the roof is suffering from over-stressed members or deflection is a material alteration, and will be treated as such and will have to meet the Building Regulation requirements. In the case of an historic roof format consisting of trusses and pole rafters this can cause many complications, including the need to have the rafters overdrawn with construction having not less than 30 minutes fire resistance together with the fitting of smoke alarms in the roof space. It is common sense to fit smoke detectors into thatched roofs, but the other measures can be problematic for thatched roofs, demanding the roof be completely stripped off, with the consequent loss of historic thatch layers.

The situation in Photo 6.36 where the roof construction became a veritable forest of timbers, somewhat defeating the objective of maintaining the integrity of the ancient roof form, is a prime illustration. The result also generated considerable discussion with building control, who

Photo 6.36 A traditional A-frame roof under construction (photo © Rob Buckley)

were adamant that it was a new roof and as such should have met the demands set out above. Fortunately a more holistic view was eventually taken.

In addition, there are restrictions on the minimum distance from boundaries for new build with new thatch roofs, anything less than 6 m (6½ yd) not being acceptable. For slate and clay tile, distances of less than 6 m (6½ yd) are acceptable. For flat roofs and pitched roofs using lead, or metal sheeting (aluminium, zinc, or copper) less than 6 m (6½ yd) is also acceptable.

It is worth noting that roofing materials have performance ratings, with clay tiles having a Class 0 rating as a non-combustible material, whereas wood shingles and thatch will be Class 3. Other materials that could be used on flat roofs are timber hardboard, wood particle board, and plywood with a density more than 400 kg/mall of which are Class 3.

Roofs – conservation of fuel and power (Approved Document L)

Thermal standards for elements measured in W/m^2K for roofs are the key to providing good sustainable new build and for upgrading the thermal qualities of existing dwellings without compromising their integrity. A repair of an existing roof will frequently involve the need to remove the existing roof cladding due to nail sickness or worm infested battens. This is the ideal time to introduce extra insulation, and all roof voids are capable of being upgraded with insulation at ceiling level. The standards are 0.16 for underside insulation, and if between the rafter 0.20. Flat roofs need 0.25. Roof windows/rooflights should have a value of no less than 3.3 (all in W/m^2K). Glazed elements in the roofs of conservatories need to have higher standards for dwellings of 2.2W/m2K. Retained thermal elements have to be upgraded to meet these values.

Chimneys and fireplaces (Approved Document J combustion appliance and fuel storage)

Chimneys and fireplaces are features that are at the heart of every existing traditional dwelling, and also feature at least once in most sustainable new build to service a wood burner, wood being regarded as a renewable fuel via coppicing, short rotation willow, etc. There are, however, some very stringent regulations to be observed to satisfy the Building Act 1984 Section 73. New chimneys adjacent to buildings which already have chimneys, joined by a party wall or being less than 1,800 mm (6ft) away, will have to match in height, and if taller the builder will be required to raise the height of adjacent chimneys. The regulations provide for access to be provided for this purpose. All this can seem rather unreasonable but the logic is presumably to prevent noxious fumes from being sucked down into adjacent dwellings via the lower chimneys. The remainder of the regulations are common sense, involving the avoidance of hidden voids through which smoke could travel, visible instructions by the builder to any installer of appliances regarding performance capabilities of the chimney construction, and the tying of chimneys into the construction. Sensible precautions are taken regarding height so that this should not exceed 4.5 times the width, this being the least horizontal dimension, all dimensions taken from the point of intersection with the roof, unless additional restraint is provided. The issue of flues passing through compartment walls and floors is a clear cause for concern, with a requirement that the flue wall should have a fire resistance of at least half that of the wall or floor. Fire stopping at all junctions, particularly between fire-separating elements and pipes and ducts, needs to be in cement, gypsum plaster, or a mix of these with perlite or vermiculite, glass fibre, or intumescent mastics. The latter expand to completely seal any gaps on meeting heat.

Surprisingly there are few directives about hearths, which are so essential in separating fire from people and carpets, but there is a requirement for a projection of at least 500 mm (20 in) for an open fire and 840 mm (33 in) for a freestanding stove, non-combustible materials, and a change in level. Fire recesses designed for an open fire must have a canopy, whilst a simple square opening is suitable for a closed stove. Existing chimneys in flats will require additional fire protection to assist with compartmentation. Of paramount concern with any flue is the avoidance of areas where sulphurous-rich condensate can congregate. Such areas lead to the conversion of cement or lime joints to sulphoaluminanates or calcium sulphate respectively, which expand and cause disruption of joints and structures. Bends should not therefore have an angle greater than 45 degrees. Bends of 90 degree need a cleaning access. The old adage of flues not having openings into more than one room is strictly enforced, as flue gases can be very dangerous. Flues should be high enough to ensure sufficient draught to clear the products of combustion. Flue outlet clearance to *thatched or wood shingle roofs* needs to be increased, according to where they are situated on the roof, in accordance with tables supplied in Approved Document J, wherein there are numerous illustrations of flue types, to be adhered to for particular situations. *Users of wood-burning stoves in buildings with such roofs should be aware of the danger of hot spots in that part of the chimney adjacent to the ridge, caused by hot flue gases. This can cause the thatch to self-combust, and many thatch roof fires have started in this way.*

Sustainable builders will often be more concerned with the construction of flue-pipes rather than chimneys, and these should be of cast iron, stainless steel, or vitreous enamelled steel. The renovation, repair, or refurbishment of existing flues by the installation of a flue liner, whether by insertion of flexible or rigid components, or cast in-situ liners, must satisfy the Building Regulations. Oversized flues, such as those which accompany the traditional ingle fireplace, are generally regarded as unsafe and in need of a liner. Inspection openings need to be considered to allow for the sweeping of the whole flue.

Gas appliances are a field of their own and Gas Safety (Installation and Use) Regulations 1998 (GS(IU)R 98) should be consulted. There is a strict need for such appliances to be installed by a Gas Safe Registered engineer.

Stairs

Airborne sound, the need to meet laboratory sound insulation values as set out in the regulation, and fire protection has already been discussed. Nonetheless stairs are the primary means of escape, and all habitable rooms in the upper storey need to be provided with a window, or external door, or direct access to a protected stairway. Basements need to be provided with a separate staircase. The very essence of an escape staircase is the protected corridor at all levels, particularly if the stair is serving a storey that is higher than 18 m (20 yd). An escape stair may be open provided that it does not connect more than two storeys and reaches the ground storey not more than 3 m (3 ft) from the final exit, or there is another protected stairway. Single stairs in small premises can also qualify provided that the floor area of any storey does not exceed 90 m² (107 yd²). In flats there are restrictions on the distance allowed from the flat entrance door to the common stair, which are 7.5 m (8 ft) in one direction only, and 30 m (33 ft) for escape in more than one direction. In all events, in flats all corridors should be protected shafts, with the wall between each flat and the corridor being a compartment wall. Smoke vents, sprinkler systems, and smoke detection systems are an essential part of common staircase areas in flats, and most rigorously so in conversions.

The conversion of a loft space in a habitable area of an existing dwelling will always generate the need for the staircase to become a protected shaft, even if it already exists, and this in turn generates the need for all doors onto this staircase to become fire doors. This has meant the demise of sound traditional doors, which is contrary to sustainable building construction. Doors on escape routes have very rigorous standards. Every dead-end corridor exceeding 4.5 m (5 yd) in length should be separated by self-closing fire doors from any part of the corridor that provides two directions of escape, or continues past one storey exit to another. Dwelling houses with more than one floor 4.5 m (5 yd) above ground level usually need a protected stairway that extends right to the final exit. This also means fire doors at ground level. If the top floor can be separated from the lower storeys by compartmentation and be provided with its own alternative escape route, this is a much better means of preserving existing doors and doorcases within the body of the building. Such external staircases need to be provided with a weather-proof covering if the vertical extent is more than 6 m (6½ yd). Another factor which is particularly pertinent in buildings of historic quality where the existing staircase is of a quality wood like oak and is highly decorative, is that the flights and landings of escape stairs should be of limited combustibility. Discussions will inevitably arise regarding the merits of the existing staircase to meet this requirement. The answer may lie in the coating of the wood with intumescent paint, but this is sometimes not conducive to maintaining its visual quality, and as with all of the above discussion one should seriously question whether conversion of, for example, large country houses into flats is indeed in the best interests of the building. Also there is a need to provide separate ventilation systems for such protected stairways, and although rare in the sort of conversion discussed above the requirement nonetheless exists. Another popular introduction into conversions is the mezzanine or gallery into tall rooms to enable a studio flat to be contrived. It should be remembered that any cooking facilities within a room containing a gallery should be enclosed with fire-resisting construction or be remote from the stairs to the gallery. Basement conversions are also popular but need to be provided with their own protected stairway leading from the basement to the final exit, or if of the Georgian town house variety, leading onto a paved area with its own access up to street level, and be provided with an external door or tall window suitable for egress.

New blocks of flats also have to be provided with fire-fighting stairs, constructed of material of limited combustibility, and equipped with fire mains having outlet connections on every storey. In addition, main stairs have to be provided with refuges, which are safe waiting areas for short periods for disabled people, on each floor of a protected stairway.

Stair dimensions for new build are also rigorously set out, with the rise being between 155 mm (6 in) and 220 mm (8½ in) and the going between 245 mm (9½ in) and 260 mm (10 in), with a maximum pitch of 42 degrees. The equation that governs the relationship between the rise R and the going G is that 2R+G should be between 550 mm (21 in) and 700 mm (27½ in). Stairs with open risers were popular in the mid 20th century and thought to be very fashionable. In reality they are very dangerous unless the gap between the risers is less than 100 mm (4 in) and the steps overlap each other by at least 16 mm (¾ in). Many traditional staircases of the newel type which change direction meet the regulations unless the tapered end is less than 50 mm (2 in). Handrails should never be stinted upon, and two handrails are needed if the stairs are more than 1 m (3¼ ft) wide or either side of a fixed ladder to loft conversions. Spiral stairs, also favoured in the mid 20th century and still used in some instances to create an architectural statement, or in a conversion of say a warehouse to apartments, need to be designed in accordance with BS 5395.

Conversions generally provide a challenge when inserting staircases as the headroom on the access between levels should be not less than 2 m (6½ ft), and landings provided at the top and bottom of every flight. Another problem which besets such conversions is the need for a clear space of 400 mm (16 in) across the full width of the flight in front of door openings, be they entrances to rooms or cupboards. Guarding should be well above the height of a 5-year-old child, and be at least 900 mm (35 in) high or at least 1,100 mm (43 in) high if acting as a balcony or balustrade on a roof, and the spacing between balusters less than 100 mm (4 in). Approved Document M provides many instructions for those needing to consider the construction of staircases for disabled people, and is so rigorous that it needs separate consideration.

Ramps

Ramps are now a feature of many buildings designed for disabled access. The slope needs to be no more than 1 in 12, with handrails of height between 900 mm (36 in) and 1,000 mm (39 in), and provided with a landing. Headroom throughout should be at least 2 m (2 yd).

Windows

The chief requirement of windows is to provide adequate ventilation for people in buildings, and to limit heat gains and losses. In respect of the latter it is the responsibility of the contractor to achieve compliance with Approved Document L and provide a certificate of compliance. An aspect which impinges on the conversion or re-use of existing buildings is the need for all habitable rooms on the ground floor, and upper storeys served by only one stair, to be provided with an emergency egress window. Such windows need to be at least 450 mm (18 in) high and 450 mm (18 in) wide and have an openable area of at least 0.33 m² (0.39 yd²). The bottom of the opening area should not be more than 1,100 mm (43 in) above the floor. Inner rooms such as bathrooms and dressing rooms, and galleries below 4.5 m (15 ft) in height, whose only escape route is through another room, also need to be provided with egress windows. In existing traditional buildings equipped with perhaps only small two-light casements, this can be a major disruption to an elevational treatment and the loss of an existing window.

Window dimensions are also concerned with ventilation, and the area of the opening part of hinged pivot windows designed to open more than 30 degrees, or sash windows should be at least ¹⁄₂₀th of the floor area of the room. Those designed to open less than 30 degrees should be at least ¹/₁₀th the floor area of the room. Rooms with more than one window can score here, as the areas of all the opening parts can be added together.

Ventilation (Associated Document F) and the sealing of buildings in order to eliminate heat loss (Associated Document L) appear to be in diametrically opposing corners and this causes endless confusion when there is a change of use of existing traditional buildings or new sustainable construction. The whole issue of pressure testing existing traditional buildings is a vexed one. The very nature of traditional building materials, particularly lime mortars, renders, and plasters, stone and brick/tile is that they are porous materials through which ventilation of the building occurs naturally. This aspect is not even considered in the Building Regulations. The three types of ventilation that are considered are:

- **Extract ventilation:** by means of extract fans used in rooms where there is most water vapour or pollutants, such as bathrooms, kitchens, and photocopying areas (in fact, very few organisations respect this), the aim being to minimise the spread of vapour or pollutants to the rest of the building. This is not a sustainable method, because it uses electricity and in sustainable buildings is often substituted with passive stack ventilation.
- **Whole building ventilation:** via trickle ventilators is often inserted when the pressure test has been completed. The aim is to provide for the dilution and dispersal of water vapour and pollutants not already removed by extract ventilation. It is readily accepted that furniture and carpets can produce these pollutants, and that occupants themselves emit water vapour every time they breathe out.
- **Purge ventilation:** (previously called 'rapid ventilation') seems to have a low priority in the Building Regulations, being regarded as an intermittent medium from which the occasional concentration of pollutants such as food odours can be released. Most sustainable builders would view it in a very different light, as being the only sound means of improving thermal comfort, particularly at night in high summer. This is best combined with passive stack ventilation, which depends on the pressure differential between the inside and outside of the building due to the temperature difference between indoors and outdoors, and pressure of wind passing over the roof, for areas such as kitchens and bathrooms. The regulations provide for internal duct diameter and internal cross-sectional area, and emission at the ridge or at ridge height.

Replacement windows are a difficult issue. There should be more emphasis on upgrading windows by adding secondary glazing rather than wasteful replacement of perfectly good windows. There is far too little choice of secondary glazing, and the standard response from homeowners who have been seduced by the lure of the UPVC window is that secondary glazing is ugly. It need not be, and what is required is an equally seductive form of secondary glazing, and recognition in the minds of building inspectors that this can produce a draught-proof unit, suitable for extensions. Area weightings of 1.8 W/m2K for new fittings provided in an extension, and 2.0 W/m2K for replacement fittings in an existing dwelling, need to be calculated with individual U values. Approved Document L does at least provide for the mitigation of high internal temperatures caused by solar gains being minimised by a combination of window size and orientation, shading, ventilation, and high thermal capacity. These are all important issues in sustainable building construction, and the latter in particular is a joy; if only the regulations paid more attention to thermal capacity, life would be easier for sustainable building. On the opposite side of the coin is a concern that the windows taken out of one dwelling, or from a reclamation yard, appear not to able to be used in an extension or new dwelling. The major concern seems to be the possibility of gaps in the external insulating envelope caused by joints between the windows and the wall structure, and yet this joint can be packed tightly. There is of course a requirement that thermal bridges should be avoided, but such concerns militate against the re-use of materials which is the very essence of sustainable construction (see Photos 6.3 and 6.4).

There is also a limiting factor on the introduction of glass into extensions, which also creates a problem for the sustainable builder intent on maximising solar gain to be absorbed into a clay floor tile or similar surface and then reflected back into the extension during the night. The area of windows, roof windows, and doors in an extension should not exceed the sum of 25% of the floor area of the extension plus the area of those windows, which, as a result of the extension, no longer exist or are no longer exposed. The alternative is to regard the extension as a new building and meet the requirements accordingly. Conservatories are frequently used as extensions, and thermal elements should have U values no worse than 3.3W/m2K, together with windows between the conservatory and the building being insulated and weather-stripped. Material change of use and extensions to buildings with a total floor area greater than 1,000 m² (1,196 yd²), or changes to the building services of the latter, mean that all existing windows with a U value worse than 3.3W/m²K should be replaced and have trickle ventilation.

Glazing does of course present a potential hazard due to impact or collision. Such scenarios are frequently discussed when the need arises to replace glass in sash windows. Traditionally this would have a thickness of 3 or 4 mm (⅛ in), but is required to be 6 mm (¼ in) in replacement, particularly in the critical area of 800 mm (31½ in)from the floor level. Such sashes have very thin glazing bars, this being the height of fashion in mid 18th/early 19th-century England, and were not designed to hold this excessive thickness. Steps therefore need to be taken to limit the risk of contact with the glazing in order to limit potential damage, as a way round this tricky problem. In new build it makes sense to ensure that glazing, if damaged, produces harmless particles, is of toughened glass, and preferably is in small panes. Large areas of glass forming part of internal or external walls needs to incorporate imagery which indicates that it is glass, and be safely accessible for cleaning, as well as having accessible controls for opening and closing. Safe means of access is to be provided for cleaning both sides of glazed surfaces where this is danger of falling more than 2 m (2 yd). Modern sash windows need a mechanism for allowing the window to be reversed and held in this position to enable cleaning (Approved Document N).

Doors

Requirements for doors are very similar to windows in respect of the conservation of fuel and power (Approved Document L), ventilation (Approved Document F), fire safety (Approved Document B), and disabled access (Approved Document M) all apply. Approved Document F demands an undercut of 10 mm (½ in) above the floor finish for standard doors of 760 mm (30 in) width (equivalent to the old 2 ft 6 in) to ensure good transfer of air throughout the dwelling. The height times the width of an external door should be at least ¹⁄₂₀th of the floor area of the room. Approved Document B determines that in extensions the area of doors should not exceed 25% of the floor area of the extension, and there is the usual requirement for U value of external doors to be not in excess of 3.3 W/m2K with an area weighted dwelling average of 2.2 W/m²K.

Doors become an important issue in material change of use, such as conversion of farm or industrial buildings, important for the archaeology/past way of life that they illustrate. Fortunately it is usually possible to keep existing traditional joinery as a shutter to an internally fixed thermal element that meets the need for a U value of 3.3 W/m2K. This also avoids the problems associated with ensuring that there is no thermal bridge between the door-jamb and the wall structure. Conversions of country houses to apartments, etc. will demand fire doors in protected shafts and need to meet the requirements of Table B1 of Appendix B to Approved Document B (fire safety). Again the main problem is the value and quality of existing doors, which by their very nature are dateable elements within buildings of quality and historic interest. Although often capable of upgrading with an 'intumescent paper' if of substantial frame and panel, i.e. capable of meeting the minimum fire resistance (usually 30 minutes), there are other factors which mitigate against their use. This can include opening so they block a corridor, or are not fitted with simple fastenings readily operated from the side from which escape is required. Nonetheless, to stop the wanton wastage of historic and sustainable doors, more thought should be given as to whether they can be modified, or whether in fact a fire door can be fitted in the same opening, as wall thickness is normally sufficient to allow this. Thus together the upgraded historic door and fire door can provide the required fire resistance, as the sum of their individual resistances. Double doors on openings have an historical precedent in the country house, originally designed to prevent eavesdropping by servants. English Heritage has produced detailed guidance on the whole issue of upgrading fire doors (English Heritage, 1997).

It should be noted that conversion of existing traditional buildings to care homes provides particular problems with historic doors, as every bedroom needs to be enclosed in fire-resisting construction (a facet that frequently affects ceiling decoration as well). All doors must be fire resisting, and every corridor serving bedrooms must be a protected corridor. A suggestion that there should be double doors on every room where frail people are involved may well be met with disbelief. This is altogether a very difficult subject, and despite English Heritage advice in the past, it is still very difficult to resolve.

(The above text adapted from: Tricker and Algar (2007); The Building Regulation Approved Documents; Workmanship on Building Sites – Codes of Practice.)

7

THE PATHOLOGY OF
TRADITIONAL BUILDINGS

On of the most inhibiting factors when using traditional materials and construction is the fear of major repair problems. Older buildings in particular generate this concern. In reality all building construction and materials generate repair problems if they do not receive regular maintenance, as this is the key to every potential malady. Older buildings are often seen as an obstruction to progress; develop an expectation that they will shortly be removed or their components completely replaced, and hence are offered no maintenance or minor works. There are numerous wooden windows all over the country left to rot in the expectation that they will shortly be replaced with UPVC.

Modern construction, all sparkly and new, generates an expectation that it will always be so, so to add to the conundrum it also receives infrequent maintenance, with the result that it goes downhill a lot faster than its more elderly cousins. The materials are by and large less robust (fast-grown wood that rots out, steel that corrodes, etc.); this alone is an instant indicator that older traditional buildings are more eminently sustainable.

It would appear, however, that as early as the 18th century, regular maintenance of timber framed buildings, such as renewal of the wattle and daub infill panels, had started to take a back seat. Brick noggin (panels of brickwork laid on bed or on the diagonal) was inserted in many instances. Likewise, a rotted-out frame might be replaced with a large area of brickwork or stone, or the whole building clad with such. Towns such as Sturminster Newton and Wimborne in Dorset, and even the famous Ludlow in Shropshire, certainly display this tendency, although brick skins may have had as much to do with fashion as lack of maintenance. In Dorset much of the framing was replaced with stonework, possibly because it was seen as a more durable material. In areas where good clay abounded, thatched roofs were replaced with tile, possibly seen as less troublesome than a thatched roof that required regular beefing up on ridge and chimney apron, although most farm labourers could thatch because thatching a hayrick was a fundamental part of gathering in the harvest. In Kent boarding was used as a form of rain-screen cladding. Render was applied to elevations in an attempt to disguise decay of timber frames, or hide the poor man's rubble work, whilst at the same time the lining out of these elevations to simulate ashlar stonework took place. This practice became popular in the late 18th and early 19th centuries.

Cob buildings continued to be occupied, and re-rendered in lime, until the mid 20th century when disastrously they were rendered in cement, and any mass cob construction had generally ceased to be done. In general the substitution of cement for lime has been devastating for cob buildings, holding in damp so that the cob started to take on a porridge-like consistency, leading to inevitable collapse.

It is in fact the use of deleterious modern materials which inhibit breathing walls that is one of the main reasons why traditional buildings have gained a bad reputation. Such materials

301

Photo 7.1 Wooden window left to rot in the expectation of replacement with UPVC

have caused problems such as damp and crumbling plaster, or woodworm and fungal decay that are rampant but unnecessary. Some buildings have literally crumbled away.

It is the vision of scenarios such as these that have given traditional building material and construction a poor reputation without a thought to the unsuitable materials used, in this case undoubtedly cement render. Given the correct and proper care they will last indefinitely.

Maintenance and minor works

Care starts with maintenance and minor works, most of which could and should be done during the summer, or at the very least by late summer. This includes painting of joinery, this being an ideal time as the late summer sun is rarely so hot that it can result in expansion of materials

Photo 7.2 A cob building barely standing and crumbling away for want of attention

and cracking of paint films. The paint film gets a chance to really harden off before the next summer. Render expansion cracks should be pointed up in a lime putty mix, even if the render is cement, as the cracks are often due to the expansion of the substrate due to heavy rain or hot sun. A lime putty mix will gently accommodate any further movement.

November is the time for clearing of leaves from rainwater goods (gutters and downpipes), valleys, and parapet gutters, although sadly this is rarely done. It is assumed that the wind will clear the leaves, but failure to do so can result in blocked hoppers, causing water-falling down the building, or blocked gullies causing foundations to flood. Snow should be anticipated by the use of heated tapes in parapet gutters and lead flats. Again the risk of build-up pushing itself under slates, and the blocking of exits with ice, is a constant hazard for traditional large houses with valleys or flat roof areas.

Photo 7.3 The clearing of gullies and checking of downpipes for blockages is fundamental for all types of construction; here ivy is also a problem and needs to be kept under control

Minor works include:

- The patching of render, or the minor re-pointing of any open joints, all in a lime putty mix, but no later than the first week in October when the first frosts start. Autumn is also to be avoided for such work because of the heavy rain that can wash a lime-based mix from the wall.
- The replacing of any loose tiles or slates, or the addition of a new ridge and apron on thatched roofs.
- Lead flashings, particularly around chimneys, should be carefully checked, and repair effected to any lengths not lying flat that can be whisked away by fierce winds.
- The re-pointing of chimneys where loose joints are evident, or even the total rebuilding, using the same bricks if possible, of chimneys that are showing the characteristic curve on a south or west elevation. This is due to constant wetting and drying cycles mobilising sulphates already in the brick, or in rainwater due to acidic gases in the air, and sulphur from the open fire/stove. The lime- or cement-based mortar turns to calcium sulphate or ettringite, respectively, and expands in the joint and causes the chimney to bend over to accommodate the increase in height of the joint.
- The creation of French drains around particularly troublesome areas of buildings where water tends to pond, or where there is obvious evidence of rising damp inside the building. If this cannot be done, areas of porous paving/gravel should be created to assist with water removal away from the base of the building. Porous paving is undoubtedly the great saviour of all buildings.
- The cleaning out of all gutters, downpipes, and gullies that may have become blocked, so that they flow freely to cope with the heavy rainfall/melted snow in winter.
- Repair of any damaged joinery which can let water into a joint or even onto internal cills, accelerating the decay of the window or door.

Our forefathers knew the importance of all of these things, and by and large they put them into operation.

Photo 7.4a Former pigsty in West Yorkshire has suffered many years of neglect and variable minor works, just enough to ensure its survival

Photo 7.4b Some former minor works are deleterious, notably the turnerising of this roof with bitumen impregnated hessian, rather than a full roof repair as in the adjacent regular slate roof

Photo 7.4c Ironwork is incidental to most traditional buildings but needs to be coated in oil or oil-based paint and retained

Photo 7.4d Years of neglect of a chimney flashing for want of minor works, evident in the poor condition of this chimney breast

Photo 7.4e Minor
works in cement will
only result in further
decay to the masonry

Photo 7.4f Vegetation
at gutter level is a sure
sign of major repair
problems in the future

Stonework – decay mechanisms and repair strategies

Stone is a remarkably resilient material, or at least it was until the Industrial Revolution started to affect the levels of acidic gases in the air. Decay is a natural response to changing conditions, and can either be slow in response to weathering, the simple effect of wind-borne particles abrading the face of the stone, or fast, as the result of decay due to a number of mechanisms. The first of the fast-acting decay mechanism is salt crystallisation, which is tied up with acidic gases in the atmosphere, as well as thermal differentials and the constant wetting and drying cycles, characteristic of the British climate. The next, frost action, is often blamed when it is innocent and salts are the real problem, and wind erosion is almost certainly only a major culprit when the beds of stone are weak, or the stone is heavily infested with salts. Other aspects like the effect of living organisms, such as bacteria, lichens, and algae, have only a minor effect and should really be classed as acceptable slow change, whilst errors due to incorrect bedding of stones in tandem with poor design and construction can be fast or slow, but in all events are definitively avoidable in new build. The chemistry of stonework is complex, and decay often arises from misunderstandings related to the use of reagents applied to the stone for cleaning, the use of unsuitable mortars which react with the stone, and the use of other materials such as iron set within stone which is incompatible, especially when the iron itself decays. Finally, stone suffers when it is subjected to excessive heating during cataclysmic events, such as house fires. Not surprisingly limestone can turn to lime putty if the heat is great enough, and the fireman's hose produces enough water to set off the slaking action.

The following is intended only to give a brief indication of the nature of these problems. There are many weighty texts devoted to this subject alone, and research is always ongoing. In addition, there is much debate and controversy about what causes stone decay, particularly amongst academics.

Photo 7.5 Lichen is concentrated in the middle wet zone whilst decay is concentrated in the lower zone due to rising and mobilised salts

Photo 7.6 Moss is a sign of continuously wet masonry, the drier zone above it is showing calcium sulphate erosion

Salt crystallisation and acidic gases in the atmosphere

Acidic gases are rampant in the atmosphere. We are all now attuned to the presence of carbon dioxide emissions, which are said to be eating a hole in the ozone layer, as a result of which temperatures are said to be rising, ice melting, and the resultant global warming causing massive changes in weather conditions. In fact, the atmosphere contains a whole range of acidic gases, an aerosol laden not only with carbon dioxide but sulphurous and nitrogenous gases, which when combined with oxygen form sulphur dioxide and nitrogen dioxide. In addition, there is an abundance of carbon particulates and tar acids. Also, there have in the past been major design defects inherent since day one. For example, much of the City of Bath was placed in a hollow: at the time the chemistry of stone was unknown and the potential reaction of acid-rich condensate, itself a function of the Industrial Revolution, a complete mystery to the builders of this beautiful creation.

Whilst modern life is capable of producing carbonic, nitrogenous, and sulphurous gas via central heating vents and vehicle exhausts that are difficult to quantify, there may be some justification in saying that these have not reached the zenith of the Industrial Revolution in Victorian England. At that juncture every domestic and factory chimney was producing a vitriolic aerosol that was so dense that the resulting smog, deleterious to human and building alike, is notable in most contemporary accounts of life in Victorian England. It is thus important to stress that the situation in the past with gaseous emissions may have been considerably worse in the British Isles than it is at present, and so it should not necessarily be assumed that new stone buildings will suffer the same degree of malaise.

The chemistry of these acidic gasses is not difficult but requires some knowledge of chemical equations. The relevant equations are set out below, together with a rather more digestible discussion of the chemical reactions for the benefit of non-chemists.

The formation of calcium bicarbonate as a result of the attack of carbonic gas on limestone (calcium carbonate)

Carbon dioxide dissolves in rainwater and forms carbonic acid. The effect of this on limestone is to combine with the calcium carbonate of the limestone to form calcium bicarbonate, which is soluble in water and washes away. Thus rain-washed areas of stonework gradually dissolve on their surface, and the surface left is very clean but slightly powdery, awaiting the next fall of rainwater to wash it away. The stone gradually melts away on its surface. The equations make this clear. Carbon dioxide dissolves in rainwater in the following way:

$$CO_2 + H_2O = H_2CO_3 \text{ (i.e. forms carbonic acid)}$$

The effect of carbonic acid on limestone is:

$$H_2CO_3 + C_aCO_3 = C_a(HCO_3)_2$$

This is calcium hydrogen carbonate, or using its more popular name, 'calcium bicarbonate'. This is water soluble and washes away in rainwater, thus removing the surface of the limestone.

The formation of calcium sulphate as a result of the action of sulphurous gases reacting with rainwater, which in turn reacts with the calcium carbonate of limestone.

Sulphur dioxide is soluble in water, forming sulphurous acid. The equation for this is:

$$SO_2 + H_2O = H_2SO_3$$

There are two possible reactions.

Reaction 1: sulphurous acid in the air acts directly upon limestone

If this sulphurous acid directly affects the limestone then the result is calcium sulphite (only three oxygen atoms), but this readily combines with further oxygen and in all events forms calcium sulphate, which together with water is washed away and carbon dioxide, a gas, is also removed. This means that a large percentage of the limestone turns to calcium sulphate, which is also water-soluble and washes away. As above, the rain-washed areas are washed away, and the face of the residual stonework appears always bright and shiny new, but in reality is slowly diminishing as it powders away in rainwater. More insidiously, those areas that are sheltered under ledges or overhangs are prone to attack from acid-laden condensate. This slower attack first forms patches of calcium sulphate, resembling miniature volcanoes which erupt, spilling powder down the face of the building. The whole starts to resemble the surface of the moon. The effect is particularly noticeable on churches, which are blessed with numerous small overhangs, be they roll mouldings, cornice projections, or string courses. Underneath each overhang, acid-laden droplets of condensate lingers in autumn, winter, or early spring, its lethal abilities to corrode the stone working away unseen.

This condensate-affected stone behaves in the following way. In some areas of the stone, calcium sulphate reacts with water, which then becomes chemically bound to calcium sulphate, forms calcium sulphate di-hydrate, the stone then becoming tantamount to gypsum. This is more stable although still very prone to water solubility. It probably explains why some shaded areas of stone retain a solid black form, with carbon particulates bound to the gypsum surface, a sure sign that the carbon particulates have been fixed to the stone when the calcium sulphate crystallises, trapping them intact. Other areas erupt like volcanoes. The white areas of eruption make a stark contrast to the black areas on the average Victorian church in a city or town. In rural areas the stone assumes a pale brown colour, having formerly been a cream-coloured limestone. The equation for sulphurous acid directly attacking the limestone is:

$$H_2SO_3 + C_aCO_3 = C_aSO_3 + H_2O + CO_2$$

This converts the limestone to calcium sulphite plus water which runs away and carbon dioxide gas which emits into the atmosphere. The calcium sulphite combines with oxygen:

$$C_aSO_3 + O = C_aSO_4$$

This completes the conversion of limestone to calcium sulphate, which is water soluble and powders away in rainfall.

Reaction 2: the sulphurous acid joins forces with oxygen in the air to form sulphuric acid,
which then acts on the limestone

Sulphurous acid (three oxygen atoms) reacts with oxygen to form sulphuric acid (four oxygen atoms). The equation for sulphuric acid is:

$$H_2SO_3 + O = H_2SO_4$$

Sulphuric acid plus calcium carbonate (limestone) plus carbon dioxide (a gas which dissipates into the atmosphere) and water (hydrogen times two plus oxygen) which washes away leaving the soluble and friable calcium sulphate to powder away in rainfall. This leaves a brand new but powdery surface, free of all carbon particulates but continuously melting away.

$$\text{Equation is } H_2SO_4 + C_aCO_3 = C_aSO_4 + CO_2 + H_2O$$

The water and carbon dioxide are removed, leaving calcium sulphate C_aSO_4.
In fog (condensate) the situation is:

$$C_aSO_4 + 2H_2O = C_aSO_4 \cdot 2H_2O$$

This shows calcium sulphate becoming chemically bound with water forming a more stable calcium sulphate, similar to the actual gypsum mineral.

Pore density, discussed below, has an influence on whether the black or brown sulphate skin adheres to the stone. The sulphate is a salt, and as such it can fill small pores to capacity and over, and cause eruption. Limestone with low pore density or with many large pores will remain relatively inert. Limestone with many small pores and a high density of them will simply not be able to cope and will erupt in sheltered areas on a grand scale, or similarly wash away very fast. The situation is not helped by the fact that the calcium sulphate heats up on contact with water in condensate, although on a very small scale, and the blisters experience thermal movement, erupting and leaving cavities.

Photo 7.7a Calcium sulphate decay – the erosion beneath an overhang has all the signs of being caused by acid-laden condensate

Photo 7.7b Calcium sulphate decay in a much larger sheltered area

Photo 7.7c Calcium sulphate decay – the cement pointing is accelerating decay by forcing more wetting and drying cycles to take place in the stone

Photo 7.7d Crypto florescence is occurring adjacent to areas showing the typical brown discolouration of calcium sulphate decay

Photo 7.7e Advanced calcium sulphate decay in limestone

Photo 7.7f The whole of this church tower is coated in brown calcium sulphate decay

The effect of nitrogen in the atmosphere on limestone

Nitrogen dioxide dissolves in water, which then acts upon the limestone to form calcium nitrate, a white deliquescent compound that is very soluble in water. This is roughly the way nitrogenous fertiliser is made.

The equation for nitrogen dissolving in water, of which the HNO_3 acts as the nitric acid is, is:

$$(NO_2)_2 + H_2O = N_2H_2O_5$$

The effect of the nitrogen dissolved in water on limestone is seen in the following equation. It shows the formation of calcium nitrate and the carbon dioxide gas which dissipates into the atmosphere. It can be seen that making nitrogen fertilisers causes carbon dioxide emissions, albeit on a relatively small scale.

$$N_2H_2O_5 + C_aCO_3 = C_a(NO_3)_2 + CO_2$$

From the chemistry discussed above it can be concluded that carbonic acid, sulphuric acid, and nitric acid all act upon limestone to form a salt which, trapped within the upper surface of the limestone, acts in a manner likely to cause decomposition of the face. The constant washing of surfaces exposed to rainfall results in surface removal, as the salts trapped within the pores are water soluble. In sheltered areas, particularly with sulphuric acid, there are miniature eruptions of the salt-laden pores, deep within the surface (crypto-florescence).

SALT CRYSTALLISATION

From the above it can be seen how acidic gases are responsible for the formation of salts.

Salts crystals grow in size in a drying cycle, within the pore structure of stone. In the next wetting cycle, more 'water of crystallisation' (or 'wc') containing a salt solution will be fed into the large pores from the small pores/capillaries. This solution adds to the ability of the crystals to grow even larger in a drying cycle. The larger crystals start to push against the internal wall structure of the pores, causing heave similar to that which occurs when water freezes within a pore structure. The wall of the pore crumbles under the onslaught and the stone face starts to disrupt from beneath the surface, called 'crypto-florescence'.

Even in a wetting cycle the problem is relentless because the salt-laden 'wc' dissolves the existing solid crystals of salt, and together the total volume of 'wc' also causes heave within the pores and hence disruption of the stone structure beneath the surface.

This problem can be abated if there are many large pores and fewer small pores, as the larger pores can sometimes cope with dry crystal growth, and the feeding solution from the small pores is diminished.

If, however, there are few large pores and many small pores, then the problem can be severe as the growing crystals are continuously fed 'wc' from the small pores and there are fewer of them to cope with expansion, forcing this expansion into the small pores. The result can be enhanced disruption of the stone beneath its surface.

Efflorescence on the surface, the deposition of a white powdery salt film, is a sure sign that all is not well in the stone (similar situations arise in render, mortar, and brick), and such efflorescence should be dry-brushed away as soon as it appears, on a dry day. Under no circumstances should water be applied as this will simply mobilise the salts. Poultices, applied wet, consisting of a weak lime putty or clay, or acid free paper, are designed to mobilise the

313

Diagram 7.1 Salt behaviour in pore structure (adapted from Torraca, 1988: 33)

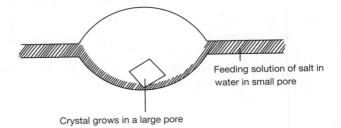

Feeding solution of salt in water in small pore

Crystal grows in a large pore

Option 1 – Many large pores, few small pores

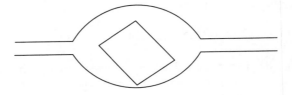

No more growth as no more feeding solution – small pores dry

Option 2 – Many small pores, few large pores

Water of crystallisation continues to act as a feeding solution and creates more pressure even if the crystal starts to break up in dehydration

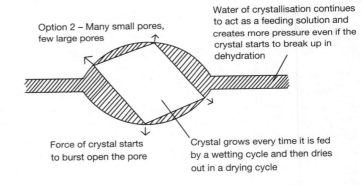

Force of crystal starts to burst open the pore

Crystal grows every time it is fed by a wetting cycle and then dries out in a drying cycle

Photo 7.8 Ice in a container can be likened to what happens in pore structure: as the ice or salt crystal expands, it fills the container and bursts open the sides

salts to the surface of the stone and then trap them as the poultice dries out. They can be effective for very problematical areas.

It is clear that pore size, and pore density in stone (or brick, render, or mortar), is crucial to how it behaves. Large pores cope with an ever-increasing salt-laden 'wc', whereas small pores do not. Few pores, be they large or small, do not cope with the stress, but a superabundance of large pores over small pores is beneficial. In fact, the size of pores is of greater importance than pore density. The porosity of stone thus needs to be gauged, particularly for any new build, and most essentially for any weathering surfaces directed upwards to the onslaught of downward precipitation.

Porosity = volume of pore space/bulk volume

Some stones have very limited porosity, less than 1%, notably marble and granite, although both of these are subject to a problem on their immediate surface as heat from the sun causes expansion of the individual mineral components and movement in the crystal boundaries, and cooling by rain causes contraction, creating cracking and fissures on the surface. These then becomes the repository for acid rain, which causes marble in particular to convert to calcium sulphate within these fissures (called 'sugaring' as a white deposit becomes obvious), leading to yet more degradation and opening up of the fissures. The cycle so precipitated then continues unabated until much of the immediate surface is in decay, and the expanding calcium sulphate in the crevices can even cause a marble headstone to bend. This is a similar problem to lime/cement mortar joints in chimneys, expanding and causing chimneys to bend. In granite the problem is exacerbated if its feldspar component is converted to kaolin clay, which is hydrophilic (water loving). Expansion of the clay takes place, with the resultant opening-up of mineral boundaries on the surface.

It can readily be seen that even in a stone with low porosity, repeated wetting and drying cycles do the most damage. Unfortunately this is the very nature of the British climate.

The hydration of salt crystals can be readily seen by illustrating the behaviour of common salt, sodium chloride, in Diagram 7.2. The H_2O molecules surround the NaCl molecule, dividing it into its component ions Na+ and Cl–. The sodium ion being positive is attracted to the negative oxygen of the water molecule, and the chloride ion being negative is attracted to the positive hydrogen, as per the diagram. The interior of pore surfaces are negative oxygen rich, so it can be seen that these surfaces want to attract the positive hydrogen of water and with it comes the negative chloride. This is how salt ions are attracted directly into the pore structure; the narrow and smaller the pore, the faster the attraction is. In the same way the positive hydrogen of the

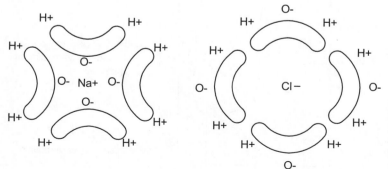

Diagram 7.2 Salt ionisation-hydration of the salt crystal sodium chloride forming two ions surrounded by water (Weaver, 1992)

water molecule is attracted to the negative oxygen of the water molecule, and with this comes the positive sodium. Once inside the pore the positive and negative sodium and chlorine join together again to form 'wc', which in turn forms salt crystals in a drying cycle.

This discussion, tortuous though it may sound, also explains why salts are hygroscopic, that is, they readily attract moisture from say warm moist air, warm air holding more moisture than cold air. Thus salts within a pore structure will attract 'air-bound moisture' and walls appear to 'wet up' as if by magic. Every type of salt has its own equilibrium relative humidity (EQRH, the relative humidity at which it will attract moisture from the air), the relative humidity itself depending on temperature. Thus not only rain but also fog on autumn mornings in 'the season of mists and mellow fruitfulness' is attracted into the pore structure of stone (or brick, render, and mortar). If the RH within the pores is low, then water will also migrate to it because it will try to boost the humidity, and with it will come salts. This certainly happens with stone window mullions. The salts or their 'wc' migrate to the area of lowest relative humidity, this being on the inner face of the mullion, and hence exfoliation of stone occurs here, rather than on the outside as would be expected. This possibly explains also why pillars holding up a porch suffer greater salt decay on their inner face, i.e. facing towards the sheltered area. The wind will be sucked into a vortex past this inner face, and hence water-bound salts will be attracted to this dry area as they try to crystallise, making the stone more friable, to be whisked away by yet more wind.

To summarise this important subject, the rate of decay by salt crystallisation depends on the:

- type of salt
- temperature
- relative humidity
- pore size
- pore density
- frequency of wetting and drying cycles.

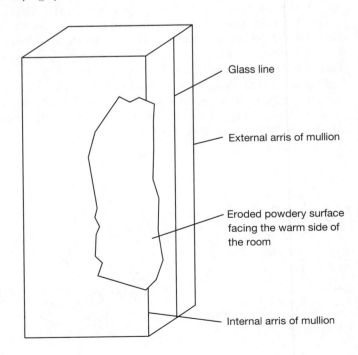

Glass line

External arris of mullion

Eroded powdery surface facing the warm side of the room

Internal arris of mullion

Diagram 7.3 Exfoliation of a stone mullion due to salt crystallisation (after Ashurst and Ashurst, 1988)

316

Photo 7.9a Salt crystallisation – wind-driven salt erosion is worse in softer stone beds

Photo 7.9c Salt crystallisation – corners of stonework are particularly prone to wind-driven salt erosion as they are evaporation zones for water

Photo 7.9b Salt crystallisation – soft oolitic limestone in Malta is particularly prone to wind-driven salt erosion

Photo 7.9d This stone pillar is graphically illustrating salt erosion in an exposed location

Sandstones, granite, and slate

Sandstone has a totally different structure to limestone unless it is calcareous sandstone (see below). It is basically grains of pure quartz, crystalline silica SiO_2 set within a cement of non-crystalline silica. There are also iron oxides present in most types of sandstone, which gives them their pale pink or buff, down to a rich orange-red colour. Also present are grains of feldspar and mica.

317

Sandstone is more resistant to acid rain, and sulphurous acid in particular, but actually looks as if it is behaving the same way as sheltered areas of limestone in that it often becomes very black with carbon particulates. In reality it is the rain-washed surfaces of sandstone that collect the particulates as they become bound to the quartz grains. Sheltered areas of sandstone stay clean. Thus the appearance is opposite, but for different reasons.

The use of lime mortars with sandstone and granite is a hotly debated subject. It is thought possible that lime migrates into the pore structure of these stones, and if it congregates in a whole raft of pores beneath the surface, and the lime changes to calcium sulphate as a result of acid rain, the resultant area can expand, causing the surface of the sandstone/granite to scale of – often referred as 'contour scaling'. This is because the appearance of such patches resembles contours on a map.

A further problem arises from iron-rich sandstone, the iron compounds being mobilised by acid rain, which then combine with the lime in the mortar lower down the face. This 'lime-rich precipitate' then crystallises, encapsulating within its crystal structure the iron oxides, which then become insoluble. The concentration of iron is often revealed during cleaning.

Calcereous sandstones, are grains of quartz set within a cement of calcium carbonate, often referred to as calcite. Acid rain dissolves the cement, releasing a whole shower of sand grains so that heavily drenched surfaces gradually wash away in a similar fashion to pure limestones. In areas which are less prone to constant rain-washing, the calcite will be converted to soluble gypsum. This will be drawn to the surface in drying cycles and re-deposited in the external pore structure. Thus the layer immediately below the surface becomes a weakened layer. The calcium sulphate will expand, and possibly exfoliate this weakened layer. Again a form of contour scaling will occur.

Calciferous sandstones have calcium carbonate, but not as the cement for the silica grains, so the sandstone structure is not so weakened by acid rain.

Dolomitic sandstones are about 45% magnesite and 55% calcite. As there is less calcite there is less risk of attack from acid rain, but magnesium can also converted to magnesium sulphate, though this does not penetrate as far into sandstone as calcium sulphate to cause expansion and hence exfoliation.

Argillaceous sandstones are very problematic as they contain clay as the cement for the quartz grains, and clay is prone to absorbing water between the layers of mullite crystals, of

Photo 7.10 Contour scaling on sandstone

which clay is composed. As it does so it expands, particularly near the surface, so the quartz grains are constantly moving about, and the whole structure of individual stone building blocks is weakened. Large numbers of traditional farmhouses and cottages in South Shropshire are built from this local stone, and whilst still robust enough to survive, many have surface decay problems.

Slate is a mudstone or clay, moulded by heat and pressure in a geological age so that its bed is turned through 90 degrees. It is as rich as silica and aluminates, but also contains calcium carbonate, and if one is very unlucky, pyrites (iron sulphide). The calcium carbonate can convert to calcium sulphate, particularly in that area formed by the lap between slates above and below (the area of the angle of creep of rainwater caused by capillarity). The calcium sulphate crystallises and dissolves on numerous occasions, gradually expanding this area of slate, so that it then laminates in the overlap. A similar problem can occur just beneath nail holes, where water can also accumulate. If lamination occurs the nail holes get larger and the slates slip. Pyrites is much more serious, with actual rust corrosion forming and causing holes in the slates. Many imported slates regrettably have this feature, which is why Welsh slate is infinitely preferable even though it is more expensive.

Granite is a dense combination of quartz, mica, and feldspar. Heating and cooling can cause movement around the boundaries of these crystals. The feldspar is an aluminosilicate and can disintegrate into a kaolin-type clay, but this is rare, and in general granites stand up to attack from acid rain provided that their surfaces are not over-heated so as to cause fissures. If this occurs, lime from mortars can wash into the fissures, form calcium sulphate and expand, causing more fissures and thereby setting up a vicious cycle of decay on the surface.

Magnesium limestone has a dominance of the mineral magnesite or magnesium carbonate. When converted to magnesium sulphate by acid rain it can penetrate more deeply than calcium sulphate, causing a much more extensive problem in deep cavernous pits. Again pore structure and its density are key issues.

Frost action

In many respects this is a lot easier to understand than salt infestation, as both result in an expansion within the pore structure which creates havoc, although they can never be confused. Frost action always produces a clean break because freezing water in pore structures tends to happen in a uniform selection of pores in the same vicinity, notably on weathering surfaces (copings and sills being prime examples) and projections of all kinds, such as cornices, string courses, and window hoods. Areas close to ground level are often badly affected where saturation lingers. Again the problem is not confined to stone and is often worse in brick.

The crux of the issue is that water expands by one-tenth of its water volume, when it freezes and becomes ice. Thus a pore structure is saturated then the action of the contents becoming ice will burst the pore wall. Even if the saturation is only just over 90%, considerable disruption will occur. Again pore size and density will be crucial as many large pores may well accommodate the expansion, to at least a limited extent. Pore distribution is crucial, as pores all of the same size in the same vicinity will be universally affected, hence the tendency for lens-like pieces of the stone (or brick) to break away. The break is always at right angles to the direction of the greatest heat loss (the thermal gradient). A less common phenomenon is a whole series of breaks radiating out in all directions, so that a whole piece of stone shatters. Very thin parts of a structure, such as an external balustrade or a coping stone, which are affected by clear night sky radiation or wind-chill factor on all sides, may be so affected in extreme cold conditions.

Diagram 7.4 Frost action on weathering surfaces (after Ashurst and Ashurst, 1988)

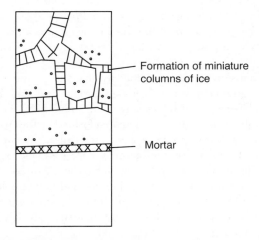

Formation of miniature columns of ice

Mortar

Universal frost shattering on the summit of a parapet
(after Ashurst and Ashurst, 1988)

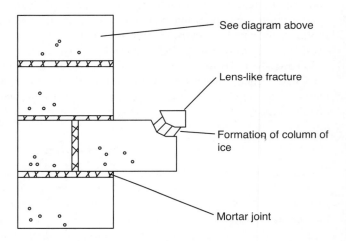

See diagram above

Lens-like fracture

Formation of column of ice

Mortar joint

Frost action on a projecting weathering surface

Wind erosion

Very few sandstone or limestone strata are so weak that when converted to a vertical pillar supporting a porch, portico, or cloister support, the weaker beds start to dissolve in wind-driven rain and pressure alone. The weaker beds of limestone will, however, be more prone to conversion of limestone to calcium sulphate, which expands within the pore structure and is water-soluble, so that wind-driven rain will start to wash it away. Similarly, weaker beds of sandstone, especially calcareous sandstone, can suffer conversion of their calcite cement to calcium sulphate, which blocks pores and loosens the grains of quartz. Sandstone is also prone to importation of chloride via wind-driven rain in coastal areas, and this also gets lodged in the pores, disrupting them and loosening the grains of quartz. The key is the constant wetting and drying cycles to which these pillars are prone on all sides, although frequently it is internal faces

320

of the pillars that are most at risk. This causes constant crystallisation of salts versus water of crystallisation, imparting extra stress. The narrower the gap which exists between pillar and wall, the greater the pressure, creating a characteristic furrowing or ridging effect.

Thermal differentials

It has already been seen that excessive heating can have an effect on marble, causing fissures which then suffer conversion to calcium sulphate. Marble fireplaces are particularly at risk, if in constant use. White Carrara marble was highly prized for such features in the 18th century, and grained marble in the 19th century. Sugaring of surfaces nearest to the fire (the source of sulphur) is frequently seen. The situation is not helped by the over-zealous use of cleaning agents containing inherent salts.

Granites, as already demonstrate, can behave in a similar way, largely due to the fact that the three minerals, quartz, feldspar, and mica, have different coefficients of thermal expansion. The boundaries of these minerals open up, providing an ingress for salts or frozen water.

Heating of existing calcium sulphate blisters will cause them to fall away faster, so in fact decay will become more evident on a hot summer's day. Expansion and contraction of sandstone with pores blocked with salts may produce contour scaling.

Thermal expansion is a key issue in building construction generally and is the reason why many materials such as render do fail. Cement renders in particular cannot hope to contain the expansion which occurs in stone or brick buildings, and consequently crack quite severely. A rise in moisture content will also cause expansion, although in theory the material will cool down, which should counteract expansion as cooling causes contraction. The same applies to expansion by heat, although this should to some extent be counteracted by contraction due to dryness. Most of the time the counter-action occurs, otherwise buildings would be in grave danger of collapse, but climate change is clearly producing long intensive periods of saturation. This will cause expansion and salt mobilisation on a scale never before experienced.

Living organisms – bacteria, algae, fungi, and lichens

Bacteria are present on the surface of every type of building stone, and can oxidise sulphur dioxide to sulphuric acid, or convert ammonia in the air to nitric acid. Both will initiate limestone conversion to salts, as has already been seen. In addition, bacteria produce organic acids with the power to dissolve the silicate cement of sandstone, loosening quartz grains.

Algae are present on stone that stays saturated for long periods of time. Such surfaces are commonly found on the north side of buildings, sheltered from the wind, and away from any solar gain. Many churches have a green wall on the north side. Algae is not in itself dangerous, but it is a sign of extreme damp being present, and thus an indication of potential dry rot for any timber penetrating or abutting such a wall. In addition, algae is food for the rather more destructive fungi.

Fungi, as the name suggests, are parasites living off algae, bacteria, and decaying humic matter (dead leaves) and bird droppings. Fungi produce oxalic and citric acid, both of which can promote limestone decay, as all acids plus a base, such as limestone, make a salt.

Lichens are a subject of hot debate amongst academics interested in stone decay. They have a symbiotic relationship with fungi, which produce water and salts utilised by lichen, and also with algae which produce food by photosynthesis also utilised by lichens. Thus the presence of lichens on a building, beautiful though they might be, is an indication of algae (damp present) and fungi (acid production). Thus, whilst lichens are fairly benign in their own right, being thought to be no more sinister than creating tension in the face of the stone, when they dry out they are in fact an indicator of the rather more sinister fungi.

Errors in design and specification

Design and specification errors are numerous in existing traditional buildings and still occur in new build, although every effort must be made to ensure that they do not.

At a very basic level, the base of every stone building must be allowed to be water permeable, so that trapped water does not mobilise salts in the lower courses. More buildings in town centres and villages than one would care to count are facing a limited lifespan because they are surrounding by a sea of impermeable tarmac, which grows higher by the year as more and more over-zealous government or locally funded resurfacing schemes are put into operation. Meanwhile, the bases of the buildings fronting onto roads and streets melt away, defenceless against the onslaught as salts created by acidic gases from vehicle exhausts which are spewed directly in their base, whilst trapped water mobilises these salts to run amok. The same havoc occurs internally on plastered finishes, causing vast annoyance to occupiers. Chimneys are always subject to calcium sulphate decay, so stone employed in the outer wall of a flue needs to have a particularly resilient pore structure.

Building materials are still placed next to other materials with which they not compatible. Water running over limestone situated above brickwork or sandstone will precipitate lime, which becomes calcium sulphate, blocking the pore structure of the materials below and causing exfoliation. Similarly, water running over brickwork situated above limestone or sandstone will mobilise the salts that are already in the brickwork, in particular sulphates, and import an extra salt loading into the pore structure of the stone below.

Iron and stone are largely incompatible. When faced with the dilemma of how to get a heavy shop sign to swing nicely on its iron bracket, the temptation has always been to drive bolts through the plate holding the bracket into the wall. Not only will water flowing down the building cause any ironwork in contact with the stone to rust, staining the stonework below, but the bolts etc. will expand as they rust, driving a wedge deep within the stone until it crumbles under the force of the ever-increasing girth of the bolt. Iron cramps set in lead were traditionally used for copings or parapets, but experience now shows that stainless steel or copper is more suited to the job.

Photo 7.11 Rising damp reaching an evaporation zone affects internal finishes; an externally sited French drain can do much to alleviate this problem

Photo 7.12 Calcium sulphate decay always occurs at the base of chimneys where hot sulphur-laden gases impregnate the stonework, altering the chemical state of the stone; rendering in cement exacerbates the problem, holding in water to further mobilise salts

Photo 7.13 The salts in the base of this internal brick wall need to be poulticed out

Stone needs to be specified for its location. Weathering surfaces, such as copings, wall capping, cornices, and string courses, need a stone that has many large pores and few small pores that can cope with the extra water mobilisation of salts that will occur. Alternatively, a stone with a very low pore density should be used. Walls in direct contact with wind velocity need a stone that is uniform in its bed structure, and has no weak planes that can furrow out. An understanding of the bedding planes versus the proposed location of the stone is fundamental, as compression forces should be at right angles to the plane of the bed. Thus in a wall, bedding plans should always be horizontal (natural bedding), whilst in a cornice, or stones laid to form a ring beam, where the force is horizontal, stones can be laid on their edge (edge bedding) so that the tension force is again at right angles to the bed. Stone should hardly

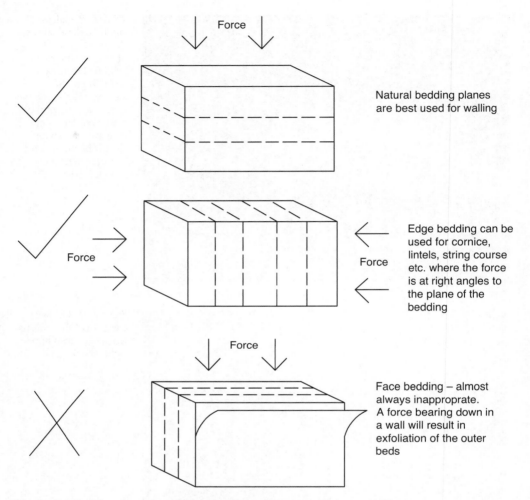

Force

Natural bedding planes are best used for walling

Force

Force

Edge bedding can be used for cornice, lintels, string course etc. where the force is at right angles to the plane of the bedding

Force

Face bedding – almost always inappropriate. A force bearing down in a wall will result in exfoliation of the outer beds

Diagram 7.5 Bedding planes in stone masonry (after Ashurst and Ashurst, 1988)

ever be laid so that the beds are travelling back into the core of the wall (face bedding), as there is a danger that the external face will simply shale off. Vents consisting of a different mineral than the main body of the stone are to be avoided, and calcite veins in sandstone can be disfiguring, although not actually structurally problematic.

Finally, and possibly the most fundamental error, but also one so easily avoidable, is the use of the wrong mortar. Mortar based on OPC (Ordinary Portland Cement) is alien to all limestone construction. This is only common sense, as when lime mortar sets it becomes a linear ribbon of limestone in the joints of the limestone building blocks, and the two are entirely compatible. Cement mortars contain an abundance of sulphates from the gypsum that is part of their manufacture, as well as alkali salts, which wash into sandstone or limestone, causing a white staining. More insidiously, Portland cement and even hydraulic limes contain calcium aluminates when set. These can react with calcium sulphate to form calcium sulphoaluminate (ettringite), which expands at an even faster rate than calcium sulphate, and the movement in the joints can be so substantial that actual cracking of stone building blocks occurs. If calcium sulphate or ettringite expansion is taking place underneath a render overcoat, then the render will develop map cracking – a series of cracks shooting in all directions. This is frequently seen on cement-mortared and cement-rendered buildings. A lime render will be more accommodating to the expansion of cement mortar joints.

This is all aside from the fact that once cement mortar is in place, it simply does not have the pore structure of the stone, so that it forces all water mobilisations to go on in the stone, front loading the salts problem from the very beginning. Stone can stay saturated for longer, and gathers fungi, lichens, and algae at a faster rate on the north side of buildings. Academic debate still rages about the best mortar to use with sandstones and granites, but whilst there can be problems with calcium sulphate washing into the pores of the sandstone from lime mortar joints, this is a small problem compared with the total inability to dismantle a sandstone/granite building at the end of its life if it is bonded with cement. If it is bonded with lime, then at least the stone can be re-used and it issuch a precious and sustainable material that this is fundamental. If builders do nothing else as a nod towards sustainability, then they will have made gigantic strides if they learn to construct stone walls/cladding in lime mortar so that the material can be used again and again. (The same applies to brick buildings.)

Fire

Hopefully fire will be a problem for few buildings, but major fires have occurred in virtually irreplaceable buildings like Windsor Castle. At its most limited in terms of impact, blackening will occur, which will then require cleaning. There is insufficient space here to discuss all the manifest problems associated with reagents, such as acids on the pore structure of buildings. It has already been demonstrated that acids react with limestone to create salts, so it would be extremely unwise to clean limestone with an acid. Cleaning sandstone and granite with hydrofluoric acid has been attempted, with disastrous results. The acid melts the silica cement, and quartz grains cascade down the building. Iron staining is also a grave problem, with natural iron oxides being readily mobilised by acids. Even water can be regarded as a weak acid, and it has already been seen that this mobilises salts, so cleaning blackened stones with something seemingly as innocuous as water would be highly detrimental. Carbon particulates that are a result of the Industrial Revolution are equally stubborn, and cleaning methods just as deleterious.

One of the greatest problems is the use of the fireman's hose, as water causes shattering, mobilisation of salts, and in cases of extreme heat, decomposition of the stone into calcium hydroxide (slaked lime) and the building simply melts away. Sandstone and granite suffer major disruption of their crystalline structure and any iron oxides turn pink.

General repair strategies (applicable for all materials)

All repair strategies must be guided by what conservators refer to as the 'ethics of conservation', although it stands to reason that these will be more rigorously applied the more precious and rare the building is. It must also be stressed that these ethics apply whatever the nature of the materials and construction of a traditional building. Briefly summarised, they are:

1   The building development and its phases must be fully understood.
2   The condition of the building and its materials must be fully understood and documented before any intervention is contemplated.

3   The repair strategy must be fully documented, with all methodologies and materials considered, even those that are rejected, and why they are not feasible.
4   Work should only be undertaken by those experienced to do so.
5   The repair should be the minimum necessary and retain the maximum amount of building material of an historically important nature.
6   The repair should be reversible if technically possible, and should not prejudice another form of repair technique that might be used at a future date.
7   The repair should respect the character of the building, and be aesthetically suitable.

There are further quite stringent ethics for buildings that are of important historical quality which deal with the vexed issue of repairs being noticeable to the eye of a trained conservator not necessarily to a lay member of the public. In addition, there is often a conflict between repairs that are aesthetically satisfactory but which remove historic fabric. In short, ethics are always a matter of debate, and generate many a sleepless night. Nonetheless, some attempt must be made with getting to grips with basic concepts such as minimal intervention. This is the hallmark of the British approach to the repair of historic buildings, and in essence, every building constructed before the Second World War is probably best regarded as an historic building. There will undoubtedly be different approaches, ranging from true restoration of the original design concept, down to very basic refurbishment. These approaches will be driven by a whole raft of standpoints, such as:

•   emotional attachment to the building, for example, identification with the builders or occupiers in the past;
•   cultural values, such as the architectural importance of the building;
•   use values, such as whether it is economically viable to do the repair, and will the finished building serve a social need for housing. In the new 'sustainability-driven' world of today, use values are going to assume an ever-increasing importance.

Repair strategies for stone

These are complex issues and can only be touched on here. Serious stone repair issues on important historic/listed buildings requires extensive research, and consultation with a wide body of experts. There is extensive literature on the subject, which should be consulted, and research continues apace. The following is guidance for the majority of existing humble traditional buildings only, and if any large areas of decay exist an architect or surveyor should be employed to advise, and detailed negotiation with the conservation officer for the local authority embarked upon.

Indentation or cutting out damaged areas of stone, which have been prone to salt infestation for example, and are exfoliating, is not to be undertaken lightly or with gay abandon over the whole façade. A diagram needs to be made of every stone in the façade under consideration and carefully annotated. Only after the most detailed of consideration should indentation be attempted, and only then by a fully qualified stone mason. Generally, decisions are based on age and scarcity value, the replacement of ashlar stone being of less concern than the replacement of carved detail, which should be nursed along until the bitter end. Weathering stones and structural stones, upon which the survival of other stone elements depend, also makes them likely candidates for replacement, if absolutely necessary. The provision of other forms of weathering devices, such as simple lead flashing over a string course or other projection, must also take precedence to cutting out and indentation. For methodologies of cutting out and indentation, and stone selection, further reference to established texts need to be made.

Re-dressing surfaces requires a similar approach. The depth of decay is always going to remain an unknown until the pressure of the chisel reveals it. When faced with such a proposal on a Grade 1 listed church (Photo 7.17) with obvious calcium sulphate decay extensive to the whole of the tower, the re-dressing to be done by operatives whose primary skills was abseiling, the advice from the author (as the area Conservation Officer) was a firm steer in another direction. This was to scaffold the tower and do a detailed survey, mapping out a whole raft of possible options, which may well have included some re-dressing, or indentation, or any of the following. This would avoid 'on the hoof' decision making whilst dangling from a rope.

Plastic stone repair is a misleading term, as the uninitiated might imagine that some treatment with plastic is implied. The term simply refers to making up a lime mortar (lime putty or hydraulic lime) together with the required colour combination of sands, plus stone dust. The sands and the stone dust determine the colour and texture. It would be better described as a dentistry repair, as the decision making resembles that which a dentist has to make about whether to fill a tooth or remove it. It is not an easy option, as only extensive experience and skill will determine the aggregate mix, and weather conditions will be more crucial than ever, the higher up the building the repair required. In reality, such techniques are only possible for very small areas, or where removal of stones for indentation will result in major disruption to adjacent areas, or where the historic fabric is so precious that any other method will involve major loss of adjoining fabric. Even mortar repair requires a ledge to be cut to support the repair and some adequate key to the backing, all of which remove stone, but not perhaps as much as if total indentation was to be contemplated. If done badly on a purely economic basis, the end result will almost certainly not be satisfactory, as trowel finishes very rarely emulate the appearance of natural stone and weathering patterns are dissimilar to those for the stone itself. Colour and texture remain the ultimate challenge, being dependent almost entirely on the aggregate and stone dust incorporated, in addition to over-zealous dubbing out (failure to apply thin layers) resulting in wholesale removal by weathering.

Mixes are many and varied and depend on whether the substrate is sandstone or limestone. Lime can be problematical with sandstone, particularly around the edge of the repair where it could form calcium sulphate and block the pore structure of the sandstone. Aggregate grading is fundamental and can include a whole raft of aggregate types, such as sharp sand, staining sand (soft red or yellow sands), and brick dust, the exact composition of which needs to be determined by trial and error during the making up of samples, with an overall proportion of one part lime putty or hydraulic lime, depending on the season, and three parts of the aggregate. For sandstone it is sometimes safer to use masonry cement, which has an air-entraining agent encouraging a wide-ranging pore structure, although it also includes kaolin mineral powder for plasticity, which can expand when wet. To create a suitably plastic mix which can emulate the pore structure of stone, the ratio of sand needs to be increased to, say five parts to one of the binder.

Grouting involves filling voids in a structure with liquid slurry. It is a major undertaking and a hazardous business. It requires a great deal of attention to detail, removing some joints all down the building to check the flow of the grout, but blocking the same, so that the grout does not run out (temporary stopping up of open joints, traditionally done with tow). Special hydraulic lime mixes (can induce alkali staining) must be employed, as carbon dioxide required to set lime is in short supply inside a wall, plus bentonite clay to ensure flow. Other lime putty slurry mixes utilise a pozzolan, such as pulverised fuel ash, to ensure a set within the core of the wall. The most pressing problem is the weight of the slurry and whether it will cause collapse in that very part of the wall required to be stabilised, or whether it will even reach that part of the wall. In reality, the whole exercise should not be attempted unless dealing with a building of major importance with massively thick walls incorporating a matrix core of mortar and rubble that defy dismantling and rebuilding without major disruption to the whole building. Only personnel with appropriate experience should be employed.

Shelter coating of limestone buildings simply involves rubbing into the stone a lime putty mix, in the proportion of one part of lime putty to two parts fine sand and stone dust to make a slurry-like render the consistency of thin cream. Casein (a protein from sour milk) can be added to give it an extra bonding boost (formation of calcium caseinate). It is then rubbed well in with hessian or a soft brush, without leaving any lumpy residue. The objective is to provide a sacrificial lime-based surface which will weather away or turn to calcium sulphate in preference to the actual stone behind. The treatment needs the same nurturing as lime render (misting and protection by hessian), and this sacrificial surface will need regular renewal.

Limewater is another well-known panacea for decay of limestone by acidic gases. Calcium hydroxide $C_a(OH)_2$ is the basic chemical constituent of the water that rises to the surfaces of a batch of lime putty or lime-wash. As soon as it is applied in very thin washes (letting each wash take on a set by taking in carbon dioxide), it starts to form a new film of calcium carbonate. The reaction which is taking place is

$$C_a(OH)_2 + CO_2 = C_aCO_3$$

The chemistry works, and indeed the methodology appears to work, giving the face of stone building blocks a new lease of life.

Lime render can be used to completely coat a degraded limestone surface. In the case of very ancient buildings, such as churches, there is ample evidence that such a practice was initiated from the very beginning, particularly for rubble stonework: first, the presence of so many mortar joints was a weakness in terms of weathering, and second, because the stonework was lined out, often in red ochre, to imitate ashlar stonework. It is the passing of time, and the fashionable insistence of the Victorians with their 'scrape' methodologies, by which external renders and internal plasters were removed to reveal the structure (truth to material), that have

Photo 7.14 Cracks in cement render due to expansion of the stone underneath – it needs replacing with lime render, a more permanent but still sacrificial overcoat

whisked away the evidence (see Chapter 1, the Victorians, p. 39). Just as an overcoat protects human beings from the cold and wet, so a lime render on a limestone building will protect it from further decay. It has to be realised that when such a treatment is undertaken on a building heavily infested with salts, the render will act as a poultice, the wet render drawing out the salts and lodging them in the render as it dries. Successive wetting and drying cycles will produce the same action, so that at some point in the future the render will become so heavily infested itself that its pore structure will break down and the render start to crumble. This is acceptable, as such a treatment can only ever be regarded as a sacrificial overcoat, to be replaced again and again, when deliberately used as a treatment. Very ancient renders assume a different role, and possibly need to be protected in their own right by multiple coats of limewash. Such renders are often characterised by swirling patterns of pitting, where fines (fine particles of sand) have been eroded out, leaving the more robust aggregate behind.

Photo 7.15 Battens for internal lime plaster on ceilings

Photo 7.16
Unfortunate cement repair of previous salt-eroded stonework on this precious mullion window is actually causing further erosion

Lime mortar repairs (plastic stone repair) needs to avoid any use of cement, as this will only further encourage decay by trapping water behind the repaired area.

A comment on sandstone

Sandstone cannot be protected with lime treatments. They are more likely to cause further decay by blocking the pore structure with calcium sulphate as a result of acidic gases. This aside, the author saw little problem resulting from the use of lime mortar joints in sandstone buildings during two decades of experience on the Welsh border. A trick discovered during this time was to encourage a new skin to form on a building block of sandstone by rubbing it with another lump of sandstone. This was possibly successful because it removed any loose material that may have had calcium sulphate blocking the outer pore structure, and consolidated the silica cement on the new face.

Note: this discussion has deliberately avoided the use of non-reversible chemical consolidants such ethyl silicate or epoxy resins, as being outside the scope of this text. As with holistic medicine for humans, it is better to run with natural treatments in tune with the original material. The advent of the use of silanes may well have revolutionised the consolidation of stone in some of the major British cathedrals, and bought more time, but no one really knows what its long-term benefits will be as only time will tell. The consolidation of limestone with limewater may well have an equally beneficial effect.

Timber – decay mechanisms and repair strategies

Traditional timber framing still has a strong following, and numerous examples of existing timber frame buildings abound. Timber has also been used from time immemorial for all manner of external and internal doors and windows, internal flooring, floor joists, wall cladding, etc. Up until the 18th century most timber was grown and sourced locally, as well as being processed locally. It was the practice of charcoal-burning as well as the demands of the British navy for best quality timber that caused a serious depletion of the massive oak forests of medieval England, not to mention wholesale burning to create agricultural land. From this point onwards it became necessary to purchase long straight softwood trees from the very cold regions of the Baltic state. This was not really suitable for external carpentry, but was used extensively for all external and internal joinery.

With care and attention to weathering, external timber can last for hundreds of years, and its existence in the clay regions of England proves this without a shadow of doubt. The same degree of care, ensuring that damp is kept at bay, will produce the same result for internal joinery. It is the failure to deal with that arch enemy, water penetration, which dictates otherwise.

The anatomy of timber

Understanding decay mechanisms demands an understanding of how timber grows and develops. It is tempting to ignore this aspect in the belief that timber is everywhere about us, and its structure is often visible, particularly in modern fast-grown softwoods. The visible structure is, however, often confusing as it is frequently viewed as a tangential cut through the length of the wood, and not a horizontal cut through its core.

It is also such a magical material that it is sometimes difficult to believe that a mighty oak can grow from a 'whip' or twig that has sprouted from an acorn buried in the detritus from decaying leaves. Generally for all wood, the twig develops a ring of vascular bundles, joined together by a ring of miniscule cambium cells, that also run through the centre of the bundles and which transport minerals from the ground and carbon dioxide from the atmosphere. As a result of the meeting of these chemicals, the process of photosynthesis takes place, whereby cell walls are formed that are composed of a sugar, a form of glucose. This arranges itself in a polymer chain which depends on hydrogen bonding between each glucose molecule. This linear cellular structure, often likened to a collection of drinking straws, is bonded together with hemi-cellulose and is protected by two structures: first the bast, then the external bark.

The analogy of paper drinking straws is a useful one as the cell walls hold water, as does the centre of the cells or tubes, just as paper drinking straws hold liquid, and if left in the surrounding liquid long enough the paper start to absorb it, soften and swell. Also if denied moisture the centre of the tube will dry out first, then the walls of the straws, whereupon they will contract and possibly if left dry long enough will go brittle. The cellular structure of timber behaves exactly as same, the common link being the cellulose of which both the paper drinking straw and wood are constructed. Timber, however, has another key extra ingredient, lignin, which exists in the cell walls and penetrates the cavities between the cells and glues them together. (This natural glue can be re-utilised when timber is processed into a board made of woodchip, the heated lignin once again acting as glue in the compressed board.)

As the cells multiply they form the much larger cell structure of the wood, and define the structure that generally demarcates the difference between softwood with thin cell walls and large cell cavities, and hardwood with thick cell walls and small cell cavities. The original twig becomes the pith, and the older cells around it, laid down year by year as growth rings, and near the centre, are termed the 'heartwood' and contain larger deposits of lignin, and tannins, hence their darker colour. The newer growth laid down in yearly deposits is more likely to contain sapwood, and is more vulnerable to all manner of decay from fungi and beetle infestation.

Softwood contains about 90% of the long tubular cells, called 'tracheids', which perform the main function of conducting nutrients via the springwood, which has larger cell cavities and thinner cell walls but also gives structural strength, plus the summerwood, which has smaller cell cavities and thicker walls. In hardwoods, such as oak, there are larger cells called 'vessels'

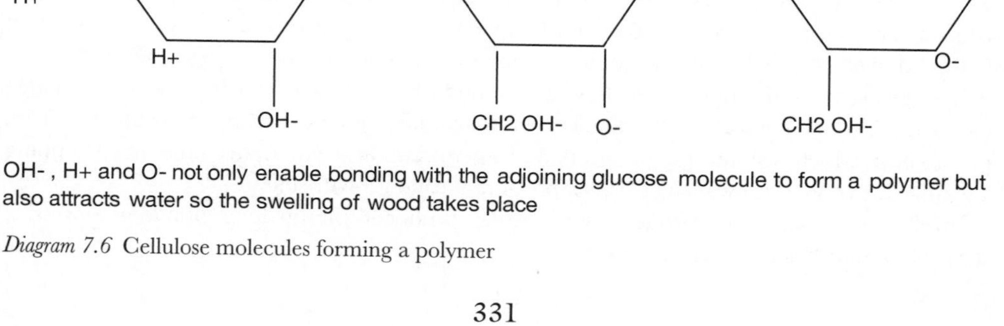

OH- , H+ and O- not only enable bonding with the adjoining glucose molecule to form a polymer but also attracts water so the swelling of wood takes place

Diagram 7.6 Cellulose molecules forming a polymer

Photo 7.17 Cross-section of Canadian slow-growing softwood to be used for carving a totem pole, needing longevity

that conduct the nutrients, which appear as pores, and hardwoods generally also have rays across which it is easy to split the wood using a wedge or froe, and cleave it across the grain. This grain is formed by the arrangement of the fibres of the wood, formed by the tubular cells.

Wood is described as 'anisotropic' because it has the ability to change its appearance depending on which angle the grain is cut through. Cutting horizontally across a section will reveal the growth rings and in hardwood, the medullary rays. Cutting vertically reveals the growth rings in a vertical format, and cutting tangentially through a vertical plane produces a 'figure', highly prized by furniture makers. This latter requires quite a lot of imagination when studying the following diagram.

The cell structure has many uses. As well as absorbing water, the cell walls and cell tubes hold glue very efficiently, meaning that two pieces of timber can be well bonded together. The use of steam will soften the linear cell walls, meaning that they can be bent into curved shapes (hence the bentwood chair), and the linear cell structure also imparts considerable strength in tension, so that timber could be very effectively used for tie-beams, collars, ridge purlins, and side purlins. It also has disadvantages, depending on the direction of any cut. Wood can swell quite happily when it is in the round as it has plenty of room to spread over the cross-section of the tree trunk, but when cut into a floor board and accidentally saturated in situ, implying that each board is constrained by its immediate neighbour, the cellular structure will swell (with water in the tube and in the cell walls) and rebel against the constraints of its neighbours. It will rise upwards, creating a cupping effect. Conversely, joists cut out of the round, and then seasoned, will experience shrinkage in the opposite direction to the pattern of growth rings, with cupping towards what would have been the pith (see Diagram 7.9 showing shrinkage patterns). This is in contrast to a radial sawn board, i.e. cut across all the growth rings from pith to bast, which will universally shrink on its perimeter and experience little or no cupping because shrinkage is universally across a portion of each growth ring.

Swelling with water also produces weakness in timber so that its longitudinal compressive strength diminishes as it wets up.

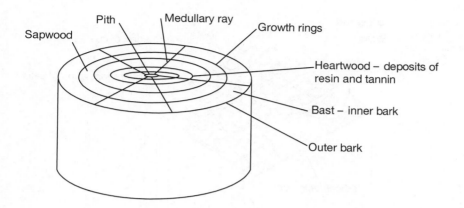

Horizontal section through a tree

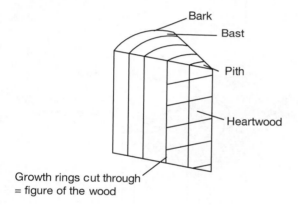

A wedge shaped section cleaved
along the meduliary rays

Diagram 7.7 Tree sections
(adapted from Desch, 1947)

Decay mechanisms: dry rot versus wet rot and fungal decay

The basis of this form of dry rot (serpula lachrymans) decay is a concerted attack on the cellulose cellular structure of wood. The process is essentially a reverse process to photosynthesis, whereby the cellulose and hemi-cellulose are broken down into their original sugar components. These in turn react with oxygen to produce carbon-dioxide, water, and energy for growth of the fungi. Lignin is not affected, and because it is dark in colour it starts to predominate, giving the wood its characteristic dark appearance following an attack by the dry rot fungus. The latter is one of a serious of so-called brown rots, so named because of this effect. In addition, the wood becomes much lighter in weight, loses strength, and a very characteristic pattern occurs consisting of large cubicular cracking. This will often be picked up underneath a paint film covering, with the horizontal cracking appearing to predominate.

The onset of dry rot is initiated by water penetrating wood, usually from a localised source such as a leaking gutter or down pipe saturating masonry against which or into which wood is

Diagram 7.8 Cell
structure of softwood
versus hardwood
(adapted from Desch,
1947)

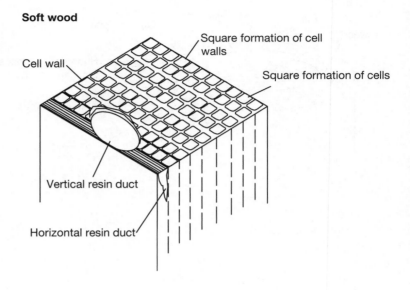

Soft wood

Square formation of cell walls

Cell wall

Square formation of cells

Vertical resin duct

Horizontal resin duct

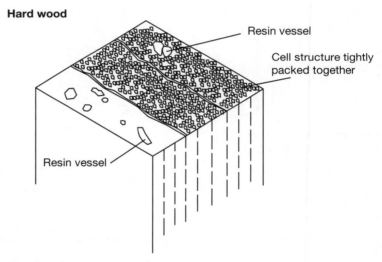

Hard wood

Resin vessel

Cell structure tightly packed together

Resin vessel

situated. The saturated wood is host to bacteria and micro-fungi which start to break down the surface of the wood, thus allowing even more water penetration. It is then that a minute spore of dry rot can germinate, rather similarl to bean sprouts. The first sprout is tube-like and penetrates the wood, only to subdivide into filaments called 'hyphae'. These produce enzymes which break down the invaded wood even further, making it even more receptive to water penetration, and increasing the water content by virtue of the breakdown of the sugars, of which water is a by-product. The hyphae eventually form a fluffy white cottonwool-like mass of mycelium, which upon closer examination can be seen to emanate tear-drop like globules of water, hence the name Merulius lacrymans given to dry rot, 'lacrymans' being the Latin for 'weeping'.

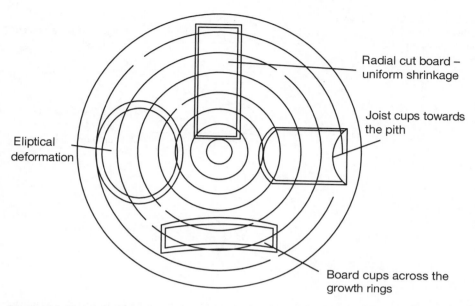

Diagram 7.9 Shrinkage format – determined by growth rings after drying (adapted from Desch, 1947)

Photo 7.18 Dead mycelium emanating from a timber block in brickwork in a Victorian school

Photo 7.19 Desiccated dry rot on a cill in a Georgian building

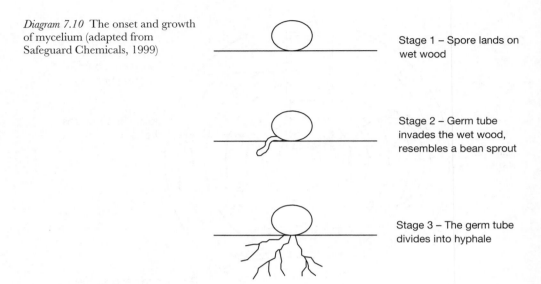

Diagram 7.10 The onset and growth of mycelium (adapted from Safeguard Chemicals, 1999)

Stage 1 – Spore lands on wet wood

Stage 2 – Germ tube invades the wet wood, resembles a bean sprout

Stage 3 – The germ tube divides into hyphale

In drier conditions mycelium adopts a very sinister grey skin-like appearance stretched across areas of timber, like something out of a horror film. In fact this appearance is to be welcomed because it means that the mycelium has encountered fresh air and started to shrink backwards, ventilation being the most efficient form of dry rot treatment. The larger hyphae, called rhizomorphs, which form part of the mycelium in a wet zone, continue to invade even mortar joints in the persistent desire to find the next timber food source. The author has seen mycelium climb chimneys via the mortar joints in an attempt to discover a source of food. Cunningly, the hyphae carry water and nourishment with them until they reach their next destination, thus potentially being capable of softening the next timber victim. In addition, the calcium in mortar is utilised by the fungus as an alkali to neutralise the acid produced by the breakdown of the cellulose. It is this ability to travel through mortar, distances in excess of 2 m (2 yd) and produce an annual growth rate of up to 4 m (4 yd) per annum, which makes dry rot so very destructive and likened to cancer in its behaviour. A 20% growth rate alone can be achieved on food reserves held in the mycelium, Amazingly dry rot favours both the acidic conditions of wood and the alkali conditions of mortar. (Data from publications by Safeguard Chemicals, 1999.)

The next phase is the growth of a fruiting body, the actual fungus itself. This is occasioned by what the fungus deems to be a stress situation whereby the mycelium encounter drier wood, or where two different strands of mycelium meet, and occasionally when light is encountered, itself a stressful scenario for the mycelium and yet another hint as to what can be used as a cure. The fungus resembles a ginger-brown/orange plate but with deep craters, folds, and pits. The spores form on what is termed the 'basidium', the spores only being released by virtue of pressure from a droplet of water that forms a sphere and thus ejects the spores into the surrounding air. In a particular severe situation, a pile of reddish brown dust will be seen coating surfaces below the fruiting body. The spores can live up to 3 years, at say 20°C, waiting for the right conditions in which to germinate, and can remain in damp masonry without a suitable host for around 12 months. At even lower temperatures, such as those encountered in a cellar or in floor voids, the survival rate can be extensive, up to 10 years. It is thus essential that in these areas, as with all infected masonry, existing and new timber is isolated from any masonry, using inert materials such as plastics, even though there is a tendency for the latter to harbour condensation. This problem indicates how difficult dry rot is to combat, as the condensation formed at the junction between wood and plastic will soften wood, starting the cycle of bacteria and micro-fungi all over again, even though the plastic might protect the wood from major catastrophe from well-advanced mycelium.

Continued growth and survival of the spores is highly dependent on the quantity of water, the exact percentage of which is the basis of much argument that rages amongst professionals. Clearly a good saturation is needed for the spore to thrive and sprout, say 28–30%, but the mycelium really only need a moisture content in excess of say 22% in stagnant conditions. Typical areas are behind skirting and window reveals, to the rear of lath and plaster on studwork walls used to clad masonry in Georgian buildings, timber lintels over windows, beneath floor-boards – a favourite place for carpenters to leave their timber shavings – under window boards, and in cellars, to mention just a few places. Thus to be on the safe side, moisture content of timber should be 18% or under.

Wet rot

In comparison, wet rot is remarkably innocuous. Called by its Latin name of Coniophora Puteana, it requires a higher moisture content to be sustained over a long period. It can be dangerous in that as the moisture content dries out to, say 22% and the wood has been considerably softened by the wet rot, then dry rot can be stimulated into growth, the spores finding a ready host in the softened timber. This can often be witnessed where a localised outbreak of wet rot in say, a window lining is accompanied by an outbreak of dry rot further into the building at the extremity of the moisture penetration. Wet rot produces a mass of white hyphae also, but has an effect on the lignin and so tends to reveal a white tinge on the wood, which is the cellulose; hence it is termed a 'white rot'. Wet rot is much more complex in many respects, as there are many varieties. The cubicular cracking, itself a product of differential drying due to the robbing out of the moisture content, is smaller in size than that for dry rot, and this factor alone is very useful in distinguishing between dry rot and wet rot. The green leathery flat plate of the fungus, with a nobbly surface but with no craters or pits, is another telling factor.

Other types of fungal decay

Various other fungi exist, but they do not have the unabated destructive effect of the brown rots. They include the blue, green, or yellow wood-staining that causes paint surfaces to flake

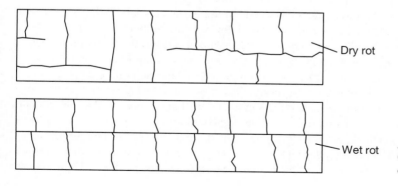

Dry rot

Wet rot

Diagram 7.11
Cubicular cracking –
dry rot versus wet rot

away, but do not lead to any appreciable loss of strength. Gilli fungi, manifesting as black straggling strands and emanating from saturated laths, are a particularly frightening sight emerging through ceiling plaster.

Repair strategies for fungal decay

The repair strategy is both simple and yet complicated because its success cannot be guaranteed due to the ability of dry rot to travel through masonry at an alarming rate. Thus what has started as a localised ingress of water, which by common sense can be terminated by simply removing the supply of water (mending the gutter, downpipe, slate roof, or chimney flashing, etc.), thus changing the environmental conditions, can be effective as long as the outbreak has not spread to all surrounding timber areas. In some cases the outbreak can be halted in its tracks simply by opening all windows on bright and breezy days, although not completely eliminated as it will re-commence if the water source has not been curtailed as well. In a previous era, notably after the Second World War, when the problem first started to manifest itself on a large scale due to the saturation of buildings following fire damage during the Blitz, the most effective means was to cut off the food source by removing all timber around an outbreak. The distance varied according to the practitioner, and could be anything from 2 m (2 yd) upwards. This obviously resulted in the loss of much historic fabric, and it is not a practice which is championed today. A variety of other techniques can be used for areas of building that can be isolated, such as raising the temperature extensively, and for smaller artefacts, freezing to lower the temperature dramatically. None of these are practical on a large scale. In fact, the only methodology which is successful revolves around completely altering the environmental conditions within the affected building, having first effectively curtailed any possible source of water. This involves great attention to all possible areas of ingress, constant monitoring of former areas of water ingress (see below), the use of dehumidifiers to remove excess water, and a copious amount of ventilation in appropriate weather conditions (dry and windy). Certainly all affected timber needs to be removed, but only up to the next piece of sound timber. Pasting all suspect remaining timber with a boron-based paste is fundamental. Boron is an alkali metal, which is only triggered by contact with water into producing a strong alkali, which fungus cannot cope with. The products most commonly used are: Boracol, for flooding walls by injection (a complicated and not always successful procedure because penetration is unknown); boron paste, by far the most effective for suspect timber; boron rods, inserted in particular into the end grain of timber, such as the ends of rafters and joists in masonry walls; sodium borate, otherwise known as borax, is the boron-based compound for these treatments. These treatments have by and large, and very thankfully, replaced the use of pentachlorophenols mixed with sodium hydroxide in a solvent solution, all highly noxious and dangerous but still in use in the very late 20th century. Zinc oxychloride paint was also still in use in this period, for isolating timber from masonry. Very traditional techniques utilised wood creosote (i.e. Stockholm tar, formed by rendering wood to release its natural resins and tannins), coal tar creosote or solignum (a tar oil from coal), and copper and arsenic salts, the former still a component of tanalised timber.

One of the biggest problems is itself the result of traditional building construction: the huge thickness of a stone or even brick wall, and the length of time that this takes to dry out, thought to be around 25 mm (1 in) per month in good drying conditions. For walls which can be as much as 600 mm (2 ft) in thickness, this drying out can take up to two years. Timber specialists use timber dowels bored into the wall, which have been weighed beforehand, and then weighed again after the elapse of, say 1 week to calculate the moisture content of the wall and map its diminishing course. Dogs trained to sniff out the metabolites produced by the fungus can be a useful aid in locating outbreaks, and endoscopes/boroscopes, a fibre optic illumination on the end of a rod which can be inserted behind linings, are a useful device but can take some degree of practice to achieve success. Cobwebs can look remarkably like hyphae in the bright light and the large turning circle of the light somewhat disorientating as to what exactly is being viewed.

Another long-term and in fact essential component of monitoring, post treatment, is the siting of electronic sensors in key points of former water ingress or areas of potential water ingress, which are wirelessly connected to a central data logger, and can be downloaded onto computer-generated charts. Any rapid moisture ingress will trigger an alarm system, remotely if necessary, to the human contact so that preventative action can be taken to literally save the life of the building. Unfortunately many contracts, having reached their culmination with a massive deficiency in the contingency figure, all too readily dispose of this laudable aim.

Thankfully, most major historic houses in the care of major bodies, such as the National Trust, use individual data loggers in each room to detect any sudden changes in temperature and moisture content.

Photo 7.20 Plastic studs for holding a steel mesh, designed to resist a further attack by dry rot in an already infested building

Insect decay

Beetle infestation of timber is again a function of the fact that trees are a factory for making sugars. Cellulose is formed as a combination of water, carbon dioxide, and sunlight. It is an ideal food for timber beetles, in addition to which they utilise the nitrogen in the sapwood for making protein. This is why fungus infected timber is so attractive to beetles, as the fungus produces nitrogen. In addition, the timber is softened by the action of the degradation of the cellulose, so that the beetles, who use their head structure to push through the timber, find that it offers less resistance. In the same way softwoods offer less resistance than hardwoods, and thus are more affected, and sapwood is easier to traverse than heartwood, so the outer part of a timber member is normally more affected than the inner core. This was one of the reasons why timber beams in ceiling frames were chamfered, as this removed the majority of the sapwood, and once chamfered the ends had to be stopped, providing a valuable dating criteria from the many styles that were adopted from the Middle Ages to the 18th century when visible beams became unpopular. Ceiling frames (the lid over the ground floor) were actually rarely affected by beetle infestation as the timber was often dry from the constant fire, and beetles abhor dry timber with a moisture content of less than 10%. A much more likely location is the ends of beams and joists buried in masonry walls, and roof timbers that have from time to time been affected by the elements due to leaks in the roof cladding. It is not unusual to see the sapwood on roof timber components badly affected by furniture and death-watch beetle, the tell-tale signs being the size of the flight holes, small and almost pin-pricks for the former and large up to 2 mm ($\frac{1}{16}$th in) across for the dreaded death-watch beetle. Beetles produce flight holes in the period April–July when they emerge to mate and lay eggs in cracks and crevices of softened wood. In the process of making these flight holes bore dust is generated, which in the case of the furniture beetle is very fine, but in the case of the death-watch beetle is in the form of bun-shaped tiny pellets, so-formed so that they can be kicked behind them so as not to block the passage forward. Traces of this brown dust or pellets, freshly generated, is a warning sign that should not be ignored. A worrying saga I experienced many years ago was an invitation to lecture at an Irish Castle, which had a newly restored medieval style roof atop a previous roofless ruin. The chairs for the delegates were peppered in brown dust, a sure sign that the walls were still saturated, and were wetting up the new roof components to the evident delight of a resident beetle population.

An examination of the boundary of flight holes always pays dividends. If they are sharp and creamy in colour they are fresh, and if grey and dirty they are redolent of inactive beetle.

The furniture beetle is far more common than the house longhorn beetle, and both are more common than the death-watch beetle, but the latter is more active in the south-west of the British Isles, where the rainfall is highest. Oak is the most resistant timber for the furniture beetle, which as its name suggests favours small items of softwood, chairs and table legs being the all-time favourites. Unfortunately they can rapidly spread into the main softwood structural timbers and claddings, skirtings being a prime target, and the outer areas of structural beams. These rapidly resemble a crumbling mass of chewed wood, called frass. Many old beams show scoops where such frass has been cut away. The death watch beetles does attack oak and other hardwoods, and can be very destructive. The eggs hatch into larvae, which proceed to burrow channels along the grain ranging up to the entire length of extremely long timbers, such as purlins and wall plates. As each female lays eggs in the hundreds, the subsequent larvae can create havoc, particularly for joints, for example the scarf joint in a purlin. The death-watch beetle strikes a particular fear due to the knocking sound it makes when constructing the flight hole to enable its emergence for mating purposes.

Photo 7.21 Timber joint attacked by beetle – the area of affected timber can be cut out and replaced with sound timber

Photo 7.22 Ancient floorboards affected by death watch and furniture beetle demand careful decision making regarding the cutting out and replacement with similar widths of board

Repair strategies for timber

Much depends on the scantling of the wood, ass major roof and floor timbers will rarely be so attacked as to have lost all integrity, whilst floorboards may look sound on the surface but in reality have the strength of paper. Many a surveyor has known that sinking feeling as they disappear through what appears to be a sound area of floorboards. Sadly, wood attacked by death-watch beetle in particular is generally irretrievable, particularly on its outer areas where frass will have formed. Frass is now rarely cut away as it is thought to form a poultice if soaked in a solvent-based timber treatment. In many instances, however, it will be better to cut away any frass and thoroughly paste any remaining wood with an insecticide treatment. It is sensible to eschew brush-on and spray methods, as they rarely penetrate more than say 2 mm ($^{1}/_{16}$th in), and sapwood can certainly exceed this, leaving resident beetles happily ensconced and ready for the mating season next spring. April to July is obviously a good period to attempt treatment, as the beetles will be making their way to the surface to emerge. Traditional treatments involved the use of sulphur candles, sadly as fatal to humans as to the beetle, and more pertinently spiders, probably the most efficient natural predator and are thus to be encouraged rather than killed off. As with fungi, the most fundamental treatment is to remove any possible sources of water ingress than can soften wood, and provide lots of ventilation in dry breezy periods so that timber can dry out and remain dry. Fortunately eaves ventilation is now a requirement of the Building Regulations. Unfortunately heat is another requirement for drying out timber, but feeding heat into the roof is contrary to the Building Regulations. Few householders are going to want to heat the loft, and indeed most roof insulation is designed to produce a cold roof by the laying of insulation across the top of upper-floor ceiling joists. Warm moist air is of course taboo, and it may be one of the reasons why so many traditional buildings have started to suffer enhanced beetle infestation, because of the gargantuan quantities of condensation produced by constant showers, baths, and boiling of kettles. Beetles also dislike strong alkalis and so the traditional treatment of ceiling frames in particular, with layer upon layer of limewash, was an ideal deterrent. The Victorian fashion for blackening timbers was unhelpful, unless it was with Stockholm tar, i.e. wood-based or coal-based creosote, and the modern fashion for baring all was equally unhelpful in inhibiting beetle infestation. Coal tar used in creosote contains tar acids such as phenol, xylol, and cresol, together with pyridine, chinoline, and acridine, and as such was possibly a useful way of preserving timber, all these substances being noxious to beetles and fungi alike, so the Victorians may have been correct in their use of it for preservation, but in all events tar oil distillation is no longer in operation, and the substance deemed hazardous to humans. In existing buildings where roof timbers are prolific and locked away out of sight, as so many roofs are, limewashing should once again be given serious consideration. Boron compounds also possess insecticide properties and have the added value of being flame retardant. This treatment is going to be more effective on damp wood. Boron rods are particularly useful when inserted into the ends of timber adjacent to masonry which can be wetted by contact with the latter. Joists, beam endings, and window linings are also ideal candidates.

Fired earth, brick – mineral composition, decay, and repair

The chemical constituents of bricks are equally, if not more complex, than those of stone. This is due to the nature of the clays from which they are fired. Clays are formed from weathered old rocks, usually granites or gneiss, deposited in still water to form sediment, which gradually shrinks as the water dries out. The water can have gathered in estuaries, lagoons, or a deep abyss in ancient rock formations. Thus the chemical structure of clay reflects that of the rock from which it is formed, usually the igneous rocks, such as silicon dioxide (silica), aluminium oxide (alumina), potassium, sodium, iron, magnesium, carbon, calcium, vanadium, chromium, and titanium. If pressure has been applied in a geological phase, then a shale or mudstone is formed, and this is the basis of many argillaceous sandstones. Clays thus contain a range of minerals, metals, and salts. If a deposit of sediment has been moved by a glacier and then redistributed by wind to re-deposit, it forms brickearth, a sandy clay which is excellent for brick-making, especially for very finely sculpted bricks such as those used for so-called gauged brickwork or moulded features (see the discussion on rubbed bricks). The weathering produced hexagonal crystals consisting of layers of silicon oxide SiO_2 and aluminium oxide Al_2O_2 where the aluminium and silicon are positively charged and the oxygen is negatively charged. The crystals are very tiny, less than 2 microns, with each crystal being formed of several hundred wafers, each composed of silicon oxide and aluminium oxide alternating according to the clay type. Illite or montmorillonite clays have wafers composed of two layers of silicon dioxide and one layer of aluminium oxide. The oxygen makes the individual wafers very attractive to water, via the hydrogen bond, which is why clay has such a propensity for swelling. The presence of iron oxides in clays FeO_2 also enhances the attraction of water. Some clays swell less than others, notably the illite clays which also contain calcium, is positively charged and has little attraction for water, meaning less swelling and less shrinkage. Kaolin is a very pure clay, with alternating layers of silicon dioxide (silica) and aluminium dioxide (alumina), but with no iron so swelling and shrinkage is more limited. Some clays contain sodium (Na) and potassium (K), which with silica and alumina are all present in igneous rocks. Other minerals present include: calcium carbonate, which fires white or cream; organic vegetable matter, which produces black cores; manganese, which produces a brown hue; chromium, a pink hue; vanadium and copper, a green tinge; cobalt, the blue or black bricks so beloved by the Victorians; and antimony, which produces yellow bricks. Iron produces black or brown bricks. These distinctively coloured bricks are not just a product of their chemical constituents, but also of how much oxygen is available in the kiln, a high concentration producing the characteristic red tinge, and a reduction firing producing the greys and buffs.

Clay was traditionally quarried in the autumn, to allow frost action to break down the clay before it was transferred to a pug mill to tease out the stones and larger particles. This opportunist natural weathering was not as easy as it sounds because it involved considerable labour in treading, raking, and spreading the newly won clay. Small piles of clay were then delivered to a bench where a man scooped the clay into a portion resembling a 'butter curl' – a clay warp – whereupon he threw it with great velocity into a sanded mould box, cut off the top with a wire, and placed it on a wooden stock board ready for delivery to a drying stack. The tell-tale linear bulges of clay to be found on traditional bricks betray not only the diagonal stacking to allow maximum warm air to penetrate the stacks, but also the sides of bricks display the wrinkles formed in the production of the clay warp. This drying was absolutely essential: water trapped in the clay layers would boil once placed in the kiln, and the energy so formed would blow the bricks apart.

The firing process at temperatures between 850 and 1,200 °C fused clay and sand together, but more pertinently fused aluminium silicate together, the combination being called 'mullite' and dispersed amongst the quartz. If potassium and sodium were present, and the temperature high enough, the quartz would form an amorphous mass of glass containing the mullite.

This process underlines the understanding of decay mechanisms, or put quite simply, the more glass the less chance of pore structure, which means less decay by invading and inherent salts, acids, and alkalis, and frost crystal expansion. This explains why some early bricks, such

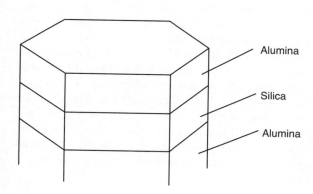

Alumina

Silica

Alumina

Diagram 7.12 Clay – a series of hexagonal crystals, each with several wafers of aluminium and silicone dioxide (after Ashurst and Ashurst, 1988)

Crystalline aluminium silicate 'mullite' and quartz with porosity

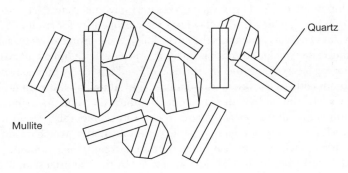

Non-crystalline mass of glass around the quartz with minimal porosity after intense firing (vitrification)

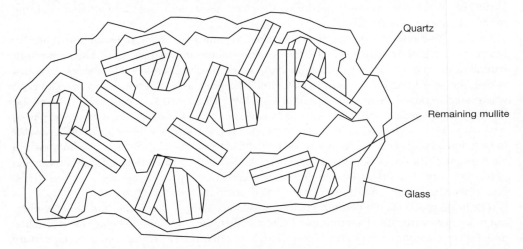

Diagram 7.13 Brick after firing (after Ashurst and Ashurst, 1988)

as those used in major buildings like Hampton Court, which were often intensively fired at least on the face, are so resilient to decay. In contrast, many 18th- and 19th-century bricks made in local brickyards, of which there were numerous throughout all clay areas, were often poorly fired. These bricks as a consequence of being 'green' have an unburnt clay content liable to swelling, and many small pores. If the clays also contain insufficient sand, and lack potassium or sodium, this negates the possibility of glass formation around the mullite. As a result, the bricks have a mass of tiny pores which cannot hope to cope with salt crystal growth, the salts being inherent in the clay or from external sources. Much also depended upon the quality of the kiln, so that by the 19th century the development of down-draught kilns, which fed hot air up the bag walls of the kiln and down into the stack standing on a dished floor, were generally more efficient than their brick clamp predecessors. There were other improvements, such as the introduction of the Archimedean screw in the pug mill process, and wire-cut bricks. It was the invention of the Hoffman Kiln in the late 19th century, a long sausage-shaped bunker with many fireholes, which produced even more efficient firing.

Photo 7.23 The Hoffman Kiln at Pant, Shropshire: this example was used for lime burning, but other similar kilns were used for brick firing

Photo 7.24 Interior of the Hoffman Kiln showing the fire holes

Porosity is the main enemy of brick survival, as the majority of bricks have a volume of pore space up to 80% of the total volume, and the majority of these are interconnecting small pores. Intensive firing and the production of glass around the mullite (crystalline aluminium silicate) not only resists salts but also attack from acids and alkalis, and frost. The latter is a persistent problem on weathering surfaces and brick projections on buildings.

Dimensional stability of new bricks is an issue for all building construction projects. Bricks freshly out of the kiln instantly start to take up moisture within the pore structure and expand. This expansion, very rapid in the first two days, does not cease until three to four months have elapsed. When lime mortar was the order of the day, this expansion was not a problem even when bricks were used in their expanding state, but with cement mortar this can produce disastrous results.

The decay of brickwork

This has already been shown to be a function of porosity, and with most early bricks having limited firing, unless placed very close to the heat source, then expansion within the pore structure of salt or frost crystals is inevitable if incorrect mortars are used. It is thus fatal to use such bricks with anything other than lime mortar, which will cope with the water mobilisation before the bricks have to, and hence salt or frost crystals form in the porous mortar in deference to the bricks. It is most certainly a waste of time re-using handmade and under-fired bricks for secondary features such as garden walls or garden paths. The latter in particular will disintegrate into mush within a short space of time, being no match for soil-bound salts.

Terracotta has had much more controlled firing available since the late 19th century, with more refined kilns and thus generally performs better with the primary problem generally being the corrosion of the steel or iron frame that acts as the supporting structure for terracotta.

Another major factor in brick decay lies in external sources such as unequal loading patterns and more pertinently thermal expansion. Brick is particularly prone to this, hence the need for

Photo 7.25 Frost action on garden path exacerbated by ground-driven salts

Photo 7.26 Frost action in garden wall, also showing salt decay from the ground behind

expansion joints in long areas of brick boundary walls. Thermal expansion is often noticeable as cracks running from the cill of windows in an upper range, to the lintels of windows in a lower range, because of expansion in a long unbroken area of brickwork between the two ranges of windows. Wrapping brick buildings securely in a cement corset will produce major cracking, the brick and the cement having differential thermal expansion coefficients. The use of cement for pointing or renders is taboo for anything other than engineering brick, which would not in any event need a render. Well-fired modern bricks will sustain cement pointing but not a cement render, which will crack upon expansion of even modern bricks.

Repair strategies for brickwork

Unsurprisingly, the key ingredient of all repairs to existing brick structures is lime mortar. Joints should be raked out only if the mortar is obviously soft and crumbly, and then to a depth of twice the width of the joint. Most vernacular buildings do not demand detailed analysis of mortar mixes, although buildings of key significance, and of great age and scarcity value, most certainly will. Generally, a standard mix of one part lime to three parts of well-graded sand is sufficient for most vernacular buildings. Georgian buildings may have at least one elevation in gauged brickwork which has deliberately very fine joints, or at the very least has rubbed brick arches, with similarly fine joints that will demand a different mix and technique (see below). Joints need to be well rinsed out with water to remove all loose material and to remove any suction, and well packed with the new mortar, applied gradually so that there is a continuous build up, having allowed the preceding applications to take on a set. Application in brick joints is always best achieved by the use of a pointing iron, specifically made for the width of the joint. Trowel application is unsatisfactory for narrow brickwork joints, as the mortar cannot be packed well back into the space left by raking out. The balance between water in the mix evaporating and carbon dioxide entering needs to be carefully monitored, with the work being well protected with hessian sheeting, damped down in very dry conditions, or used dry in wet conditions. The sheeting needs to allow for at least a 200 mm (8 in) gap for the taking in of carbon dioxide. After it is clear that an initial set has taken place, the joints need to be well tamped with a churn brush to bring the aggregate to the surface. This removes the 'fat' lime and provides a better weathering surface, ensuring a longer lasting repair.

Photo 7.27 Patching old stone walls with cement is a recipe for further decay

Photo 7.28 Cement render can rarely be totally removed

Mass repointing of existing brick elevations is rarely necessary. Some mortar joints or renders of very ancient vintage showing a characteristic swirling pattern due to the fines being weathered out of the mix by wind erosion, reflecting the pattern of mixing (similar in characteristics to the wind erosion effect on stonework affecting the weaker beds). Often the residue of the weathering action is very robust and should be left intact, particularly if it is thought to be 17th century or earlier.

Render should generally be treated in much the same way, with only selected areas being cut out and replaced with the same one part lime to three parts well-graded sand. Hydraulic limes are now more generally used on the basis that such a large exposed area will be subject to intense weathering conditions. There is still a case for lime putty based renders when the work can be done at temperatures and humidity levels that are perfect, such as a still cloudy day with a reasonably high relative humidity, and a temperature of around 18 °C. Providing that the required set is achieved, and the lime is not entirely pure but contains some natural pozzolanic ingredients as a result of firing with coke, then the resultant render should be enduring. Gas-fired limes with a high calcium carbonate content will almost invariably mean some glitches in the end result of render or mortar joint repair.

Gauged brickwork

Gauged brickwork is a practice which has long since died out, so any remaining examples are particularly precious and require special techniques in terms of repointing or cutting out and indentation of replacement bricks. The bricks themselves are now difficult to obtain, so each and every remaining example needs to be carefully husbanded. If replacements are required, then specials are normally made from a prepared wooden mould. Fortunately they rarely decay from salt or frost action, although differential loading of an arch may cause cracking. Made from the finest brick earth, a clay with a high sand content, traditionally obtained from Kent, Berkshire, or Sussex, they were fired to a point just short of total vitrification. They were then literally rubbed on a 'rubbing stone' – usually a flat area of reasonably course limestone – to

Photo 7.29 Cement pointing on gauged brickwork has fortunately not extended to the brick arches, or as yet caused much decay overall

ensure the sides were straight and square. This involved much use of the set square to ensure absolute precision.

Special shapes required for, say a cornice or wedge-shaped bricks required for an arch, were cut using a frame holding a two metal wires twisted together, and following a template held against the secured brick, the latter by means of a strut and bridge, clamping the brick firmly to the base board. A great deal of extraneous matter in the form of sand was produced in the process. Whole elevations of gauged brickwork with miniscule joints required only the rubbing process, but rubbed brick arches continued to be used until the First World War using this laborious process.

Mortaring the bricks was quite unlike the normal process. The bricks were dipped into a pail of water to ensure that all the air in the pore structure was replaced with water, thus negating any possibility of drying out the thin film of mortar to be utilised. In addition, the silica was encouraged to come to the surface, and this achieved a form of case hardening which also accounts for the longevity of these bricks. This was achieved by dipping the brick in a box containing little more than lime putty, sometimes with some very fine sand added. The dipping motion depended on the arm resting on a batten laid across the box to ensure only the surface was coated. Shellac (a sticky exudation from the lac beetle) and white lead was sometimes used, especially if more modification of the brick shape was required to be done in situ. Sliding the bricks into place often results in a fine bead of mortar being exuded, which needs to be cut away with a sharp knife. Eighteenth- and early 19th-century bricklayers would have finished the work with a finely pored piece of carborundum (silica carbide – made by heating silica with carbon in an electric furnace) to ensure that all the joints were flush. The process accounts for the extremely fine joints in rubbed brick arches in particular, which now defy standard raking out and repointing. In fact, it is rare to have to do so unless the joints have been infested with masonry bee. Such an exercise now involves various contortions, such as using cake nozzles or laying a limerich mortar strip (made with fine silver sand) between two sheets of greaseproof paper and offering the whole up to the joint. Whatever ingenious method that can be devised needs to be employed.

Other aspects of existing brickwork that demand appropriate recognition and correct treatment are ruled and also tuck pointing. Much Georgian brickwork was done at a very fast speed, using a feeble hydraulic lime, often produced because of using wood or coke for firing. In addition, the burnt lime was often used raw on the site, and incorporated with wet sand to form what is termed a 'hot mix'. The joints would be scraped back whilst still green, and repointed with a superior mix using the finest lime putty and fine sand, to give the illusion of gauged brickwork. The illusion of regularity was further enhanced by the ruling of these joints, either down the centre or top and bottom of the joint. The ruling would have taken place once an initial set had taken place, to ensure that the build-up of a bead of mortar was minimal. Such mortars used for ruling might employ a half measure of hydraulic lime to one part of lime putty and six parts of sand, of which at least three parts would be a fine silver sand.

Further uniformity could be achieved by colour-washing with a thin limewash coloured with ochre or iron oxide. Such practices were common in the late 18th and early 19th centuries, but continued up to around 1900. The even colour so achieved was highly prized. Unfortunately it is rarely recognised for its significance today, and in many cases has also faded. Any surviving examples are thus extremely precious.

The practice was also used where tuck pointing was utilised, the general mortar in the bricks, referred to as 'stopping' being coloured, again with oxides to match the bricks, or colour washed, before being incised to make a gap for a thin ribbon of white mortar to be set within. This was an extremely skilled task using a very unusual a high-calcium lime in a 'hot mix', that is, with the burnt lime added to a fine wet sand on the site. The resultant mix was very dry in order to encourage it not to slump out of the incision in the 'stopping mix' intended to house it. The stopping mix was similar to that used for ruled joints. There was very little water in the tuck pointing mix, but what little there was laced with glue size (animal glue), or cream of tartar, or keratin (animal protein), in order to retard the setting of the lime by introducing a carbon hydrogen compound to interrupt the crystal formation. Bricklayers would be occupied for the whole of their 12-hour day on the task of tuck-pointing alone. The practice continued up to the Victorian period, when black and red mortars with a white ribbon of 'tuck' were popular, as was red stopping with a black ribbon of 'tuck'. Black became a popular colour to honour the demise of Prince Albert.

It is very difficult to even contemplate wholesale repair of such characteristics, but minor repair using a very fine pointing iron, almost a knife edge, with a suitably dry mix is possible. It is more essential to recognise such characteristics in the first instance and respect them in any repair regime, even if it is only conserving the last traces, and not obliterating the features with mass repointing.

Unbaked earth – decay and repair

This is probably the most traditional and most sustainable material of all, and its use was widespread from very early periods, about which little is known, to the late 19th century in south-west England, East Anglia and parts of the Midlands. Its unique feature of producing houses that were warm in winter and cool in summer, farm buildings that produced a warm even temperature for animals, and garden walls that absorbed heat from the sun to protect adjoining plants from frost, was well known and understood. Unfortunately the intervention of mass-produced building materials, such as bricks and latterly cement, and the intervention of the First and Second World Wars resulting in the loss of essential skills, as well the speedier production of concrete, have nearly eliminated its use. Existing examples suffer from a whole raft of maladies due to the fact that the essential requirement of a good hat (roof), overcoat (lime or cob slurry render), and boots (a sturdy flint, brick, or stone plinth) are not maintained. Houses fare better because there is more attention to the overcoat and the hat, usually a thatch roof, but boundary walls and farm building suffer dreadfully from lack of attention, so much so that it is normal to read the construction of the building, via the stratified layers of the material, when it should not be possible to do so.

A basic understanding of the nature of the material itself is essential. Variations depend on the chalk, clay, sand, or gravel content of the local soil. All subsoils are to some extent stable in themselves, depending on the proportions of the granular components of sand and gravel, and the small particles of silt and clay. It is the latter which absorb water and encourage the larger particles to bind together via the hydrogen bond (the negative oxygen and positive hydrogen attraction). Each particle of is surrounded by a miniscule film of water whose thickness determines the binding power. Instability is a product of wetting and drying. In reality a subsoil which has a high sand content is more stable than that which has a high clay content, because the latter is more prone to swelling and shrinkage and sand particles do not absorb moisture. Thus a good cob mix has only sufficient clay and silt to bind the sand and gravel together. The film of water surrounding the clay particles becomes thicker, forming a type of lubrication and increasing the plasticity of the material (Van Der Steen, 1964).

Thus if a subsoil is to be used for a new cob building and its true provenance is unknown, then some soil analysis will be essential. This is achieved by taking a quantity of the selected soil, crushed, and placed in a dish to enable it to be heated in an oven to extract moisture. An exact amount is then put into a measure, transferred to a shallow dish, and repeatedly washed in water until the water runs clear of silt and clay particles. The remaining material is sand and gravel, which is then dried and placed in the measure. The amount of silt and clay removed in washing can then be gauged against the amount of sand and gravel remaining, so that it can be decided whether the aggregate content is greater than the clay and silt content.

Needless to say, existing cob buildings did not involve any of these technical processes, but rather an experiential understanding, innate and passed on by word of mouth through endless generations regarding which subsoils were suitable. This means that all existing cob buildings are very important to conserve as they encapsulate this fountain of innate knowledge.

The construction of cob took various forms, but two forms predominate at least in the south-west of England. By far the most popular is the slow process of a 'lift' at a time, around 450 mm (18 in) of a stiff slurry of subsoil with chopped straw, and covered with a layer of straw. The chopped straw acted as an internal corset and prevented cracking upon drying out. The whole was allowed to take on a set before the next lift was attempted. This implies that when building a whole township, such as Milton Abbas in Dorset, constructed in the 1780s as a model village, the build process must have taken many months, and considerable labour. In this area of Dorset cob is largely chalk based. The calcium carbonate has an equally high attraction to the positive hydrogen in water via the negative oxygen, so again particles of chalk are surrounded by a layer of water and for this reason the particles bond together.

A quick process is also noted in some areas, often clad with brick. The thinner layers were laid more rapidly and tended to slump, taking with them the brick courses. In Buckinghamshire, where the material is known as 'wychert', whole villages were again constructed of the material, undoubtedly of a greater timescale – from the medieval period to the 19th century – the intensive work required representing the embodied energy of many generations, deserving due recognition and every attempt at preservation.

Clay lump is not common in Britain, apart from areas of Norfolk, Suffolk, and Cambridgeshire, and yet it is the nearest relative to world-wide use of adobe, itself characteristic of ancient civilisations. One of the great unanswered questions must surely be why this ancient technique was superseded by a mass walling technique of compacted clay or chalk. Clay lump, or rather mud blocks, has many advantages, in that soil of a higher clay content can be used (as much as 80%, which would be untenable for cob), as initial shrinkage will take place in the

Photo 7.30 A dramatic grouping of cob buildings at Haddenham, Buckinghamshire

drying phase before the mud blocks are utilised, and the main concentration of energy is in the puddling of the earth to remove major pebbles. Any shrinkage cracks on the surface will act as a key for the lime or clay slurry render. New adobe buildings in commonwealth countries have incorporated a damp course, which is probably acceptable for pre-dried clay blocks were moisture content is dealt with via the clay jointing material. They are not so dependent on a homogeneous moisture content within all of the material, as solid cob walls definitively are. The insertion of a damp course in a cob wall will diminish the hydrogen bond in the clay and silt surrounding the aggregate of a cob wall, and it will literally turn to powder. Too much water from a leaking roof or downpipe will alternatively render the cob into a porridge-like consistency. With both these unfortunate conditions that are the result of no maintenance or inadvisable damp courses, the danger of collapse is imminent.

Clay block is thus the preferred material for the repair of cob walls where they have failed due to inadequate maintenance of the hat, boots, and more often than not, the overcoat. It is a simple matter to scoop out a degraded area of cob, making a flat shelf upon which the mud blocks are to sit, and bind them with clay, as was done in all traditional adobe work. They are laid in much the same way as bricks with staggered joints. Very small areas do not even require this treatment, needing only a twig to act as an armature, a ledge, and then a portion of cob material thrown at the area with great velocity. Cracks in cob can be repaired in a similar way, by removing a scoop either side of the crack, with uniform sides and base, and then utilising the same technique.

Cob in farm buildings can be affected by rats tunnelling. The only effective shield for this is a close wire mesh of stainless steel, extended up the outside of the walls to a suitable height, above plinth level and slightly overlapping it. Doubtless a technique of grouting with a cob slurry could be utilised for the repair of this defect, at least at obvious entrances and exits, and the lower areas protected with stainless steel wire mesh, but regrettably very little effort is made to do so and remaining cob farm buildings are disappearing at an alarming rate.

Complete areas of collapsed cob can be effectively dealt with by simply remixing the material with chopped straw and water, and on a suitable footing, equivalent in loading to that of a solid masonry wall of the same thickness, overlaid with a plinth at least 225 mm (9 in) above the ground level or floodplain, re-erect the material in the traditional manner in lifts, or make blocks of the material and bind them with a slurry of the material. Walls need to be at least 300 mm

(12 in) minimum, but in an existing building it is advisable to match the existing wall thickness as the whole needs to act as a thermal store, regulating temperature and humidity. Cob walls butting up against more solid materials like brick again appear to cause consternation in the minds of structural engineers and building society surveyors unused to natural materials. The age-old ability of a soft putty lime render to embed itself in any cracks, such as those formed by the butt joint between two dissimilar materials, and effectively seal out moisture appears long forgotten but is in urgent need of resurrection.

To conclude, roofs with adequate overhangs to protect walls beneath, soft lime render overcoats and a stout plinth should effectively maintain such buildings for nigh on eternity. A bi-annual coating with limewash will also build up to a fine plaster in any small cracks and seal them against moisture. Particular attention needs to taken to ensure that any vegetation such as ivy is never allowed to encroach, and if present cut away at the base to ensure die-back before attempting to pull it away. Killing ivy at source is a dilemma never entirely solved by patented weed killers, so it is essential to monitor the situation continuously and stem any re-growth.

Invariably new occupation of cob houses raises the spectre of extension. If new cob walls cannot be contemplated for whatever reason, then timber frame or blockwork can be used, butting up to or toothed in respectively. The main problem is with the creation of new openings to allow egress to the new extension. It is possible to cut new openings with a powered saw and span them with timber, having effectively needled the area to allow support, but the work should be approached with extreme caution. The advice of a structural engineer skilled in earth construction is essential, and the size and number of such openings should be kept to an absolute minimum. Cob buildings were deliberately constructed with small openings for windows, to allow for the optimum redistribution of loading patterns. Interfering with these patterns must be done with due consideration of the possible consequences. A further consideration with extensions is the abutment of the normal solid concrete floor with damp-proof membrane up against the breathing floor of the cob building. The latter will allow the cob walls to breathe, but the new floor will force excessive damp up the abutting cob wall. For this reason alone, it is essential to make a plea for a limecrete floor for the new extension, as this is also a breathing floor and will not cause this problem.

Two further earth wall categories deserve a brief mention. Sods or turf were used for centuries in Ireland and Scotland for low social status housing. Little remains of tangible examples, and other than museum reconstructions, the most potent being seen at Arnhem, Holland, which positively reeked of damp, it seems unlikely that the technique would ever be resurrected as a new form of sustainable housing.

Pisé or rammed-earth wall construction, made by ramming suitably moist sandy loam between shutters or formwork, is not generally a British technique, although it can be used for repair provided that the subsoil used contains just enough silt and clay to bind the aggregate content together, and not enough to cause excessive shrinkage upon drying. Generally a sand content of between 65 and 75% is suitable, compared with the 40% sand needed for clay cob or blocks, and a drier mix than that used for traditional cob is essential. Ramming needs to be done in 100 mm (4 in) layers until each layer is compacted to 75 mm (3 in) in depth and a ringing sound is heard. Another layer can then be immediately placed on top and the process begun again without any waiting for drying or curing.

Lime mortars, renders, and plasters – the fundamental tools for all masonry repairs

Lime has been referred to throughout the entire discourse of this text as it is the fundamental building block of repair. Lime mortar, with a well-graded aggregate, has a pore content that

Photo 7.31a Cob repair using unbaked bricks to join an existing cob wall to a blockwork wall (photo © Rob Buckley)

Photo 7.31b A crack in a cob wall can be repaired by stitching across the deliberately created ledge with cob bricks or flinging handfuls of cob to fill up the gap

Photo 7.31c A new cob wall being constructed in lifts, re-using old cob chopped up with straw and water

allows adjoining building blocks or timber to breathe, the pore structure of the mortar dealing with the accumulated moisture and expelling it before it can mobilise salts. Lime external renders act as overcoats, taking the worst of the weather, engaging in wetting and drying cycles, and dealing with salt mobilisation in preference to the building material beneath, as well as coping with thermal expansion.

Internal lime plasters act in a similar way, as well as wicking away moist air before it can settle on cold surfaces to form condensation and its attendant black mould. In addition, the lime plaster adds to the thermal store properties of the building and creates a buffer for high relative humidity. What is not generally realised is that such plasters can only buffer moisture content and wick it away if the general internal environment of rooms is conducive to this exercise. The air so contained needs to be warmed in cold damp conditions so that it will hold the moisture and encourage it to be transferred into the lime plaster. On–off central heating systems will not fulfil this function. A low gentle heat from storage heaters or a multi-fuel stove and a working chimney creating the requisite air changes per hour will encourage the right conditions. Traditional stone or cob buildings with thick walls cannot function in any other way. Great care therefore needs to be exercised then when attempting to meet the needs of modern building regulations for the minimum number of air changes per hour. Covering over lime plasters with what is essentially an internal plastic bag, be it in the form of foam panels or a vapour membrane, or even astonishingly thick plastic vertical membranes supplied by specialist companies to resist damp, simply defeats the ability of the building to function.

A major contribution to the ability of buildings to breathe has been the development of the limecrete floor, an insulated breathing lime floor that negates the necessity for damp-proof membranes and eliminates the subsequent explosion of damp in the walls. For decades the ramifications of this latter form of treatment of existing traditional buildings has resulted in unwise attempts at silicone injection in order to stem the rising damp, complete with the standard removal of the first 1 m (39 in) plus of lime plaster and its replacement with a dubious waterproof renders. The latter has much more to do with the fear of contractors to be asked to return to deal with mobilising salts and the resultant efflorescence than provision of a performing finish. The Victorians famously disguised any trace of rising damp with match-boarded dados, having installed cement-based tile floors, and may well have started the continuing unhappy trend of ignoring the cause and patching up the symptoms.

Insulated, breathing limecrete floors

Initially developed for use in conjunction with underfloor heating systems, the pore structure of the lime holding air warmed by the hot pipes beneath, the technique has obviated many of the problems indicated above. The floor is dug out to the normal depth required for the standard treatment but the sub-base is well compacted and given a layer of geotextile to prevent any loose material from disappearing downwards. A layer of loose fill, coarse, non-permeable aggregate is then laid on top to a depth of around 200 mm (8 in). It has now become usual for this to consist of recycled foam glass from Norway, compacted. As there are no fines there are no small pores and hence there is no potential for capilliary action. This is very difficult for those who are used to laying down a layer of dense polythene to understand, but the point needs to be laboured.

The loose fill is then covered with another layer of geotextile or hessian to avoid the limecrete from slumping into any of the voids. Basic limecrete is made by substituting hydraulic lime, usually NHL 3.5 for cement in the mixer with the aggregate. The sand and gravel used for the aggregate can be made more insulating by the incorporation of leca, or lightweight pumice, or chopped hemp (hempcrete). Leca is made by injecting droplets of liquid clay into a kiln at a temperature in excess of 1,000 °C. Moisture in the clay turns to steam, expanding the clay and forming numerous air bubbles which become entrained in the firing. It is manufactured in sizes ranging from 1 mm (¼ in) to 20 mm (¾ in) and can thus be combined in these various sizes to make a well-graded aggregate for the usual proportion of one of lime and three of aggregate. A slab of around 150 mm (6 in) is normal. Leca can be used as the loose fill, but there is a danger of crushing it and introducing fines whilst walking across the geotextile covering to lay the limecrete. Hempcrete could, in theory, be used instead of the limecrete. It is now becoming much more widely available in Britain, as it is slowly being realised that it is not the narcotic

Photo 7.32a Limecrete floor technology – a model made up by Rob Buckley showing a leca sub-base, leca and lime, and limecrete

Photo 7.32b Limecrete floor technology – a newly laid foamed glass sub-base

variety and can be purchased in ready made bags rather than combining the chopped material, which resembles chaff in the mixer with the hydraulic lime, although this is considerably less expensive.

The whole is then made suitable for underfloor heating by the application of 50 mm (2 in) of lime screed (or lime mixed with crushed glass, termed 'glaster'), which then encapsulates the heating pipes or wires, and the whole overlaid with flagstones or quarry tiles if available or desired.

A leca insulated lime floor slab has a U value of ranging from 0.45 to 0.1, and a hempcrete slab has a U value of around 0.11. Thus in theory there is no reason why the technology cannot be universal for existing and new-build traditional buildings.

Photo 7.32c Limecrete floor technology – lightweight pumice and lime ready to be laid on to the sub-base of foamed glass

Photo 7.33d Limecrete floor technology – model made by Nigel Jervis of Ty Mawr Lime Ltd showing the foamed glass sub-base and the next layer of lime with lightweight pumice

BIBLIOGRAPHY

Allback, S. and Fredlund, B. (2004) Windowcraft – Part One, *Journal of Architectural Conservation*, No 1, Vol 10, 53–66.

Allin, S. (2005) *Building with Hemp*. SEED Press, Co. Kerry.

Anderson, W. (2006) *Diary of an Eco-builder*. Green Books, Totnes.

Anderson, W. (2007) *Green Up*. Green Books, Totnes.

Anink, D., Boonstra, C. and Mak, J. (1996) *Handbook of Sustainable Building: An Environmental Preference Method for Selection of Materials for Use in Construction and Refurbishment*. James and James, London.

Ashurst, J. and Ashurst, N. (1988) *Practical Building Conservation. English Heritage Technical Handbook, Volumes 1–5*. Gower Technical, Aldershot.

Ashurst, J. and Dimes, F.G. (1984) *Conservation of Building and Decorative Stone*. Butterworth-Heinemann, London.

Banister Fletcher, Sir (1921) *A History of Architecture on the Comparative Method*. Batsford, London (republished 1996 with Dan Cruickshank as co-author and editor. Architectural Press, Oxford).

Bankart, G.P. (1908) *The Art of the Plasterer*. Batsford, London.

Barley, M. (1980) *Houses and History*. Council for British Archaeology, York.

Beresford, J. (1935) *James Woodforde, The Diary of a Country Parson 1758–1802*. Oxford University Press, Oxford.

Berge, B. (2000) *The Ecology of Building Materials* (translated from the Norwegian by Filip Henley with Henry Liddell) Architectural Press, Oxford.

Billington, M.J., Bright, K. and Waters, J.R. (2004) *The Building Regulations Explained and Illustrated, 13th edition*. Blackwell, Oxford.

Borer, P. and Harris, C. (1998/2005) *The Whole House Book: Ecological Building Design and Materials*. Centre for Alternative Technology Publications, Powys.

Bowyer, J. (1981) *Handbook of Building Crafts in Conservation*. Hutchinson, London (reprint of *The New Practical Builder and Workman's Companion* by Peter Nicholson, 1823).

BREEAM (2009) *BREEAM: The Environment Assessment Method for Buildings Around The World*. Available at www.breeam.org.

Brereton, C. (1991) *The Repair of Historic Buildings: Advice on Principles and Methods*. English Heritage, London.

British Standard Codes of Practice (1990) *BS 8000 (Workmanship on Building Sites*. BSI, London.

Brunskill, R.W. (1970) *Illustrated Handbook of Vernacular Architecture*. Faber and Faber, London.

Brunskill, R.W. (1985) *Timber Buildings in Britain*. Victor Gollanz in association with Peter Crawley, London.

Brunskill, R.W. (1990) *Brick Building in Britain*. Victor Gollanz in association with Peter Crawley, London.

Building Regulations – Approved Documents A, B, C, D, E, F, G, H, J, K, L1A/L2A, L1B/L2B, M, N, P. Available at www.planningportal.gov.uk

Building Research Establishment (1992) *Recognising Wood Rot and Insect Damage in Buildings*. Building Research Establishment, Watford.

Building Research Establishment (2007) *The Green Guide to Housing Specification*. Building Research Establishment, Watford.

Calloway, S. (1991) *The Elements of Style: An Encyclopaedia of Domestic Architectural Details*. Mitchell Beazley, London.

Charles, F. (1984) *The Conservation of Timber Buildings*. Hutchinson, London

Clifton-Taylor, A. (1987) *The Pattern of English Building, 4th Edition*. Faber & Faber, London.

Code for Sustainable Homes: www.communities.gov.uk/publications/planning.

Collins, J. (2002) *Old House Care and Repair*. Donhead, Shaftsbury.

Cunnington, P. (2002) *Caring for Old Houses*. Marston House, Yeovil.

Daintith, J. (1996) *Oxford Dictionary of Chemistry*. Oxford University Press, Oxford.

Davey, A., Heath, B., Ketchin, M. and Milne, R. (1995) *The Care and Conservation of Georgian Houses*. Butterworth-Heinemann, Oxford.

Desch, H.E. (1947) *Timber: Its Structure and Properties*. Macmillan, London.

Doyle, P., Hughes, T. and Thomas, I. (2008) *England's Heritage in Stone: Proceedings of a Conference in 2005*. English Stone Forum, Folkestone.

Ellis, M., Hutchinson, B. and Barton, J. (1975) *Maintenance and Repair of Buildings*. Newnes-Butterworths, London.

English Heritage (1994) *The Smeaton Project: Factors Affecting the Properties of Lime Based Mortars*. English Heritage, Collaborative Research in association with ICCROM and Bournemouth University, London.

English Heritage (1997) *Timber Panelled Doors and Fire (Upgrading the Fire Resistance): An English Heritage Technical Guidance Note*. English Heritage, London.

English Heritage (2002) *Building Regulations and Historic Buildings: Balancing the Need for Energy Conservation with Building Conservation*. English Heritage, London.

Fielden, B. (1982) *Conservation of Historic Buildings*. Architectural Press, Oxford.

Georgian Group (n.d.) Advisory leaflets. Available at www.georgiangroup.org.uk.

GHEU (2007) *Cutting Down on Carbon: Improving the Energy Efficiency of Historic Buildings*. Government Historic Estates Unit Annual Seminar, Building Research Establishment, Garston, October.

Glover, P. (2003) *Building Surveys, 5th Edition*. Butterworth-Heinemann, Oxford.

Graham, P. (2003) *Building Ecology: First Principles for a Sustainable Built Environment*. Blackwell, Oxford.

Griffiths, N. (2007) *Eco-House Manual*. Haynes, Yeovil.

Gwilt, G. (1891) *An Encyclopedia of Architecture*. Longman, London.

Hall, K. (2006) *The Green Building Bible, Volume 1*. Green Building Press. Available at www.greenbuilding press.co.uk.

Harris, C. and Borer, P. (2005) *The Whole House Book: Ecological Building Design and Materials*, 2nd edn. Centre for Alternative Technology Publications, Powys.

Harris, R. (1993) *Timber Framed Buildings*. Shire Publications, Princes Risborough.

Hinks, J. and Cook, G. (1997) *The Technology of Building Defects*. Spon, London.

Historic Scotland (n.d.) Technical literature. Available at www.historic-scotland.gov.uk.

Hudson, K.H. (1972) *Building Materials*. Industrial Archaeology Series, Longman, London.

Hughes, P. (1986) *The Need for Old Buildings to Breathe*. Technical Pamphlet. Society for the Protection of Ancient Buildings, London.

Innocent, C.F. (1916/1999) *The Development of English Building Construction*. Donhead, Shaftbury.

Insall, D. (1975) *The Care of Old Buildings Today: A Practical Guide*. Architectural Press, London.

Jackson, A. and Day, D. (1998) *Collins Care and Repair of Period Houses*. Harper Collins, London.

Jones, B. (2002) *Building with Straw Bales: A Practical Guide*. Green Books, Totnes.

Kelly, R. (1999) *Latent Heat Storage in Building Materials*. Building Services Engineering Diploma Dissertation, Dublin Institute of Technology.

Lander, H. (1986/1992) *The House Restorer's Guide*. David and Charles, Newton Abbot.

Lawrence, M., Heath, A. and Walker, P. (2009) Determining Moisture Levels in Straw Bale Construction, *Construction and Building Materials*, Vol. 23, No. 8, 2763–2768.

Lee, R. (1988) *Building Maintenance Management*. BSP Professional Books, Blackwell Science, London.

Lever, J. and Harris, J. (1966) *Illustrated Dictionary of Architecture 800–1914*. Faber and Faber, London.

Lynch, G. (1990) *Brickwork, History, Technology and Practice*. Donhead, Shaftsbury.

Machin, R. (1994) Lecture to the Winter Meeting of the Vernacular Architecture Group, London, December. See www.vag.org.uk.

Marley Eternit Ltd (2006) *Pure Roofing.* Marley Eternit, Burton upon Trent.

Massey, J. and Maxwell, S. (1995) *Arts and Crafts.* Archetype Press, London.

May, N. (2005) *Breathability, the Key to Building Performance.* Natural Building Technologies, Oakley.

Melville, I.A. and Gordon, I.A. (1974) *Structural Survey of Dwelling Houses.* Estates Gazette, London.

Melville, I.A. and Gordon, I.A. (1997) *The Repair and Maintenance of Houses.* Estates Gazette, London.

Mercer, E. (1975) *The English Vernacular House: A Study of Traditional Farmhouses and Cottages.* HM Stationery Office, London.

Mills, E. (1994) *Building Maintenance and Preservation: A Guide to Design and Management.* Butterworth-Heinemann, Oxford.

Minke, G. (2000) *Earth Construction Handbook.* WIT Press, Southhampton.

Nash, G. (2003) *Renovating Old Houses.* Taunton Press, Newtown, CT.

Natural Building Technologies (2010) www.natural-building.co.uk, accessed 5.8.10.

Newbold, H.B. (1923) *House and Cottage Construction, Vols 1–3.* Caxton, London.

Nicholls, R. and Hall, K. (2006) *The Green Building Bible, Volume 2*, Green Building Press. Available at www.greenbuildingpress.co.uk.

Oliver, A. (1997a) *Dampness in Buildings, 2nd Edition.* Blackwell Science, Oxford.

Oliver, P. (1997b) *Encyclopedia of Vernacular Architecture of the World.* Cambridge University Press, Cambridge.

Oxley, R. (2003) *Survey and Repair of Traditional Buildings: A Sustainable Approach.* Donhead, Shaftsbury.

Pearson, D. (1989) *The Natural House Book.* Gaia–Conran Octopus, London.

Pearson, G.T. (1992) *Conservation of Clay and Chalk Buildings.* Donhead, Shaftesbury.

Pevsner, N. (1985) *The Sources of Modern Architecture and Design.* Thames and Hudson, London.

Pilkington, B., Griffiths, R., Goodhew, S., and de Wilde, P. (2008) Thermal Probe Technology for Buildings: Transition from Laboratory to Field Measurements, *Journal of Architectural Engineering*, Vol. 14, No. 4, 111–118.

Powys, A.R. (1929/1995) *Repair of Ancient Buildings.* Society for the Protection of Ancient Buildings, London.

Ridout, B. (2000) *Timber Decay in Buildings.* Spon, London.

Roaf, S., Fuentes, M. and Thomas, S. (2003) *Ecohouse 2: A Design Guide.* Architectural Press, Oxford.

Robson, P. (1999) *Structural Repair of Traditional Buildings.* Donhead, Shaftesbury.

Rutherford, J., Gettens, J. and Stout, G.L. (1942/1966) *Painting Materials: A Short Encyclopaedia.* Dover Publications, New York.

Safeguard Chemicals (1999) *Dry Rot and Its Control: A Guide to the Biology and Control of Dry Rot.* Available at www.safeguardeurope.com.

Saunders, M. (1987) *The Historic Home Owners Companion.* Batsford, London.

Schofield, J. (1995) *Lime in Building: A Practical Guide.* Black Dog Press, Crediton.

Seeley, I.H. (1987) *Building Maintenance.* Macmillan Building and Surveying Series, Basingstoke.

Smeaton, J. (1791) *A Narrative of the Building of the Eddystone Lighthouse, and a Description of Its Construction.* Hughes, London.

Snell, C. and Callahan, T. (2005) *Building Green: A Complete How-To Guide to Alternative Building Methods.* Lark Books, New York.

Society for the Protection of Ancient Buildings (SPAB) (n.d.) All technical pamphlets and information sheets: www.spab.org.uk.

Stroud-Foster, J. and Greeno, R. (2007) *Structure and Fabric, Part 1, 5th Edition*, Mitchells Building Series. Pearson Education, Harlow.

Taylor, C.D. (2000) *Materials in Construction: An Introduction.* Pearson Education, Harlow.

Teutonico, J.M. (1988) *Architectural Laboratory Manual.* ICCROM, Rome.

Torraca, G. (1988) *Porous Building Materials.* ICCROM, Rome.

Thompson, F. (1945) *Lark Rise to Candleford.* Penguin, London.

Tricker, R. and Algar, R. (2007) *Building Regulations in Brief, 5th Edition.* Butterworth-Heinemann, Oxford.

Tutton, M., Hirst, E., Louw, J. and Pearce, J. (2007) *Windows: History, Repair and Conservation.* Donhead, Shaftesbury.

Ty-Mawr Lime Ltd – Limecrete Floors: www.lime.org.uk.

Vale, B. and Vale, R. (2000) *The New Autonomous House.* Thames and Hudson, London.

Van Der Steen, A. (1964) *Earth Wall Construction*. Commonwealth Experimental Building Station, Sydney.

Van Lengen, J. (2008) *The Barefoot Architect: A Handbook for Green Building*. Shelter, Bolinas, CA.

Victorian Society (n.d.) All advisory leaflets. Available at www.victoriansociety.org.uk.

Waterfield, P. (2006) *The Energy Efficient Home. A Complete Guide*. Crowood Press, Marlborough.

Watkin, D. (1979) *English Architecture: A Concise History*. Oxford University Press, Oxford.

Watt, D. and Swallow, P. (1996) *Surveying Historic Buildings*. Donhead, Shaftesbury.

Weaver, G. (1992) *Science for Conservators*. Routledge, London.

Weaver, M. and Matero, F.G. (1993) *Conserving Buildings: A Guide to Techniques and Materials*. Wiley, Chichester.

Wilde, E. (2002), *Eco: An Essential Sourcebook for Environmentally Friendly Design and Decoration*. Quadrille, London.

Wood, M. (1981) *The English Medieval House*. Ferndale, London.

Woolley, T. (2006) *Natural Building: A Guide to Materials and Techniques*. Crowood, Marlborough.

Woolley, T. and Bevan, R. (2009) *Constructing a Low Energy House from Hempcrete and other Natural Materials*. Proceedings of the 11th International Congress on Non-conventional Materials and Technologies, Bath. Available at www.opus.bath.ac.uk.

Woolley, T. and Kimmins, S. (2002) *Green Building Handbook, Volume 2: A Guide to Building Products and Their Impact on the Environment*. Spon Press, London.

Woolley, T., Kimmins, S., Harrison, P. and Harrison, R. (1997) *Green Building Handbook, Volume 1: A Guide to Building Products and Their Impact on the Environment*. Spon, London.

Workmanship on Building Sites – Codes of Practice. Available at www.bsigroup.com.

Wright, A. (1991) *Craft Techniques for Traditional Buildings*. Batsford, London.

Yarwood, D. (1966) *English Houses*. Batsford, London.

Yeomans, D.T. (1997) *Construction Since 1900: Materials*. Batsford, London.

INDEX